Advances in Neutron Optics

T0093457

Advances in Neutron Optics

Fundamentals and Applications in Materials Science and Biomedicine

Edited by

Maria L. Calvo and Ramón F. Álvarez-Estrada

CRC Press
Taylor & Francis Group
Boca Raton London New York

CRC Press is an imprint of the
Taylor & Francis Group, an **informa** business

CRC Press
Taylor & Francis Group
6000 Broken Sound Parkway NW, Suite 300
Boca Raton, FL 33487-2742

First issued in paperback 2023

© 2020 by Taylor & Francis Group, LLC
CRC Press is an imprint of Taylor & Francis Group, an Informa business

No claim to original U.S. Government works

ISBN-13: 978-1-138-36446-2 (hbk)
ISBN-13: 978-1-03-265356-3 (pbk)
ISBN-13: 978-0-367-81605-6 (ebk)

DOI: 10.1201/9780367816056

Library of Congress Cataloging-in-Publication Data

A catalog record for this book has been requested.

Visit the Taylor & Francis Web site at
http://www.taylorandfrancis.com

and the CRC Press Web site at
http://www.crcpress.com

Contents

SECTION I An Introductory Approach to the Foundations of Neutron Optics

SECTION II Neutron Optics-Based Technologies and Applications

Preface

This book is presented with the intention to be an updated contribution to the field of neutron optics. This interdisciplinary area among nuclear physics, optics, and quantum physics is based upon the dual particle-wave behavior and very specific properties of slow neutrons and contains, in turn, a number of multidisciplinary subjects. The slow neutrons are enjoying a noticeable expansion since they are related to many interactive areas of science, namely, major branches as material science, biomedicine and related based technologies. Moreover, there is an active development of optical (or optics-inspired) components used for neutron instrumentation, new neutron sources, and new emerging branches involving the introduction and the design of new key results created by and expanding the previous contributions. An update in neutron optics has a double focus: the revision and expansion of the theoretical formalisms and mathematical models, and the development of new revolutionary technologies. This double view leads an eventual student to a more complete understanding of this interesting and promising field. Neutron optics has such a wide scope that a number of interesting specific subjects had to be omitted from the present update, a fact for which we apologize and which may be only partially amended through the references given in the book. We can only hope that the subjects included in the book justify the present (hence, partial) update.

Although neutron optics is not yet a subject routinely included in the academic programs for physics in many universities, we are convinced that, in a near future, it will constitute a standard offer for students of masters and many specialties of physics that include materials science and biomedicine. Also, we believe that currently, in the field of nuclear engineering, neutron optics constitutes quite a challenging topic.

We hope and wish that this book will be of help and provide knowledge support to many readers interested in this fascinating area of science.

Maria L. Calvo
Ramón F. Álvarez-Estrada

Acknowledgments

We would like to thank Professor Vasudevan Lakshminarayanan, Editor of the CRC Press/Taylor & Francis Books Series in Multidisciplinary and Applied Optics, for his kind invitation to perform this quite exciting and challenging project. We also thank Dr. Marc Gutierrez, from CRC Press/Taylor & Francis Group, for all the accorded technical support and kind facilities.

Our sincere gratitude goes to all authors for their contributions without which this book could not have been edited. We emphasize here that all authors are deeply involved in research in the area of neutron optics, which provides important impact on the contents of the corresponding chapters. We also thank all authors' efforts to maintain the initial proposed timing.

About the Editors

Maria L. Calvo, Ph.D., is Emeritus professor at the Department of Optics, Faculty of Physical Sciences, Complutense University of Madrid (UCM) (Madrid, Spain). She initiated research at the CNRS (Paris, France) in the Glass Laboratory. She also founded and leads the Interdisciplinary Group for Optical Computing (GICO-UCM) created in 2001 from the former Interdisciplinary Group for Biooptics initiated earlier in 1993. Her theoretical research interests include classical formalism for light scattering, optical waveguide theory and neutron waveguides, photonic devices, and applications to particular media such as holographic gratings. In experimental physics, she has relevant contributions in the design, synthesis, and characterization of holographic photomaterials. She has been a visiting professor in other institutions as the University of California at Berkeley and the University of Saint Louis-Missouri. On all the mentioned subjects she has published as author or co-author more than 200 scientific articles. Professor Calvo has taught general and specialized courses in optics for both undergraduate and master's degree candidates in physical science at UCM. She is involved in education and professional issues, and has authored, coordinated, and edited various textbooks in Spanish and in English. She has been president of the International Commission for Optics (ICO) during the term 2008–2011.

Ramón F. Álvarez-Estrada, Ph.D., is Emeritus professor of physics at the Department of Theoretical Physics, Faculty of Physical Sciences, Complutense University of Madrid (UCM), Madrid, Spain, where he has taught undergraduate and graduate courses since 1975. He has been group leader on research projects (1986–2012), sponsored by successive Ministries of Science in Spain. He has been visiting scientist in Faculté des Sciences d' Orsay, Université de Paris, and visiting professor at Theory Division, CERN, Geneva, Switzerland and at Theoretical Physics Group and Nuclear Science Division, Lawrence Berkeley Laboratory, University of California at Berkeley, California. He has published, as author and co-author, more than 120 research papers in international scientific journals on elementary particle physics, quantum field theory, quantum physics and scattering, classical and quantum optics, non-equilibrium classical and quantum statistical mechanics, statistical mechanics of macromolecular chains and, since 1984, neutron optics. He has co-authored one research book in English, co-authored three textbooks in Spanish, and authored and co-authored contributed chapters to books, in particular, on neutron optics.

Contributors

Ramón F. Álvarez-Estrada
Emeritus Faculty of Physical Sciences
Department of Theoretical Physics
Complutense University of Madrid
Madrid, Spain

Maria L. Calvo
Emeritus Faculty of Physical Sciences
Department of Optics
Complutense University of Madrid
Madrid, Spain

Ignacio Molina de la Peña
Faculty of Physical Sciences
Department of Optics
Complutense University of Madrid
Madrid, Spain

Nikolay Pleshanov
Neutron Research Department
Petersburg Nuclear Physics Institute
of National Research Centre
«Kurchatov Institute»
Gatchina, Leningrad Region, Russia

Ignacio Porras
Faculty of Science
Department of Atomic, Molecular and
Nuclear Physics
University of Granada
Granada, Spain

Ángel S. Sanz
Faculty of Physical Sciences
Department of Optics
Complutense University of Madrid
Madrid, Spain

Markus Strobl
Neutron Imaging and Applied Materials
Group (NIAG)
Paul Scherrer Institut
Villigen, Switzerland
and
Niels Bohr Institut, University of
Copenhagen
Copenhagen, Denmark

Introduction

The very early discovery of the neutron by Chadwick in 1932 established it as a subatomic particle and, therefore, as a constituent of the atomic nucleus. In the almost ninety years of continuous scientific research about the neutron and its interaction with matter, there has been an impressive expansion of many subjects related to it. The key properties of the neutron (its dual particle-wave nature, mass, spin, electric neutrality, strong interactions, and presence in all atomic nuclei, except those of ordinary hydrogen) have given rise to a new interdisciplinary area in physics. For the latter purpose, neutrons have to be slow. Neutron optics is that interdisciplinary area, dealing with the theory and application of the *wave behavior* of slow *neutrons*. Neutron optics involves studying the interactions of matter with a *beam of free slow neutrons*, much as *spectroscopy* represents the interaction of matter with *electromagnetic radiation*. There are two major mechanisms and artificial sources for the production of beams of free slow neutrons: (1) *fission* reactions at *nuclear reactors* and (2) collisions in *particle accelerators* (small ones, spallation sources) of *proton* beams with targets of heavy atoms, such as Tantalum and Uranium. In this book, attention will be paid to artificial neutron sources of the above two classes (including some short reference to various spallation sources geographically distributed over the world, including the new big facility, namely, the European Spallation Source in Lund, Sweden).

The chapters of this book can be broadly divided into two major areas or sections: Section I—An Introductory Approach to the Foundations of Neutron Optics and Section II—Neutron Optics-Based Technologies and Applications.

Chapter 1, co-authored by Álvarez-Estrada and Calvo, deals with fundamentals. This chapter introduces various general aspects of neutron optics. It treats fundamental concepts and tools providing a general overview and, so, allows to afford the more specific contents of the book. Significant references are included, that is, those having a historical meaning and those offering a wide perspective, which could eventually help to fill the gap corresponding to several interesting subjects and references that could not be included here due to space limitations. The authors apologize for both omissions. Readers could amplify the references given and fill the gaps generated by the omissions, according to their own interest. For instance, the subsection on inelastic neutron scattering is intended as an introduction to further, much wider and specialized references. And the (more expanded) subsections on neutron diffraction are aimed to help the reader to connect, in particular, with later chapters. Several sections and subsections offer updated information (for instance, neutron holography) and, even, new material (a new formulation of the extinction theorem, in a simple context). The total number of sections of this chapter is nine. Several developments, omitted from the main text, have been included as exercises with solutions. For a general presentation, the generic term *slow neutrons* is employed to cover the rather wide range from ultracold neutrons to epithermal ones. Upon coming to specific studies, the energy ranges involved are specified; in particular when focusing on thermal neutrons.

Sanz presents new formalisms for the study of slow neutron diffraction and inter-ferences, with an account of neutron trajectories in Chapter 2. At low energies, as any other matter wave, neutrons exhibit very distinctive features, such as diffraction, interference, and entanglement. Diffraction and interference, on the one hand, are observed in single-particle conditions, while entanglement arises in the case of spin-correlated pairs of neutrons. This chapter focuses on these fundamental aspects of neutron matter waves, providing a wide perspective on the experimental and theo-retical work developed within the areas of neutron diffraction/scattering and neu-tron interferometry, where such aspects and their applications have been extensively studied. This includes a revision of issues related to quantum measurement problem, such as Bell-type measurements and the more recent issue of weak measurements. The chapter contains five sections.

An introduction to the confined propagation of slow neutron in thin waveguides (sizes about 1 μm and somewhat smaller) is offered in Chapter 3, by Molina et al., with particular attention to possible improvements, applications, and feasibility. The total internal reflection phenomenon (studied in Chapter 1) gives the fundamentals for these waveguides with small transverse cross sections (either thin fibers or thin planar films), having air or Titanium at the core, and a clad formed by appropriate materials, as glass and silica, among others. Previous conjectures in the last decades are summarized. Two important seminal experiments are studied (which have trig-gered further experimental routes): (i) with polycapillary glass fibers (in 1992) and (ii) with thin planar films (in 1994). Recent proposals for confined propagation and possible applications, based upon thin planar films (ii) are considered; recent ongo-ing mathematical implementations and numerical simulations are discussed. This chapter has six sections.

Chapter 4, authored by Ströbl, focuses on the very active topic of neutron optical devices and the current development in various nuclear technologies. The introduc-tion provides the basic principles of neutron imaging as a technique in material sci-ence and industrial non-destructive testing applications. Focus is first on the two fundamental constituents of imaging, spatial resolution and contrast, how they are achieved and what their basics are, leading up to the information that can be pro-vided and retrieved from neutron imaging. This includes parallels and complemen-tarities with other transmission imaging methods and concludes with an overview of traditional and state-of-the-art applications. The chapter contains three sections.

Chapter 5 by Pleshanov is dedicated to quite a modern and appealing topics based upon neutron spin optics (NSO), with discussions on current perspectives. NSO is a new direction in polarized neutron instrumentation. It is based on quantum aspects of the interaction of neutrons with magnetically anisotropic layers. It opens new possibilities for spin manipulations and may lead to numerous innovations in neutron techniques. Advantages of NSO: compactness (miniaturization unlimited); zero-field option (no external fields are required, guide fields are optional); multi-functionality (handling spectrum, divergence, and spin manipulations at the same time); new possibilities and solutions, including the beam hyperpolarization and spin 3D-manipulators. All these subjects are presented and discussed in the six sections of this chapter.

The last chapter, Chapter 6 by Porras, concentrates on quite interesting and active applications of neutron optics in biomedicine, namely, the so-called Boron Neutron Capture Therapy (BNCT) and related techniques. In addition to the well-known application of neutron scattering to structural biology, here is a very relevant application of neutrons to health sciences: BNCT, a promising experimental form of radiotherapy, which is facing a new era with renewed projects and under active research. The chapter contains six sections.

All chapters have the corresponding references. The book ends with a subject index.

As a final and general comment, this book has been conceived and designed for readers having a background as scientists and technologists and with an interest in expanding their knowledge in the field of nuclear physics, by enhancing a particular attention to neutron physics, neutron optics, its fundamentals, and related technologies. The diversity in presentation illustrates the conceptual richness of neutron optics. Students in areas like physics and nuclear engineering could benefit from the reading of the chapters. This will give an opportunity to authors and readers to share the updating topics on this relevant and promising branch of physical science.

Section I

An Introductory Approach to the Foundations of Neutron Optics

Section I

1 Neutron Optics: Fundamentals

Ramón F. Álvarez-Estrada and Maria L. Calvo

CONTENTS

1.1 INTRODUCTION: SOME HISTORICAL BACKGROUND AND CONTENTS

The existence of the neutron was experimentally established by J. Chadwick in 1932 [1] (see Figure 1.1). That fundamental discovery was the culmination of other important researches (by W. Bothe and H. Becker, I. Curie, H. C. Webster, P. I. Dee, I. Curie and F. Joliot, F. Perrin, N. Feather, E. Majorana, W. Heisenberg and D. Iwanenko); see, for instance, Ref. [2] for a historical account. Experiments in 1936 provided adequate indication that crystalline materials diffracted slow neutrons: see Ref. [3] for a readable historical account. Systematic researches by E. Fermi and W. H. Zinn, started in 1944, had put neutron optics on firm grounds: their experiments, using a collimated beam of thermal neutrons from a nuclear reactor, provided evidence of mirror reflection [3–6, 25]. Since then, a whole variety of experiments have established several optical-like effects for slow neutrons (with associated de Broglie wavelength at an interval of about 1 Å). A number of researchers (G. E. Bacon, B. Brockhouse, L. M. Corliss, J. M. Hastings, W. C. Koehler, C. G. Shull, M. K. Wilkinson, E. O. Wollan, ...) in various institutions made important contributions in developing and employing neutron diffraction in order to analyze the atomic structure of matter (say, locations of the atoms and their dynamics). In particular, B. Brockhouse and C. G. Shull were awarded the 1994 Nobel Prize in Physics for their works on those subjects.

We shall not offer here a broad presentation of neutron optics: for comprehensive views about it, the reader should proceed to other sources [2, 3, 5–8]. We shall limit to treat in certain detail and for the sake of a basic understanding, some general

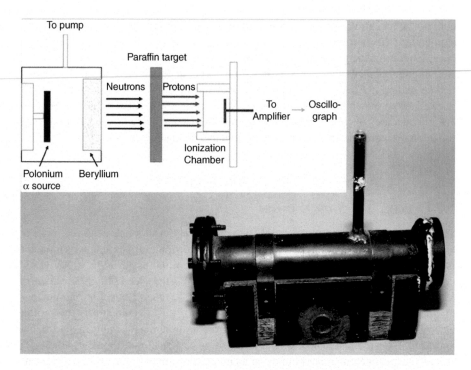

FIGURE 1.1 The experimental discovery of the neutron: Chadwick's device (Courtesy of and copyright Cavendish Laboratory, University of Cambridge, UK). Upper subfigure. A source containing Po (polonium) emits α particles which, after nuclear reactions in the Be (beryllium) target, give rise to an unknown electrically neutral radiation (solid arrow). The unknown neutral radiation goes through a paraffin target, enters an ionization chamber and generates, in the latter, further nuclear reactions which give rise to protons that, in turn, are detected. The detection of the protons was accomplished by means of an amplifier and an oscillograph. The paraffin target contains H, Li, Be, B, and N. The unknown neutral radiation emitted by the Be target was consistently interpreted as a beam of neutrons.

aspects of neutron physics and neutron optics (including some discussions of a few recent topics. The omissions of certain material will be justified, because it will be treated in other chapters of this book. See, for instance, comments in Section 1.9).

The specific contents of this chapter are as follows: Section 1.2 outlines various general aspects of neutron physics [2, 7, 9]. Section 1.3 summarizes wave propagation of slow neutrons, using quantum mechanics and yielding the basic effective optical potential acting on them. Section 1.4 deals with quantum-mechanical probabilistic issues, namely, pure states and statistical mixtures for slow neutrons. Section 1.5 discusses reflection and refraction of slow neutrons. Section 1.6 outlines some minimum essentials of magnetic interactions of slow neutrons due to their spin: the active and expanding field of neutron spin optics will be treated at considerable length in Chapter 5 (some specific aspects, like spin polarization control and spin filters, being treated, from another perspective, in Chapter 4). In search for some further completeness, Section 1.7 treats various developments with thermal neutrons, concentrating mostly on their dynamical diffraction by crystals and some applications (in particular, neutron holography). Neutron diffraction and interference, exploiting their analogies with the corresponding phenomena with light, as well as entanglement and related issues from deeper quantum-mechanical standpoints will be studied in Chapter 2. Section 1.8 deals with the extinction theorem and introduce a new approach to it in a simple context, in connection with Section 1.5. Section 1.9 contains the conclusions and some discussions. Some calculations, omitted from the main text, are presented as exercises, together with solutions for them.

1.2   GENERAL ASPECTS OF NEUTRON PHYSICS

1.2.1   PROPERTIES OF THE NEUTRON

The neutron (denoted by n) is one of the constituents of atomic nuclei. Its mass is $m_n = 939.57 \text{MeV/c}^2$, its spin is $\hbar / 2$ (thereby being half-integral), its magnetic moment is $\mu_n = -1.913 \mu_N$ and it is electrically neutral. We remind that $1\text{MeV} = 10^6 \text{eV}$ (1 eV denoting one electron-volt), $c (= 2.998 \times 10^{10} \text{cm} \cdot \text{s}^{-1})$ is the speed of light in vacuum, $\hbar (= 6.58 \times 10^{22} \text{MeV} \cdot \text{s})$ is Planck's constant and $\mu_N = \hbar |e| / (2m_p c)$ is one nuclear magneton. In turn, s denotes one second, $|e| = 1.602 \times 10^{-19}$ coulomb is the proton electric charge and $m_p = 938.27 \text{MeV/c}^2$ is the proton mass. Many experiments appear to indicate that the neutron has a vanishing electric dipole moment: more specifically, experimental upper bounds on its magnitude are of order $3.0 \times 10^{-26} |e| \cdot \text{cm}$: for a slightly improved bound $(-0.21 \pm 1.82 \times 10^{-26} |e| \cdot \text{cm})$, see Ref. [10].

The neutron is not strictly elementary—rather, it is composed by other three more elementary or fundamental entities, named quarks. Quarks have fractional electric charges. Specifically, the neutron is a bound system of one certain type of quark with electric charge $+(2/3)|e|$ (named quark u) and of two other quarks, each with electric charge $-(1/3)|e|$ (named quark d) [2, 9]. The overall electric charge of the neutron, as a bound system (due to strong interactions) of those three charged entities, vanishes consistently. All phenomena to be dealt with in this chapter take place at ranges of energies and at length scales such that each of the neutrons involved behaves as an elementary particle (its composedness not becoming manifest)—this is the point of view to be followed here. That is, we shall disregard completely and consistently the fact that the neutron be composed by other more fundamental entities.

The neutron is unstable, its lifetime being 886.7 s. All neutron phenomena (in particular, the travels of slow neutrons along various, eventually macroscopic, distances) treated in this chapter occur during time intervals shorter than its lifetime. Accordingly, one can disregard the neutron instability as a reliable approximation and, hence, one can treat it as a stable particle (as if it had an infinite lifetime).

A discussion about the influence of gravity on slow neutrons, moving on the surface of the Earth with differences of height not exceeding one m (which will always be the case here), will be given in Subsection 1.7.6.

Let us consider a nonrelativistic neutron with kinetic energy E_K and de Broglie wavelength λ_{dB}. They are related through $E_K = (2\pi\hbar)^2 / (2m_n \lambda_{dB}^2)$. In a rather strict sense, a neutron is named "thermal" if it has $E_K \simeq 0.025$ eV and $\lambda_{dB} \simeq 1.8$ Å. If a thermal neutron moves through a material medium at thermodynamic equilibrium at absolute temperature $T = 293$ K (20°C, say, room temperature), then one has, consistently, 0.025 eV $= (3/2)K_B T$, $K_B = 1.38 \times 10^{-23} \, \text{m}^2 \text{kg}^{-2} \text{K}^{-1}$ being Boltzmann's constant.

Various specific names are usually attributed to nonrelativistic neutrons according to the ranges of their kinetic energy, although there does not seem to exist unique (unanimously accepted) conventions. Thus, in a less strict sense, neutrons with E_K ranging from about 0.003 eV up to about 0.4 or 0.5 eV are also named "thermal." Optical-like phenomena with slow neutrons having E_K smaller than about 0.03 eV are also of considerable interest; such neutrons are denoted generically as cold. In a somewhat more strict sense, neutrons also receive the following more specific names: cold $(5 \times 10^{-5} \text{eV} \le E_K < 0.025 \text{ eV})$, very cold $(5 \times 10^{-5} \text{eV} > E_K > 2 \times 10^{-7} \text{eV})$ and ultra-cold $(E_K < 2 \times 10^{-7} \text{eV})$ [2]. Neutrons with E_K in the range 0.4 − 0.5 eV to 0.1 and, even, to a few tens keV are usually called epithermal (1 keV $= 10^3$ eV). In a general sense, we shall refer to ultracold, cold, thermal, and epithermal neutrons as "slow." Neutrons with E_K in the range just above the epithermal one, up to about 200 keV are named intermediate energy ones. Neutrons with kinetic energy in the range 200 keV up about 10 MeV are referred to as fast. One usually refers to neutrons with E_K larger than 10 MeV as high energy ones. Throughout this chapter, we shall have in mind thermal neutrons (in both strict and non-strict senses). However, various results for other slow neutrons (ultracold, epithermal, etc.) and even fast ones will also be mentioned.

The basic quantum-mechanical analysis for thermal neutrons, including the basic optical potential, will also hold for slow neutrons (in particular, for epithermal ones), except, at most, for a few specific details. For this reason, we shall frequently replace "thermal neutrons" by "slow neutrons" in the quantum-mechanical studies.

1.2.2 Some Important Properties of Slow Neutrons

We shall overview very briefly a few reasons supporting the interest and importance of slow neutrons [2, 7, 9].

- The de Broglie wavelength for (strictly) thermal neutrons with $E_K \simeq 0.025$ eV (namely, $\lambda_{dB} = 1.8$ Å) has the same order of magnitude as the spacings between two neighboring atoms in a typical crystal. This allows for a beam of (strictly) thermal neutrons, going through a crystal lattice, to give rise to interference and diffraction phenomena, as already displayed in researches at early stages. See Figures 1.7, 1.10, and 1.11. Then, by analyzing the resulting diffraction patterns, such thermal neutrons allow to explore interatomic spacings and structure in crystals. In this application, thermal neutrons do a job similar (or, as qualified below, complementary) to the one X-rays do. More generally, slow neutrons with $0.1 \le \lambda_{dB} \le 20$ (in Å) have proved to be quite adequate and very useful in order to study correlations among atoms and among magnetic subdomains in condensed matter.
- Slow neutron with energies E_K in the range 10^{-2} to 10^{-1} eV enable to analyze molecular vibrations, while those with E_K about 10^{-6} eV allow for exploring certain slow dynamics of polymers (their reptation).

- Slow neutrons can penetrate into various materials, without being absorbed appreciably along a certain limited depth and without producing significant modifications in the former. In comparison with slow neutrons, and on the opposite side, X-rays turn out to be more strongly absorbed and rather destructive. See Subsection 1.7.4. Then, that (relatively) nondestructive character of slow neutrons makes them quite convenient for (i) the experimental study of several properties and behaviors of biological matter, (ii) the determination of the composition of small samples of certain materials without producing appreciable damage in them (neutron activation analysis), (iii) the processes of making images of objects with neutrons (neutron imaging), which refer to a collection of nondestructive testing methods, exploiting the penetration and attenuation properties of neutrons to investigate the internal structure of the imaged objects. See Figure 3.2. Chapter 4 is devoted to neutron imaging.
- Slow neutrons which have penetrated into matter beyond some depth, could, at a later stage, be eventually absorbed through suitable nuclear reactions occurring at certain localized domains. In certain cases, those nuclear reactions can be relatively nondestructive but in other cases, they turn out to be certainly destructive. In the case of biological matter (human tissue), the damage produced by those nuclear reactions can be beneficial in order to kill tumor cells (therapies based on neutron capture). See Figure 3.18. Since that fact has enormous interest, an updated account of that active and interdisciplinary application of slow neutron physics will be given in Chapter 6.

1.2.3 DETECTION OF SLOW NEUTRONS

As the neutron is electrically neutral, its detection is not direct but it proceeds indirectly as follows [2, 9]. Arrangements are made so that the neutron to be detected gives rise to some nuclear reaction. The detection of some suitably chosen final product (a charged particle or a photon, γ) in that nuclear reaction constitutes an indirect detection of the initial neutron. See Figure 1.1.

Recall that $^{A}_{Z}X$ denotes a nucleus of chemical species X, with atomic number Z and mass number A (i.e., with Z protons and $A - Z$ neutrons). For the sake of brevity, our notations will not distinguish between the atomic nucleus and the atom which contains it: either the former or the latter will be understood and, so, easily deduced from the context if required.

The variety of neutron detectors is quite wide and, so, we shall limit ourselves here to quote some of them.

$^{6}_{3}$Li (with abundance 7.4%) captures and absorbs thermal neutrons with high probability. Thus, reasonable amounts (say, widths of about several mm) of $^{6}_{3}$Li may be sufficient to provide protection from slow neutrons. On the other hand, natural lithium (say, the mixture of isotopes of this element naturally occurring and found) has an absorption probability for thermal neutrons about an order of magnitude smaller. Natural boron (B) and, in particular, cadmium (Cd) absorb thermal neutrons strongly.

Three different (standard) detection procedures are based, respectively, upon the nuclear reactions $n + ^{10}_{5}B \rightarrow \alpha + ^{7}_{3}Li$ (in proportional counters, containing boron trifluoride), $n + ^{6}_{3}Li \rightarrow \alpha + ^{3}_{1}H$ (in spark counters) and $n + ^{3}_{2}He \rightarrow p + ^{3}_{1}H$ (neutrons bombarding helium-3, in proportional counters). $_1^3$H denotes tritium. The subsequent detections of the outgoing charged particles (the α particle or the proton, p) in the counters provide the indirect detection of the neutron which triggered the corresponding nuclear reaction.

Gas proportional detectors containing a suitable nuclide with high neutron cross section (like helium-3 or lithium-6 or boron-10 or uranium-235) allow to detect neutrons.

Another procedure, with important applications, is based on "activation" reactions induced by incoming beams of slow neutrons on suitable targets. Examples of such reactions are: $n + _{25}^{55}$ Mn (manganese) $\rightarrow \gamma + _{25}^{56}$ Mn, $n + _{79}^{197}$ Au (gold) $\rightarrow \gamma + _{79}^{198}$ Au, and $n + _{27}^{59}$ Co (cobalt) $\rightarrow \gamma + _{27}^{60}$ Co. We remind that $_{25}^{55}$Mn, $_{79}^{197}$ Au, and $_{27}^{59}$Co in the targets are the isotopes found in nature, with abundances about 100% in each case. Each of those reactions gives rise to isotopes ($_{25}^{56}$Mn, $_{79}^{198}$ Au, and $_{27}^{60}$Co, in the above examples) which are radioactive, with lifetimes ranging from minutes to days. These isotopes decay, returning to their ground states (or to less excited ones), and, in so doing, they emit various radiations (electrons, photons, etc.) which are detected. The activity $\tilde{A}$ (i.e., the number of those products emitted by the radioactive isotopes in the targets and counted by suitable detectors, per unit time) is measured. A useful approximate formula for $\tilde{A}$ (in counts per second) is [9]

$$\tilde{A} \simeq \frac{m\sigma F_0}{A}[1 - \exp(-\lambda t)]. \tag{1.1}$$

Here, F_0 is the flux of incoming slow neutrons (measured in neutrons/(cm$^2 \cdot$ s)). See Subsections 1.4.1 and 1.4.2 for discussions and interpretations of neutron fluxes. m and A are the mass (in grams) and the mass number of the element (isotopes found in nature) subject to neutron bombardment, respectively. σ is the cross section for slow neutron capture by that element. λ is the decay constant for the decay of the isotope formed by the neutron capture and t is the duration of the neutron bombardment.

In a typical application of "activation" reactions to neutron detection, such reactions allow for measuring, specifically, the flux (F_0) of the incoming neutron beam, without introducing significant perturbations in the latter. Thus, a thin target of a suitable material is exposed to the neutron beam to be measured. After some reasonable time, the irradiated thin target is removed from the beam and the γ radioactivity of the corresponding radioactive isotopes induced by the captured neutrons is analyzed subsequently in suitable counters. Specifically, the activity $\tilde{A}$ is measured. The value of F_0 follows from Eq. (1.1), if the remaining quantities in it are known. The technique is applied typically to beams of thermal and epithermal neutrons. A wide variety of materials have been employed for the thin target: in particular, $_{25}^{55}$Mn, $_{79}^{197}$Au, $_{27}^{59}$Co, $_{48}^{113}$Cd (cadmium), etc.

Certain selected elements provide further options for neutron detection through activation. For instance (in shorted notations for elements and reactions): iron (^{56}Fe$(n, p)^{56}$Mn), aluminum (^{27}Al$(n, \alpha)^{24}$Na), silicon (^{28}Si$(n, p)^{28}$Al). Those elements are suitable because they have extremely large cross sections for the capture of neutrons within a very narrow energy band.

These "activation" detections are related to the so-called neutron activation analysis to be discussed in Subsection 1.2.6, which has important applications.

1.2.4 ARTIFICIAL SOURCES FOR NEUTRONS

We shall provide a short survey of various small, medium, and large sources of slow neutron beams [2, 9].

- A source of α particles (say, the fully ionized nucleus $^4_2$He, helium), produced, for instance, in α decay of another nuclei, and a beryllium target, through the nuclear reaction $\alpha + ^9_4$Be $\to n + ^{12}_6$C, give rise to neutrons with E_K about 5 MeV. $^{12}_6$C denotes a carbon nucleus. This reaction, which gives rise to a neutron source at a laboratory scale, was precisely the one which led to the discovery of the neutron [1]. A typical α-beryllium neutron source may give rise, approximately, to 30 neutrons for every one million incoming α particles. In turn, the corresponding emission rates range from 10^6 to 10^8 neutrons per second. See Figure 1.1 [1].
- Photons (γ), previously emitted in the decay of some adequate radioisotope (like ^{124}Sb, antimony), give rise to photo-nuclear reactions like, for instance, $\gamma + ^9_4$Be $\to n + ^8_4$Be. The neutrons so produced have E_K smaller than 1 MeV.
- Californium $\left(^{252}_{98}\text{Cf}\right)$ is a source of neutrons. $^{252}_{98}$Cf suffers spontaneous nuclear fission, which gives rise to fast neutrons. The half-life of $^{252}_{98}$Cf is 2.6 years; his implies that, as a neutron source emitting in the beginning 10^7 to 10^9 neutrons per second, it may require a relatively frequent replacement. We shall comment later in Subsection 1.2.6 on one recent application of this type of source.

Just to confirm the ubiquitous interest of neutron sources, we quote shortly the following new project for a source of high-energy neutrons (to be constructed in the near future). The projected very large fusion reactor after International Thermonuclear Experimental Reactor (ITER) (namely, the future Demonstration Reactor or DEMO, which should produce electric energy) will require that its first inner wall be able to support very intense fluxes of high-energy neutrons. Materials for that first inner wall have to be tested, validated, and qualified. The latter processes, in turn, have motivated the project of a unique neutron source with energy spectrum and flux tuned to test possible materials for the first wall when that fusion reactor will be operational. This project, at the European level, of a neutron source is called DONES (DEMO Oriented Neutron Source). DONES is based on a deuteron (d) accelerator (40 MeV, 125 mA in continuous wave (CW) mode, with 5 MW beam average power). Deuterons hit, with a rectangular beam size (approximately 20 cm × 5 cm), a liquid Li (lithium) screen target flowing at 15 m/s to dissipate the beam power and generating a flux of neutrons of $10^{18}\,\mathrm{m^{-2}s^{-1}}$ with a broad peak at 14 MeV through stripping nuclear reactions, thereby reproducing the expected conditions of fusion power plants. Materials will be irradiated by the neutron beam as close as possible to the Li target to obtain damage rates. See: https://www.euro-fusion.org/programme/demo/

There are three very important artificial sources of neutrons to be treated in the following three subsections. Due to their relevance, they will also be considered in Chapter 3 (briefly), Chapter 4, and Chapter 6 (in connection with neutron capture therapy). Then, we shall limit here to some limited (and, eventually, complementary) discussions regarding them.

1.2.4.1   Nuclear Reactors

A nuclear reactor contains in its inner part (the "core") fissile material ("fuel") and other elements, to be commented as we proceed. Typical fuels are natural or enriched uranium U (containing, respectively, 0.7% of $^{235}_{92}$U or more), or $^{233}_{92}$U or $^{239}_{94}$Pu (plutonium). Neutrons propagating through the fuel are captured by the specifically fissile atomic nuclei $^{235}_{92}$U (or $^{239}_{94}$Pu or $^{233}_{92}$U), each of which suffers nuclear fission, breaks into two smaller nuclei and emits fast neutrons (having a few megaelectronvolts). The neutrons produced after a fission can give rise to new fission processes with other fissile atomic nuclei in the fuel, and so on. In many reactors, the fast neutrons after a fission require, first, to be slowed down ("moderated") from megaelectron-volt energies down to the thermal energy range, in order to produce new fission processes effectively. The process of moderation of fast neutrons (see Subsection 1.2.5) is performed by other material (the moderator), also located in the core of the reactor. In certain reactors ("fast neutron reactors") the fast neutrons, just produced after a fission and without need of moderation, give rise directly to the subsequent nuclear fission processes. For our purposes here, the most interesting feature is that beams of slow or thermal neutrons (after having suffered moderation) can be extracted from the core of the reactor, be subject to adequate physical manipulations (collimation, monochromatization), commented later, and employed at a later stage for various purposes.

Small nuclear reactors (with power a few megawatt) can give rise to stationary or constant (time-independent) neutron fluxes about 10^9 epithermal neutrons/(cm$^2 \cdot$ s).

Large nuclear reactors produce neutron beams with high constant flux. For instance, the nuclear reactor (with power about 57 MW) at the Institut Laue Langevin (Grenoble, France) produced stationary fluxes about 10^{15} neutrons/(cm$^2 \cdot$ s).

New programs and budgets for building nuclear reactors for power plants (specifically, for large-scale energy production) were strongly reduced or even eliminated in the last decades of the 20th century, in various advanced countries. Those nuclear reactors were perceived then by society as sources of pollution and potential dangers ("nuclear accidents"), so that their operations should be closed down or, at least, not renewed. In recent times, other sources of energy production at large scale have been shown either to give rise to other very important problems for society and/or to suffer from limited availability. Then, it appears that society is accepting, not without reluctance but unavoidably and progressively, the necessity of allowing and employing several different sources of large-scale energy production and, in particular, of not excluding power plants nuclear reactors but of allowing them, up to certain percentage. This change toward a nonnegative perception has also been facilitated by considerable advances and improvements, in recent decades, in performance, security, treatment of nuclear residues,..., of power plants nuclear reactors (through the development of new generations thereof). Thus, in the 21st century, one witnesses the rebirth of new budgets and programs for building nuclear reactors for power plants.

All in all, there are about 450 nuclear power reactors operating in 30 countries in the world in 2019. Moreover, in 2019 there are about 50 nuclear power reactors being constructed in various countries. A partial list of the latter, including the corresponding number of nuclear reactors under construction, is: China (11), India (7), Russia (6), South Korea (5), United Arab Emirates (4), Bangladesh, Belarus, Japan, Pakistan, Slovakia, Ukraine, and the United States (2 in each of them), and other countries (one in each of them). Consequently, one foresees an improved acceptance, within reservations, of nuclear reactors as sources of neutrons.

Nuclear reactors as sources of neutrons for various applications (therapies, condensed matter research, etc.), unrelated to energy production, should not be confused with power plants nuclear reactors, which give rise to much larger amounts of nuclear residues. Then, nuclear reactors for applications are expected to have, anyway, favorable perspectives. As an example of research reactor, with typical thermal neutron fluxes 10^9 neutrons/(cm$^2 \cdot$ s), see Ref. [11].

1.2.4.2 Small and Medium Accelerators

Small accelerators also play an important role (even if neutrons cannot be directly accelerated by externally applied electric fields). Thus, suitable nuclear reactions produced by incoming charged particles, that have been accelerated previously in small accelerators, constitute other sources of neutrons. An example of those reactions is $p +^7_3$ Li (lithium) $\rightarrow n +^7_4$ Be (with protons that have acquired energies about 2 MeV, or a bit higher, in a small or medium accelerator). See Chapters 3 and 6 in this book.

1.2.4.3 Spallation Sources

Let us consider a suitable charged particle beam (protons, α-particles, deuterons) with energies about 100 MeV or higher (coming out from some accelerator). Actually, such beams neither have constant intensities nor are stationary but, rather, constitute successive pulses. If those beams (or, better, pulsed beams) collide with a target containing some suitable heavy metal, the resulting nuclear reactions (generically named as "spallation reaction") give rise to successive pulses or pulsed beams of neutrons, copiously. For instance, let the target contain uranium $^{238}_{92}$ U. Then, the resulting fission of the latter, originated by an incoming beam of protons with energies up to about 800 MeV, may generate about 30 neutrons per proton, on the average. A pulse of neutrons, produced in a spallation source, may contain 10^{16} neutrons/s and even more [2, 12].

A complementary short account of various spallation sources for producing neutron beams will be given in Chapter 3: see Figure 3.3 and Table 3.1. Here, we shall comment briefly on one of them, namely, the European Spallation Source under construction in Lund, Sweden (with user programs planned to start by 2023). It is based upon a linear accelerator (along a tunnel 537 m long), which (due to the use of new superconducting cavities) will be able to accelerate protons until they reach 96% of the velocity of light in vacuum. The accelerated protons will impact a tungsten target (with a weight about 5 tons), cooled with helium gas. It has been estimated that the impacts of the protons with the target will give rise to neutron fluxes about 30 times higher than those generated in more recent previous spallation sources.

One of the pulses generated in a large spallation source may contain, in suitable cases and during its duration, more than $10^{17} n / (\text{cm}^2\text{s})$.

1.2.5 MODERATED, COLLIMATED AND QUASI-MONOCHROMATIC SLOW NEUTRON BEAMS

We summarize here some useful facts and items related to neutron physics [2, 7, 9].

Natural beryllium (Be) and carbon (C) have a very small probability to capture and absorb thermal neutrons. The same is true for deuterium (containing the deuteron nuclei $^2_1 H$) and for natural bismuth (Bi).

Graphite (say, C) and Bi absorb photons with large probability and, hence, contribute to eliminate γ's produced when neutron beams propagate through matter and interact with atomic nuclei in the latter.

Neutron beams coming from nuclear reactors also include, typically, fast neutrons and hard photons. Achieving an adequate degree of suppression of both fast neutrons and hard photons in the beams turns out to be very desirable, in general. The hard photon content can be reduced by employing adequate devices (filters) containing lead (Pb) or Bi. On the other hand, Bi also reduces fast neutron contaminations. Pb and Bi are relatively transparent to thermal neutrons. Outside of areas in which neutron beams are propagating, high-density concrete (say, mixed with minerals containing iron) can be employed in order to reduce hard photons.

Moderation: The atomic nuclei of certain materials (with low atomic mass) have small probability to capture and absorb slow neutrons. Then, neutrons propagating in those materials loose energy and are slowed down ("moderated"), through successive collisions with their atomic nuclei, but they are not absorbed by the latter. In particular, hydrogen is a good moderator and, hence, so are suitable (nonabsorbing) materials containing an adequate proportion of the former. Graphite, heavy water (containing deuterium), Be, and paraffin (due to its hydrogen content) are widely used moderators. Other possibilities include Al (aluminum), Al_2O_3, AlF_3, Teflon™, etc. Typically, moderators are devices located inside nuclear reactors.

The preparation of well-collimated and quasi-monochromatic slow neutron beams is a very important task, to be considered in further chapters, in particular in Chapter 6. We shall limit here to some introductory comments.

Collimation: It is very important to dispose of neutron beams formed by neutrons with momenta having approximately similar directions (although still with rather different absolute values) or, at least, with velocities not diverging much (in a transverse plane, orthogonal to some average direction of propagation). The devices producing beams in which the directions of motion of all neutrons are more or less analogous (to within some solid angle) are named collimators. Holes in the shields of nuclear reactors, walls of steel coated with Cd (which absorbs neutrons) constitute approximate or partial collimators. Those located inside the shielding reflect neutrons back into the beam. Collimators placed near the beam exit are beam delimiters, so that they should absorb rather than reflect neutrons. The magnitudes of the velocities of all neutrons in those (partially) collimated beams sweep, typically, an interval (or "spectrum"): they are said to be non-monochromatic.

Velocity selectors: In a first overall description, velocity selectors are machines that enable to transmit neutrons having velocities belonging to a narrow interval. In short, they are (mechanical) devices which, upon receiving a beam of neutrons, reject all those having velocities lying outside some selected range: the first device was inspired on Fizeau's apparatus for measuring the velocity of light. They produce approximately monochromatic beams of neutrons. In practice, they are useful only for relatively slow neutrons.

Choppers: A neutron chopper is a rotating disk or wheel, having one hole or aperture (or more). Let us consider a beam of neutrons (originating from some source) having different velocities, which arrive at the chopper (having traveled approximately the same distance) at different times. The neutrons which reach the chopper when the hole or aperture is in a position favorable (say, open) will pass though the latter: This occurs only for neutrons with certain velocities. Other neutrons, with different velocities will be absorbed by the chopper.

Monochromatization: It is very important that an (at least partially) collimated beam be formed by neutrons also with similar velocities in absolute value (say, be almost "monochromatic"). Monochromators are devices that give rise to neutron beams with all momenta in a narrow interval (in direction and magnitude), out of incoming partially collimated non-monochromatic beams. Single crystals (for instance, calcium fluoride) constitute monochromators: incoming partially collimated non-monochromatic neutron beams suffer Bragg reflections in suitably chosen crystal planes, with typical lattice spacing (distance between neighboring atoms) equal to d (a few Angstroms). We recall that the reflected neutrons with appreciable probability fulfill Bragg's law (Eq. (1.94)), θ being half of the angle between the momenta of the incoming neutron and the outgoing reflected one, λ_{dB} is de Broglie's wavelength and $n = 1, 2, \ldots$. For given d and small values of n, Bragg's law selects values of θ and λ_{dB} and, so, outgoing subbeams of neutrons, which are better collimated and are monochromatic approximately. See Subsection 1.7.3.1 for further complementary analysis.

1.2.6 SOME MISCELLANEOUS ASPECTS OF NEUTRON PHYSICS

Converters: Let us consider an initial beam, formed by neutrons having energies mostly concentrated in a certain interval. In a broad sense, a converter is a device that creates (or regenerates), out of that initial beam, another neutron beam with energy spectrum concentrated in another range, for various specific purposes (say, to improve the detection efficiency of the resulting beam, to increase the probability that the latter beam gives rise to some subsequent nuclear reaction, etc.). See Figure 1.1. A typical (fission) converter is a row of fuel elements located in the beam line, but adequately away from the reactor core. We shall limit ourselves to illustrate that concept, by describing succinctly a fission converter put into operation at the Massachusetts Institute of Technology (Boston, MA) some years ago [13]. This fission converter contains fissionable material (fuel), moderators, and filters (aluminum, Teflon, and cadmium, in this case), a suitable shield (lead) in order to absorb photons and a large collimator (with walls made up also by lead). In this converter, a (large area) thermal neutron beam, coming from the nuclear reactor at the institution, impinges upon the fissionable material. The resulting fission processes give rise to production of neutrons of higher energies, which, after moderation and filtering, generate a new beam of epithermal neutrons (about 10^{10} neutrons/(cm$^2 \cdot$ s)), with high purity. In particular, the latter epithermal neutron beam has been designed for advanced research on neutron capture therapy.

Neutron activation analysis: This is an important nondestructive technique for the determination of the composition of small samples of certain materials without producing appreciable damage in them, which has become widely used. Specifically, it is based upon the capture of slow neutrons in the nuclear reaction $n + B \rightarrow C + \gamma$. B denotes a stable nucleus (contained in the small sample), which will become radio-active by virtue of the irradiation with slow neutrons, and C is the radioactive isotope. The slow neutrons belong to an incoming beam generated in, say, a nuclear reactor, and the incoming flux F_0 is supposed to be known with adequate accuracy (contrary to what happened in the "activation" detection, considered in Subsection 1.2.3). After removing the irradiated sample from the neutron beam (say, extracting the sample from the reactor, typically), the radioactivity of C is measured. Typically, the radioactive nucleus C emits first one β -decay electron and, subsequently, one or several photons until it gives rise to some stable nucleus. The specific energies of the photons so emitted characterize the decay. Then, the precise measurement of the energies of those photons allows to determine the radioactive isotope C and, hence, its parent B. Moreover, the measurement of the activity $\tilde{A}$ enables to estimate, using Eq. (1.1), the amount m of the element B which was present in the small sample. In fact, $\tilde{A}$, the knowledge of F_0, t, and of other quantities in Eq. (1.1) yield m. There are other interesting and useful applications (forensic science, archaeological researches, etc.) of neutron activation analysis [9].

Some further applications: The number and variety of possible applications of thermal neutron beams have extended and grown along the years, and they continue to do so. We shall limit ourselves to quote the following ones, as examples.

- Their possible use in devices aimed to detect buried mines and hidden explosives (containing carbon, nitrogen, oxygen, and hydrogen) [14]. The detection device contains a small radioactive source (specifically, californium $^{252}_{98}$Cf), which produces fast neutrons, and a moderator, which thermalizes them. A beam of thermalized neutrons is sent from the device onto the inspected area in the soil. In particular, an anomalous concentration of nitrogen appears to be characteristic of most explosives. Thermal neutron are captured by nitrogen nuclei. Suitable detectors (large scintillation ones) above the inspected area allow to detect the characteristic gamma rays emitted after thermal neutron capture by nitrogen nuclei. This allows to investigate a possible anomalous concentration of chemicals containing nitrogen and, so, the possible existence of buried explosives. The overall size of a prototype [14] does not exceed some tens of centimeters.
- Neutrons, due to their much greater penetrating power and much less destructiveness than X-rays (Subsection 1.7.2), constitute invaluable tools to perform archaeology and cultural heritage studies. Neutron diffraction, small angle neutron scattering (SANS), neutron tomography, etc. have opened new perspectives

in archaeometry. The interiors (structures at microscopic scale, chemical composition, alterations, inclusions, structure bulks, etc.) of cultural heritage objects can be investigated nondestructively. Neutron activation and neutron capture analysis are also used routinely. Archaeological objects, the interiors beneath the surface of paintings, etc. can be studied. See, for instance Ref. [15].

1.3 DESCRIPTION AND QUANTUM MECHANICAL PROPERTIES OF SLOW NEUTRONS

1.3.1 SCHRÖDINGER EQUATION, COHERENT WAVE AND ITS PROPAGATION

Wave and optical phenomena for slow neutrons in three-dimensional space are based upon the following approximate Schrödinger equation ($\bar{x} = (x, y, z)$) [2, 5]:

$$\left[-\frac{\hbar^2}{2m_n} \Delta + V(\bar{x}) \right] \Psi(\bar{x};t) = i\hbar \frac{\partial \Psi(\bar{x};t)}{\partial t}, \tag{1.2}$$

where $\Psi(\bar{x};t)$ is the time (t)-dependent coherent wave function for the neutron, which describes how it propagates through either vacuum or some given material media. As usual, $\Delta = \partial^2 / \partial x^2 + \partial^2 / \partial y^2 + \partial^2 / \partial z^2$ denotes the three-dimensional Laplacian.

We recall that the range of the neutron-nucleus interaction is much smaller than (i) the separations among nuclei and (ii) the de Broglie wavelengths of slow neutrons. $V(\bar{x})$ is the approximate (optical) potential on the neutron, due to a given material medium, while $V(\bar{x}) \equiv 0$ in vacuum. The neutron is always subject to the strong interaction due to the atomic nuclei of the material medium through which the former propagates. Coulomb forces due to either electrons or atomic nuclei do not act on neutrons, as the latter are electrically neutral. The spin of the neutron also interacts with the spins of the individual nuclei. Moreover, in magnetic materials, a neutron can be subject to magnetic interactions of its magnetic moment with those of atoms (due to their electrons). There have been detailed studies about coherence and incoherence phenomena in the propagation of matter waves associated to slow neutrons through material media [6, 16–19, 21–23]. In this chapter, we shall concentrate on coherence phenomena and, mostly, on the strong neutron-nucleus interaction. Magnetic interactions due to the neutron spin will be considered shortly in Subsection 1.3.1.2 and, more specifically, in Section 1.6.

$V(\bar{x})$ provides an effective description of, at least, neutron optical phenomena arising from strong interactions with atomic nuclei. See Subsection 1.3.1.2. Since $V(\bar{x})$ is t-independent, one can simplify the quantum-mechanical description of the slow neutron by means of the t-independent or stationary wave function $\Phi = \Phi(\bar{x})$ with total energy E. Φ is an eigenfunction of the operator $\left[-\frac{\hbar^2}{2m_n} \Delta + V(\bar{x}) \right]$ with eigenvalue E. One has

$$\Psi(\bar{x};t) = \Phi(\bar{x}).\exp(-iEt / \hbar), \tag{1.3}$$

$$\left[-\frac{\hbar^2}{2m_n} \Delta + V(\bar{x}) \right] \Phi(\bar{x}) = E\Phi(\bar{x}). \tag{1.4}$$

It is very important to state that neither $\Psi(\overline{x};t)$ nor $V(\overline{x})$ nor $\Phi(\overline{x})$ depend on any specific configuration of the positions of the atoms (hence, their nuclei) forming the material medium. In other words, a suitable averaging has been carried out over all configurations of the atoms in the medium, in order to arrive at the physical description embodied in Eqs. (1.2) and (1.4). Moreover, such an averaging does not destroy wavelike properties, but allows for interference and diffraction phenomena. That is, after having performed that averaging the neutron continues to be represented by wave functions, namely, $\Psi(\overline{x};t)$ and $\Phi(\overline{x})$, which justify to refer to the latter as "coherent." Equations (1.2) and (1.4), $\Psi(\overline{x};t)$ and $\Phi(\overline{x})$ can be regarded as macroscopic equations describing (coherent) elastic scattering of the slow neutron involving no change in the microscopic state of the material medium (which behaves as a macroscopic one). $\Phi(\overline{x})$ and $V(\overline{x})$ are macroscopic quantities depending on the thermodynamic variables of the material medium (like the number of nuclei per unit volume, as explained below). We shall discuss these aspects in Subsection 1.3.1.2.

As stated in Subsection 1.2.1, the neutron spin is $\hbar/2$, so that $\Psi = \Psi(\overline{x},t)$ has two components, $\Psi(\overline{x},t)_+$ and $\Psi(\overline{x},t)_-$, which correspond, respectively, to the two spin states $|\alpha>$ with projections $\alpha = +\hbar/2$ and $\alpha = -\hbar/2$ of the neutron spin along the positive z-axis (see Section 1.6). We shall remind briefly the following general property of the two-component neutron wave function. Let us consider an axis and a complete rotation, by an angle 2π, about that axis. Then, under such rotation, the transformed wave function Ψ' equals the former one, except for an overall change of sign, that is, $\Psi(\overline{x},t) \rightarrow \Psi(\overline{x},t)' = -\Psi(\overline{x},t)$ (see Section 1.9).

Under the approximation that $V(\overline{x})$ does not influence the neutron spin, we shall treat the slow neutron as if it were spinless. Then, it suffices to regard $\Psi(\overline{x};t)$ as just one (complex) function and a unique wave function suffices, as $\Psi(\overline{x};t) = \Psi(\overline{x},t)_+ = \Psi(\overline{x},t)_-$. In Section 1.6, we shall deal with magnetic interactions, which give rise to more general $V(\overline{x})$'s influencing the spin degree of freedom. Then, we shall deal with two (in general, different) components: $\Psi(\overline{x},t)_+$ and $\Psi(\overline{x},t)_-$.

Let the material medium appear (to slow neutrons), approximately, as homogeneous and isotropic, for instance, a gas, a liquid, an amorphous solid at length scales such that its discretized atomic structure can be disregarded. Then, for slow neutrons, we can approximate $V(\overline{x})$ by the following [2, 5, 7, 8]:

$$V(\overline{x}) = \frac{2\pi\hbar^2\rho b}{m_n},\qquad(1.5)$$

where ρ is the number of nuclei per unit volume (say, per cm^3) in the homogeneous material medium. b now stands for the average (over nuclear spins and over isotopes) coherent amplitude for the low-energy purely nuclear scattering of a neutron by the nucleus of an atom in the medium. The nucleus, being more massive than the neutron, is regarded to be at rest and its recoil (due to the interaction with the neutron) is neglected, in such an approximation. Equation (1.5) is also known as the Fermi pseudopotential [25]. Equation (1.5) also holds for solid materials, upon a suitable averaging allowing to regard them, approximately, as homogeneous and isotropic.

Throughout much of this chapter, we shall work with Eq. (1.5) (and, hence, with b) which will suffice for our purposes. Equation (1.5) allows for the possibility that the neutron could be absorbed by atomic nuclei. If that absorption can be neglected, then b is real ($b = Re(b)$, Re denoting the real part). However, if nuclear absorption is not negligible, then b is complex: $b = Re(b) + iIm(b)$, with imaginary part $Im(b) < 0$. b has been experimentally measured for many naturally occurring elements in the periodic table and for many isotopes as well. Various compilations of data for b exist [2, 7, 24]. The order of magnitude of the real part $Re(b)$ is about 10^{-13} cm [7]. Then, $|Re(b)|$ is much smaller than both the separations among nuclei and slow neutrons de Broglie wavelengths. The sign of $Re(b)$ is positive for almost all nuclei, except for a few ones: $Re(b) < 0$ for the naturally occurring H, Li, Ti, V, and Mn. On the other hand, $Re(b) < 0$ holds individually for several isotopes of certain nuclei (say, without averaging over isotopes), for instance, for $^{62}_{28}$Ni and $^{64}_{28}$Ni. However, upon averaging over isotopes, one has $Re(b) > 0$ for natural Ni [2, 7]. One also has $Re(b) < 0$ for $^{152}_{62}$Sm (samarium). In other cases, $Re(b)$ has opposite signs for different isotopes, but the contribution of one isotope having $Re(b) < 0$ dominates in the (isotope averaging yielding the) naturally occurring element. This is the case for Li and Ti, where both $^{7}_{3}$Li (abundance 92.6%) and $^{48}_{22}$Ti (abundance 73.8%) have $Re(b) < 0$. On the other hand, the isotope $^{55}_{25}$Mn has an abundance about 100%, so that it coincides, in practice, with natural Mn. See Table 1.1 (for thermal neutrons) and a complementary discussion in Chapter 3.

For typical solid materials (like titanium, carbon, etc.), regarded as a first approximation as homogeneous after a suitable spatial averaging, ρ is about 10^{22} nuclei per cm^3. Then, the order of magnitude of the right-hand side of Eq. (1.5), for typical material media of physical interest, is about 10^{-8} to 10^{-7} eV. For a dilute gas, ρ is about two or three orders of magnitude smaller.

Another very important case occurs for thermal neutrons traveling through a crystal, in which the discretized (approximately, periodic) atomic structure has to be taken into account. Then, Eq. (1.5) does not hold as an approximation and $V(\overline{x})$ will have the same periodicity as the crystal and neutron diffraction by the latter will occur. See Subsections 1.3.1.2, 1.7.2, and 1.7.3.

The approximate Eq. (1.5) omits any sort of spin-dependent (or magnetic) interaction of the neutron. Upon employing Eqs. (1.5) and (1.2), both components of $\Psi(\overline{x};t)$ (corresponding to the two possible and opposite projections of the neutron spin on the z-axis) have the same behavior. That is, $\Psi(\overline{x};t)$ will be regarded as a one-component wave function (thereby, omitting the neutron spin).

1.3.1.1 Cross Sections by Individual Atomic Nuclei

The scattering and the absorption of a slow neutron by an individual atomic nucleus in the medium turn out to be relevant processes. Then, we need to consider σ_s and σ_{abs}, which are the cross sections (integrated over all angles) for those processes, respectively.

We shall deal with slow neutron propagation through a chemically pure material medium, in which the nuclei are fixed in space (as a first approximation) at various positions $\overline{x}_l$, as l varies. That chemical purity still allows for different isotopes of the same element to be present at various locations. One can describe approximately the low-energy interaction of the neutron with the nucleus at $\overline{x}_l$ through the amplitude [2, 5]:

$$b_{0,l} + c_l \overline{s} \overline{I}_l,$$ (1.6)

TABLE 1.1

Some Typical Useful Data

Nucleus	$b(10^{-12}\,\text{cm})$	$\sigma_s(10^{-24}\,\text{cm}^2)$	$\sigma_{abs}(10^{-24}\,\text{cm}^2)$
$^{1}_{1}\text{H}$	−0.378	81.5	0.19
$^{2}_{1}\text{H}$	0.65	7.6	0.0005
$^{7}_{3}\text{Li}$	−0.22	1.4	—
$^{6}_{3}\text{Li}$	0.2	—	570
Li	−0.18	1.2	40
$^{9}_{4}\text{Be}$	0.774	7.54	0.005
B	—	4.4	430
$^{12}_{6}\text{C}$	0.661	5.51	0.003
$^{14}_{7}\text{N}$	0.940	11.4	1.1
$^{16}_{8}\text{O}$	0.577	4.24	0.0001
$^{23}_{11}\text{Na}$	0.351	3.4	0.28
$^{27}_{13}\text{Al}$	0.35	1.5	0.13
Si	0.42	2.2	0.06
K	0.35	2.2	1.2
$^{48}_{22}\text{Ti}$	−0.58	—	—
Ti	−0.34	4.4	3.5
V	−0.05	5.1	2.8
Mn	−0.37	2.0	7.6
Fe	0.96	11.8	1.4
$^{62}_{28}\text{Ni}$	−0.87	—	—
Ni	1.03	18.0	2.7
Cd	0.36	—	2650
In	$0.38 + i0.12$	—	115
Sm	—	—	11700
Gd	—	—	19200
Pb	0.96	11.4	0.1
$^{238}_{92}\text{U}$	0.85	—	2.1

where $\bar{s}$ is the spin operator for the neutron; see Section 1.6. $\bar{I}_l$ and I_l are the spin operator and the spin value, respectively, for the atomic nucleus fixed at $\bar{x}_l$. Now, $b_{0,l}$ and c_l are suitable parameters describing the low-energy scattering of a neutron by the atomic nucleus at $\bar{x}_l$, due to strong interactions and to the coupling of their spins.

We allow for the possibility that different isotopes of the same stable nucleus be located at different positions $\bar{x}_l$. At the latter, there may be L nuclear isotopes, each with its own spin I_l and parameters b_l and c_l, with fractions f_l, fulfilling $f_l > 0$, $\Sigma_l f_l = 1$. One may interpret σ_s as the following average:

$$\sigma_s = 4\pi \sum_{l=1}^{L} f_l \left[\frac{I_l+1}{2I_l+1}\left|b_l + 2^{-1}c_l I_l\right|^2 + \frac{I_l}{2I_l+1}\left|b_l - 2^{-1}c_l(I_l+1)\right|^2 \right], \qquad (1.7)$$

suitable averages over nuclear spins and isotopes being performed. b in Eq. (1.5) can be regarded as a suitable average (over nuclear spins and over isotopes) over the quantities in Eq. (1.6). See Subsection 1.3.1.2.

More practically, one has the following approximations for thermal neutrons with energy E:

$$\sigma_s \simeq 4\pi|b|^2 \, , \sigma_{abs} \simeq \frac{4\pi|Imb|}{|\overline{k}|},$$ (1.8)

with $E = \left(\hbar^2|\overline{k}|^2\right)/2m_n$. The total cross section is $\sigma_{tot} = \sigma_s + \sigma_{abs}$. Table 1.1 (for thermal neutrons) summarizes some useful data [2, 7] for $b, \sigma_s, \sigma_{abs}$ for various isotopes (omitting, for simplicity, their atomic numbers and certain data in some cases) and natural elements. The order of magnitude of both σ_s and σ_{abs} is about $10^{-24}\,\mathrm{cm}^2$ (= 1 barn), with appreciable variations [2, 7].

For atomic number Z below 45, σ_s is usually (but not always) larger than σ_{abs}, while for $Z \geq 45$ the typical situation seems to be the opposite one [7]. As Table 1.1 displays, for certain very absorbing nuclei, like boron and cadmium, σ_{abs} is several orders of magnitude larger than σ_s.

The sum of the approximations for σ_s and σ_{abs} given in Eq. (1.8) could not be an adequate approximation for σ_{tot}. An approximate formula for σ_{tot}, free of the above objection, is [5]

$$\sigma_{tot} \simeq \frac{4\pi|Imb|}{|\overline{k}|} + 4\pi\left(|Re(b)|^2 - |Im(b)|^2\right),$$ (1.9)

with the same $|\overline{k}|$ as in Eq. (1.8). In turn, one has $\sigma_s \simeq \sigma_{coh} + \sigma_{inc}$, where σ_{coh} and σ_{inc} stand for the coherent and incoherent scattering cross sections, respectively, and the effects of diffuse scattering [5] have been disregarded, as a zeroth-order approximation. One has $\sigma_{coh} = 4\pi|b|^2$, b being the average coherent amplitude discussed above. σ_{inc} describes both spin and isotope incoherence. Accordingly, in many cases of physical interest, σ_{coh} is larger than σ_{inc}, but not always (as it happens for H, Li, Na, V, etc.) [2, 5].

1.3.1.2 Approximate Justification of the Slow Neutron Optical Potential $V(\overline{x})$: The Coherent Wave

We shall summarize here certain essential aspects of the important treatment in Ref. [17], with various simplifications. See Refs. [18, 19] for further, also important, improvements. Let us suppose that the medium is an extended (macroscopically large) distribution of N microscopic scatterers (atomic nuclei), in three-dimensional space at the positions $\overline{x}_j, j = 1,....,N$, N being macroscopically large. To fix the ideas, we consider first that the medium is a homogeneous isotropic random distribution of scatterers without any sort of correlations among them (say, a dilute gas).

Let us consider a specific configuration (sc) of the scatterers. Let an incoming slow neutron propagating freely (outside the medium) be represented by the plane wave $\Phi(\overline{x})_{in} = \exp[i\overline{k}_{in}\overline{x}]$ and enter into the medium. Then, it interacts successively with all scatterers and, so, experiences multiple scattering due to them. Let $\Phi(\overline{x})_{mi,sc}$ be the total wave function representing the slow neutron subject to the multiple scattering of the specific configuration sc of the microscopic (mi) scatterers inside the medium. Notice that the range of the neutron-nucleus interaction is much smaller than the internuclear separations.

It can be justified [17] that $\Phi(\overline{x})_{mi,sc}$, for the configuration considered, is given by the following approximate multiple scattering equations:

$$\Phi(\overline{x})_{mi,sc} \simeq \Phi(\overline{x})_{in} + \sum_{j=1}^{N} G_1(\overline{x} - \overline{x}_j) \frac{2\pi\hbar^2 b_j}{m_n} \Phi(\overline{x}_j)_{mi,sc,j}, \qquad (1.10)$$

$$\Phi(\overline{x}_j)_{mi,sc,j} \simeq \Phi(\overline{x}_j)_{in} + \sum_{l=1,l\neq j}^{N} G_1(\overline{x}_j - \overline{x}_l) \frac{2\pi\hbar^2 b_l}{m_n} \Phi(\overline{x}_l)_{mi,sc,l}, \qquad (1.11)$$

$$G_1(\overline{x} - \overline{x}') = -\frac{m_n \exp\left[i\left|\overline{k}_{in}\right|\left|\overline{x} - \overline{x}'\right|\right]}{2\pi\hbar^2 \left|\overline{x} - \overline{x}'\right|}. \qquad (1.12)$$

$G_1(\overline{x} - \overline{x}')$ is the standard t-independent three-dimensional Green's function in vacuum (see also Eq. (1.109), in Section 1.8) [20]. Equation (1.11) has a rather intuitive character. The crucial factor $\frac{2\pi\hbar^2 b_l}{m_n}$ represents the approximate effective amplitude for the scattering of a neutron by a nucleus at $\overline{x}_l$, at low energy. In so doing, we are treating only the purely strong interaction of the neutron with one nucleus. One may interpret $\Phi(\overline{x}_l)_{mi,sc,l}$ as the overall wave function at the l-th scatterer, describing the wave due to all multiple scattering processes. Then, the l-th scatterer produces a new scattered wave given by $\frac{2\pi\hbar^2 b_j}{m_n} \Phi(\overline{x})_{mi,sc,l}$. The latter scattered wave, being propagated by $G_1(\overline{x}_j - \overline{x}_l)$, is represented by the wave function $G_1(\overline{x}_j - \overline{x}_l) \frac{2\pi\hbar^2 b_l}{m_n} \Phi(\overline{x}_l)_{mi,sc,l}$, at the scatterer at $\overline{x}_j$ ($\neq \overline{x}_l$). The sum of all the latter wave functions, over all $l(l \neq j)$, plus $\Phi(\overline{x}_j)_{in}$ has to be equal to $\Phi(\overline{x}_j)_{mi,sc,j}$, at $\overline{x}_j$. This argument establishes Eq. (1.11). Through a similar argument for an arbitrary $\overline{x}$, the sum of all the propagated wave functions $\sum_{j=1}^{N} G_1(\overline{x} - \overline{x}_j) \frac{2\pi\hbar^2 b_j}{m_n} \Phi(\overline{x}_j)_{mi,sc,j}$ plus $\Phi(\overline{x})_{in}$ has to be equal to $\Phi(\overline{x})_{mi,sc}$, which establishes Eq. (1.10). For convenience, Eq. (1.10) will be recast, equivalently, as

$$\Phi(\overline{x})_{mi,sc} \simeq \Phi(\overline{x})_{in} + \int d^3\overline{x}' G_1(\overline{x} - \overline{x}') \sum_{j=1}^{N} \delta^{(3)}(\overline{x}' - \overline{x}_j) \frac{2\pi\hbar^2 b_j}{m_n} \Phi(\overline{x}')_{mi,sc,j}, \qquad (1.13)$$

$\delta^{(3)}$ denoting the three-dimensional delta function. Notice that the integration in Eq. (1.13) is extended over the whole medium. We shall now perform an averaging (indicated through $< \ldots >$) over all configurations of the microscopic scatterers. The averaging is carried out by using some probability distribution for the latter, which needs not be specified. We introduce the new total wave function, upon averaging $\Phi(\overline{x})_{mi,sc}$ over all configurations sc:

$$\Phi(\overline{x}) \equiv \left\langle \Phi(\overline{x})_{mi,sc} \right\rangle. \qquad (1.14)$$

Accordingly, as $\Phi(\bar{x})$ is independent on any specific sc, it appears adequate not to append to it the subscripts mi,sc. One can argue [17] that, for large N and for an homogeneous isotropic random distribution of scatterers, the contribution of $\Phi(\bar{x}')_{mi,sc,j}$ upon taking such averaging, can be approximated just by replacing it by $\Phi(\bar{x}')$. Equation (1.11) appears to support physically such an approximation, since $\Phi(\bar{x}')_{mi,sc,j}$ would differ from $\Phi(\bar{x}')$ by corrections of order $1/N$. Then, Eq. (1.13) yields

$$\Phi(\bar{x}) \simeq \Phi(\bar{x})_{in} + \int d^3\bar{x}' G_1(\bar{x} - \bar{x}') \left\langle \sum_{j=1}^{N} \delta^{(3)}(\bar{x}' - \bar{x}_j) \frac{2\pi\hbar^2 b_j}{m_n} \right\rangle \Phi(\bar{x}'). \quad (1.15)$$

Trivially, $\Phi(\bar{x})_{in}$ is unaffected by the averaging. For the actual homogeneous isotropic random distribution of scatterers, one can write for any $\bar{x}'$ in the medium:

$$\left\langle \sum_{j=1}^{N} \delta^{(3)}(\bar{x}' - \bar{x}_j) \right\rangle \simeq \rho, \quad (1.16)$$

$$\left\langle \sum_{j=1}^{N} \delta^{(3)}(\bar{x}' - \bar{x}_j) \frac{2\pi\hbar^2 b_j}{m_n} \right\rangle \simeq \frac{2\pi\hbar^2 b}{m_n} \rho. \quad (1.17)$$

In fact, $\delta^{(3)}(\bar{x}' - \bar{x}_j)$ in Eq. (1.16) can be interpreted as the inverse of the overall volume of the medium only in a microscopically small region about $\bar{x}' - \bar{x}_j$, while it vanishes outside that tiny domain. Then, Eq. (1.16) equals, approximately, N divided by that overall volume, that is, the average number of nuclei per unit volume (ρ). Accordingly, Eq. (1.17) enables to interpret b as an average neutron-nucleus scattering length. Then, Eq. (1.15) becomes

$$\Phi(\bar{x}) \simeq \Phi(\bar{x})_{in} + \int d^3\bar{x}' G_1(\bar{x} - \bar{x}') \frac{2\pi\hbar^2 b}{m_n} \rho\Phi(\bar{x}'). \quad (1.18)$$

Equations (1.16) and (1.17) hold for any $\bar{x}'$ in the medium. Then, $\frac{2\pi\hbar^2 b}{m_n}\rho \equiv V(\bar{x})$ can be regarded as an effective potential acting over the whole medium and approximately constant in it. Thus, we have justified Eq. (1.5). Accordingly, the integration in Eq. (1.18) is extended over the whole medium. It is more convenient to transform the latter into an effective Schrödinger equation. Upon applying $\left[-\frac{\hbar^2}{2m_n}\Delta - E \right] G_1(\bar{x} - \bar{x}') = \delta^{(3)}(\bar{x} - \bar{x}')$, with $E = \hbar^2 \bar{k}_{in}^2 / (2m_n)$, Eq. (1.18) becomes Eq. (1.4).

We shall now turn to the important class of cases where there are spatial correlations among the positions of the microscopic scatterers. As the spatial positions of the microscopic scatterers are not strictly fixed, certain distributions of positions of the N scatterers occur (with a probability distribution certainly different from that for a dilute gas). The physical positions of the scatterers have now some small fluctuations about some average locations. Equations (1.10) through (1.13), the above averaging over the distribution of N microscopic scatterers and Eq. (1.14) continue to hold. At this point, it is convenient to define also the so-called local field $\eta(\bar{x}')$ through

$$\eta(\bar{x}') = \frac{\left\langle \sum_{j=1}^{N} \delta^{(3)}(\bar{x}' - \bar{x}_j) \frac{2\pi\hbar^2 b_j}{m_n} \Phi(\bar{x}')_{mi,sc,j} \right\rangle}{\left\langle \sum_{j=1}^{N} \delta^{(3)}(\bar{x}' - \bar{x}_j) \frac{2\pi\hbar^2 b_j}{m_n} \right\rangle}. \tag{1.19}$$

Due to the spatial correlations among the positions of the microscopic scatterers, to approximate $\eta(\bar{x}')$ just by $\Phi(\bar{x})$ (which would have led to Eq. (1.15)) is not valid in principle. A better approximation [19] is

$$\eta(\bar{x}') \simeq c_1 \Phi(\bar{x}'), \tag{1.20}$$

with some factor c_1, depending on the structure and state of the set of scatterers. Further investigations, a posteriori, have indicated that, in all cases and conditions of practical interest, c_1 can be approximated by unity [26, 27], which will be incorporated from now on in our analysis. Then, upon averaging, Eq. (1.13) becomes

$$\Phi(\bar{x}) \simeq \Phi(\bar{x})_{in} + \int d^3\bar{x}' G_1(\bar{x} - \bar{x}') \left\langle \sum_{j=1}^{N} \delta^{(3)}(\bar{x}' - \bar{x}_j) \frac{2\pi\hbar^2 b_j}{m_n} \right\rangle \Phi(\bar{x}'). \tag{1.21}$$

A distinguishing property of the actual case with spatial correlations is that, due to the latter, Eqs. (1.16) and (1.17) no longer apply. Actually, in Eq. (1.21), we should regard

$$\left\langle \sum_{j=1}^{N} \delta^{(3)}(\bar{x}' - \bar{x}_j) \frac{2\pi\hbar^2 b_j}{m_n} \right\rangle \equiv V(\bar{x}') \tag{1.22}$$

as the adequate optical potential.

We shall deal with slow neutron propagation through a chemically pure material medium, in which the nuclei are fixed in space (as a first approximation) at the position $\bar{x}_j$. Recall Subsection 1.3.1.1 and, in particular, Eq. (1.6). Then, one can approximate each effective amplitude b_j above by the new [2, 5]:

$$b'_j = b_{0,j} + c_j \bar{s}\bar{I}_j. \tag{1.23}$$

If the atomic nuclei located at different $\bar{x}_j$ are identical (say, they are not different isotopes) and if they have nonzero spin $I_j = I \neq 0$, then there are incoherence effects in the propagation and scattering of neutrons through the material medium (spin incoherence). Moreover, the possible occurrence of different isotopes at different positions $\bar{x}_j$ in the material medium may give rise to further incoherence effects (isotope incoherence). Specifically, let us consider a material medium which is formed by a mixture of L nuclear isotopes (each with its own b_l and c_l at the position $\bar{x}_j$) with fractions f_l ($f_l > 0$, $\Sigma_{l=1}^{L} f_l = 1$). In particular, at each $\bar{x}_j$ we now replace b_l by b'_l and the overall average coherent amplitude b in Eq. (1.5) turns out to be the corresponding average of $\Sigma_{l=1}^{L} f_l b'_l$ [2, 5, 7]. Thus, Eq. (1.5) for a homogeneous medium can be regarded as some sort of spin and spatial average associated to Eq. (1.6).

We shall now turn to one important class of cases, included in those described by Eq. (1.22), namely, that in which the spatial correlations among the positions of the microscopic scatterers correspond to an approximately ordered periodic crystal. That is, the latter has some approximate spatial periodicity, on average. The physical positions of the microscopic scatterers (atomic nuclei in the crystal) are not strictly fixed but have now some small fluctuations about some average locations, which define a mathematical fixed periodic structure. $V(\bar{x})$, now given by Eq. (1.22), should have the same periodicity as the above mathematical periodic structure defined by the average positions of the scatterers. An approximate form of such $V(\bar{x})$ in Eqs. (1.21) and (1.22), adequate for crystals, is

$$V(\bar{x}) = \frac{2\pi\hbar^2 b}{m_n} \sum_{\bar{a}} \delta^{(3)}(\bar{x} - \bar{a}), \qquad (1.24)$$

where $\Sigma_{\bar{a}}$ is extended over all locations $\bar{a}$ of the nuclei, the distribution of which describes the three-dimensional periodic lattice.

This $V(\bar{x})$ in Eq. (1.24) will be employed in Subsections 1.7.2 and 1.7.3. See Eq. (1.74). Upon applying $\left[-\frac{\hbar^2}{2m_n}\Delta - E\right]G_1(\bar{x} - \bar{x}') = \delta^{(3)}(\bar{x} - \bar{x}')$ with $(E = \hbar^2 \bar{k}_{in}^2 / (2m_n))$, Eq. (1.21) becomes Eq. (1.4) as well.

To summarize the above cases, it turns out that, after such an averaging process, wavelike properties continue to exist, that is, $\Phi(\bar{x})$ gives rise to interference (in particular, between the incoming wave and scattered ones) and diffraction phenomena at a macroscopic scale. Accordingly, $\Phi(\bar{x})$ deserves the name "coherent wave" and accounts for phenomena in neutron optics. Certainly, other processes besides elastic scattering of the slow neutron may occur. Other scattering processes are referred to as diffuse scattering. For instance, the neutron could be absorbed by nuclei, in certain cases. Those processes (besides elastic scattering) could give rise to an attenuation of the coherent wave $\Phi(\bar{x})$ in the material medium; then, $V(\bar{x})$ can be expected to be complex, with negative imaginary part.

1.3.2 Linear Coefficient for Slow Neutron Absorption

The linear coefficient μ for slow neutrons in the medium is defined as $\mu \equiv \rho\sigma_{tot}$ [7]. μ characterizes the attenuation of a neutron beam propagating in a medium, due to the absorption of neutrons in the beam by the atomic nuclei in the medium. Let us consider a slab of thickness z_0, made up of certain homogeneous material with linear coefficient μ, and some neutron beam which has approached that slab from outside (along a direction orthogonal to its parallel surfaces) and entered into it through the surface at $z = 0$. Let $F_0(z)$ be the flux corresponding to that neutron beam inside the slab, at a distance z from that entrance surface ($0 \leq z \leq z_0$): z does not represent vertical height here. Discussions and interpretations of neutron fluxes will be deferred to Section 1.4. Standard arguments yield directly: $F_0(z) = F_0(z = 0)\exp[-z\mu]$ (Lambert law) [5, 7], which displays the physical interest of μ. Typical values for μ in many cases are about $10^{-1} cm^{-1}$. μ is about $5 \times 10^{-4} cm^{-1}$ for Be, C and Bi and air, but about 26, 60, and 121 cm^{-1} for Li, B, and Cd, respectively. See Figure 3.1.

1.3.3   REFRACTIVE INDEX FOR SLOW NEUTRONS

In the geometrical-optics approximation, the index of refraction (or refractive index), n, for slow neutrons with total energy E moving through a medium (and, hence, under the action of the optical potential $V(\bar{x})$ due to the latter) is defined as [2, 8]

$$n^2 = 1 - \frac{V(\bar{x})}{E}. \tag{1.25}$$

In turn, by using Eq. (1.5)

$$n^2 = 1 - \frac{2\pi b \rho \hbar^2}{m_n E}. \tag{1.26}$$

Let a neutron propagate through vacuum or a homogeneous medium, with density ρ. The stationary Schrödinger Eq. (1.4) with the constant effective potential in Eq. (1.5) has plane wave solutions with wavevector $\bar{k}$:

$$\Phi(\bar{x}) = \exp\left[i\bar{k}\bar{x}\right], \quad E = \frac{\hbar^2 \bar{k}^2}{2m_n} + \frac{2\pi\hbar^2}{m_n} b\rho. \tag{1.27}$$

Since E is real, if $Imb < 0$, then at least one component of $\bar{k}$ has to be complex. This is required for the second equation in Eq. (1.27), yielding E, be consistent.

Notice that if b is complex, so is the index of refraction introduced in Eq. (1.26). If b is approximately real and positive, one has $1 > n^2$. In this case, a typical order of magnitude estimate for $1 - n^2$ is 10^{-5}.

1.4   PURE STATES AND STATISTICAL MIXTURES FOR SLOW NEUTRONS

1.4.1   PURE STATES AND NEUTRON INTERFEROMETRY EXPERIMENTS

Let us consider a system containing more than one neutron. As the neutron has half-integral spin, the neutrons contained in that system or beam behave as fermions (described by antisymmetric wave functions), in principle; that is, their wave functions should change sign under the simultaneous exchange of the coordinates and spin projections of any two neutrons.

We shall treat a slow neutron beam, which, in general, has an energy spectrum. In typical reactor beams of interest here, the average separations among the centers of the wave packets associated to different thermal neutrons are appreciably larger than their average de Broglie wavelengths; for a simple estimate, see Subsection 1.4.2.1. Then, it is reasonable to neglect overlaps, interferences, and interactions among different neutrons in the beam, as a first approximation.

Stated into slightly different terms, in those beams the overlaps among the wave packets associated to any two or more different neutrons are negligible (again, because the typical average separations between those wave packets are appreciably larger than the average de Broglie wavelengths). Consequently, one may safely disregard that neutrons are fermions, their wave functions need not be antisymmetric and they can be treated as independent particles.

Then, the neutrons, in those beams, are regarded as independent particles, each of which propagates through and interacts with various material media. Moreover, the different neutrons in the beam can be regarded to constitute, in practice, an ensemble of identical copies of the same quantum system (formed by just one neutron) "prepared" in the same quantum state [2]. It may be adequate to discuss further some conceptual aspects regarding them.

Let us consider, as a first (eventually, idealized) possibility, the case in which relative phases be not destroyed and incoherence effects be negligible for single neutrons, say, for single copies (even if there may an energy spectrum). Then, the quantum state describing the slow neutrons is represented by one time-dependent wave function $\Psi(\overline{x};t)$ fulfilling Eq. (1.2), also referred to as a pure state. $\Psi(\overline{x};t)$ is normalized for any t ($\int d^3\overline{x}|\Psi(\overline{x};t)|^2 = 1$, the integration being extended over the whole spatial region in which the neutron could be found), by assumption.

$\Psi(\overline{x};t)$ provides a statistical description of the ensemble of all N_n neutrons in the beam. In fact, all those neutrons, being independent from one another approximately, can be regarded to be in the same quantum state $\Psi(\overline{x};t)$ and, so, to behave as N_n identical copies of one and the same system. Thus, let us imagine that N_n identical idealized experiments (independent from one another) were carried out on that ensemble of N_n neutrons. In each experiment, a measurement is performed, aimed to detect whether there is one neutron inside a very small volume $d^3\overline{x}$ about $\overline{x}$, at time t. The detections are carried out by adequate independent counters (all being entirely analogous in all measurements). Let $N_n(\overline{x},t)$ be the total number of those counts (i.e., the total number of times in which one neutron is detected). Then, the statistical interpretation of quantum mechanics [28, 29] states the following. For large N_n, $N_n(\overline{x},t) / N_n$ (namely, the probability for one neutron to have been detected, at time t) approaches $|\Psi(\overline{x};t)|^2 d^3\overline{x}$. See Figure 1.2. The wave function $\Psi(\overline{x};t)$ represents that ensemble and encodes the largest amount of information which could ever be possible about such ensemble, and this appears to suffice for practical purposes.

The probability current density vector determined by $\Psi(\overline{x};t)$ is

$$\overline{J}(\overline{x},t) = \frac{\hbar}{m_n} Re\left[\Psi(\overline{x};t)^*(-i)\nabla\Psi(\overline{x};t)\right], \qquad (1.28)$$

where Re denotes the real part and $\nabla = (\partial/\partial x, \partial/\partial y, \partial/\partial z)$ is the gradient.

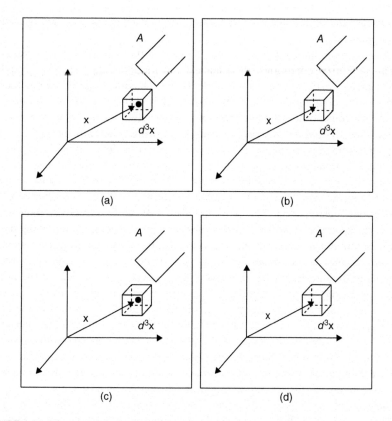

FIGURE 1.2 The system is formed by one single neutron (small black circle). $N_n = 4$ identical copies of the system, and a very small volume $d^3\bar{x}$ about $\bar{x}$ (identical, for all copies), are considered. Four identical devices A (one per copy) detect, at time t, whether the neutron in each copy is detected inside the corresponding $d^3\bar{x}$ at t or not. In cases (a) and (c), the detectors indicate the presence of one neutron inside the corresponding $d^3\bar{x}$. In cases (b) and (d), there are no detections. Then, $N_n(\bar{x}, t) = 2$.

Let F_0 be the quantum-mechanical probability flux (per unit area and unit time interval) [28, 31] of the wave function $\Psi(\bar{x}; t)$, across a small surface $dS = dxdy$ in the (x, y)-plane, about some point (x, y, z), at time t. One has

$$F_0 = \frac{\hbar}{m_n} Re\left[\Psi^*(\bar{x}; t)(-i)\frac{\partial}{\partial z} \Psi(\bar{x}; t) \right]. \qquad (1.29)$$

Experiments of neutron interferometry provide important examples in which the probabilistic interpretation discussed above is employed directly. Since several decades, procedures allowing to form almost perfect silicon (Si) crystals (actually, monocrystals) have been well established; the size of such crystals can be even up to 15 cm. Those crystals were first employed to perform experiments on X-rays diffraction and, later, have provided the experimental setups of neutron interferometry. For further discussions on neutron interferometers, see Subsection 1.7.3. Here, we shall discuss the probabilistic interpretation in the framework provided by neutron interferometers.

Let us consider an adequately prepared beam of thermal neutrons generated, all of them in similar conditions, by a nuclear reactor. The beam travels through an interferometer having typical sizes of order 10 cm. The time required for one thermal neutron to travel along the whole interferometer is about 35×10^{-6} s. The fluxes of thermal neutrons produced by different nuclear reactors for research allow that, on average, there should be only one neutron propagating coherently along the volume occupied by the interferometer. In other words, for such fluxes, it is highly unlikely that two or more neutrons could travel (or interact with each other) during the same time interval along the interferometer. Thus, even for nuclear reactors with very high flux (10^{15} neutrons $cm^{-2}s^{-1}$), the time delay, on average, between two successive thermal neutrons is estimated to be, on average, 780×10^{-6} s. This allows to interpret the corresponding experiments on neutron interferometry through the following natural simplification: during certain time interval (t_1, t_2), the whole interferometer contains only one thermal neutron from the beam. Stated equivalently, when one neutron is going along the interferometer, the next neutron is yet in the nucleus of the nuclear reactor, with very high probability. That sequence of temporal events becomes repeated a number of times equal to the number of thermal neutrons in the beam. Any of those neutrons traveling (during some interval (t_1, t_2)) along the interferometer can be regarded as one copy of the physical system under consideration. The set of N neutrons, emitted successively (in an analogous way) by the nuclear reactor (each traveling alone along the interferometer without "seeing" the preceding or the next neutron, during the corresponding time interval), constitute the set of N identical copies of the same physical system. There is no conceptual difficulty in regarding, within some suitable time scale and approximations, $t_2 - t_1$ as a (relatively) large time interval, so that one unique wave function $\Psi(\overline{x};t)$ (with $t_1 < t < t_2$) describes the set of those N copies. That is, in the present case and under certain conditions, one can accept that, approximately, N copies of one thermal neutron can be represented by one wave function (i.e., a pure state). One unique detector would then be enough to perform the successive detections of the N copies. The probabilistic interpretation of $\Psi(\overline{x};t)$ (in particular, the probability current density) plays an essential role in the analysis of those interferometry experiments.

For further and wider discussions, see Refs. [28, 33–35]. There is another interpretation of $\Psi(\overline{x};t)$, in which the wave function represents just one individual neutron (which, on the other hand, seems also a natural procedure as there is only one neutron in the interferometer, at a time). In this chapter, we have adopted the interpretation that $\Psi(\overline{x};t)$ describes an ensemble of neutrons. For a lucid comparative discussion about both interpretations of $\Psi(\overline{x};t)$, see Ref. [34] and references therein.

Let $F_{0,tot}$ be the total quantum-mechanical probability flux (per unit time) [20, 31] of $\Psi(\overline{x};t)$ across the whole (x, y)-plane, for any z. One has:

$$F_{0,tot} = \int_{-\infty}^{+\infty} dx \int_{-\infty}^{+\infty} dy \, \frac{\hbar}{m_n} Re\left[\Psi^*(\overline{x};t)(-i)\frac{\partial}{\partial z}\Psi(\overline{x};t) \right]. \qquad (1.30)$$

1.4.2   STATISTICAL MIXTURES OF SLOW NEUTRON BEAMS: DENSITY MATRIX

Typical slow neutron beams generated directly in nuclear reactors (according to some Maxwellian distribution) do not correspond to a unique wave function. In such and similar cases, relative phases are destroyed and incoherence effects (for single neutrons or copies) have to be taken into account [2, 7]. Such beams appear to be the analogue for neutrons of incoherent beams for ordinary light. As in the above case in which relative phases were not destroyed (pure case), all neutrons (say, N_n) in the actual incoherent beam, being approximately independent from one another, are also regarded as an ensemble of identical copies. The distinguishing feature now is that a statistical mixture of wave functions [2, 5, 28, 31] is what provides the statistical description of all neutrons in the beam. In the actual case, the beam is represented by an statistical mixture of the (normalized) wave functions $\Psi(\overline{x};t)_j$ with certain probabilities p_j ($p_j \geq 0$, $\Sigma_j p_j = 1$), the index (or set of indices) j varying in some discrete or continuous set. What is attributed to any individual neutron in the beam is not one single wave function $\Psi(\overline{x};t)_j$, but the whole set (statistical mixture) of wave functions, precisely due to incoherence. The ensemble of all N_n thermal neutrons in the beam (the ensemble of N_n copies) is represented by the whole statistical mixture of all wave functions $\Psi(\overline{x};t)_j$. In turn, the statistical mixture is represented by the density matrix:

$$\rho(\overline{x};\overline{x}';t) = \sum_j p_j \Psi(\overline{x};t)_j \Psi^*(\overline{x}';t)_j. \qquad (1.31)$$

Like in the above pure case, let us imagine that N_n identical idealized experiments (independent from one another) were carried out on that ensemble of N_n neutrons, aimed to detect whether one neutron has gone across the small surface $dS = dxdy$ in the (x,y)-plane, about (x,y,z), during the small time interval $(t - dt/2, t + dt/2)$. Let $F_{0,j}$ be given in Eq. (1.29), for each $\Psi(\overline{x};t)_j$. Let $N_{n,c}(\overline{x},t)$ be the number of those experiments in which the counters do detect the arrival of one neutron. In the actual incoherent case, for large N_n, $N_{n,c}(\overline{x},t)/(N_n \times dS \times dt)$ (the probability for one neutron to have reached the counter, per unit area in the (x,y)-plane about (x,y,z), per unit time) approaches $\Sigma_j p_j F_{0,j} = < F_0 >$. $< F_0 >$ is the quantum-mechanical probability flux (per unit area and unit time interval) for the statistical mixture. Compare with Figure 3.4. One may consider statistical averages for neutron beams, with probability fluxes which be non-monochromatic and time-dependent. Those statistical averages, omitted here, could possibly be adequate for describing neutron beams coming from pulsed (spallation) sources. It is instructive to compare different sources of thermal neutrons (spallation sources versus nuclear reactors) and their associated fluxes [12].

Let the average energy of a thermal neutron, in a typical beam coming from a nuclear reactor, be about 10^{-2} eV, so that the average neutron velocity v is about 10^3 ms^{-1}. Let F_0 denote some average thermal neutron flux (per unit surface and time). Typically, each neutron is represented by a statistical mixture of wave functions, so that F_0 should be interpreted as, and replaced by, $< F_0 >$. The average smallest separation between two neighboring thermal neutrons propagating in the beam is about

$d_{n-n} = (v/ < F_0 >)^{1/3}$. For the highest fluxes, $< F_0 >$ is about 10^{15} neutrons cm^{-2}s^{-1}. Then, d_{n-n} is of the order of 10^4Å to 10^5Å. The thermal neutron beam is, typically, non-monochromatic, so that a spectrum of wavelengths is met; for instance, it is ranged between 2Å and 9Å in the experiments in Ref. [36]. Since the wavelengths in such intervals are smaller than d_{n-n}, one can reasonably neglect the overlaps between the wave packets associated to different neutrons in the beam and regard each confined neutron as propagating independently.

1.4.3   APPROXIMATE FORMULAS FOR PURE STATES AND STATISTICAL MIXTURES

Let a neutron propagate through a large region $r(\Omega)$ of three-dimensional space, with volume Ω: either vacuum or a homogeneous medium, with $Imb = 0$. The neutron is in a pure state and let the corresponding wave function be approximately, for any physically interesting t, an infinite and practically continuous superposition (Ω being suitably large) of the plane waves in Eq. (1.27) (a wave packet), all components of the wavevector $\bar{k}$ being real:

$$\Psi(\bar{x};t) = \int \frac{d^3\bar{k}}{(2\pi)^{3/2}} a(\bar{k}) \exp\left[i\bar{k}\bar{x}\right] \exp\left[-iE(\bar{k})t/\hbar\right]. \qquad (1.32)$$

$a(\bar{k})$ is a complex function and $E(\bar{k}) = E$ is given in Eq. (1.27). We suppose that $a(\bar{k}) \to 0$ quickly if $|\bar{k}| \to +\infty$, in such a way that Eq. (1.32) fulfills $\int d^3\bar{x} |\Psi(\bar{x};t)|^2 = 1$. Let $a(\bar{k})$ be strongly peaked within a small region about $\bar{k}_0$, so that $\Psi(\bar{x};t) \simeq \exp\left[i\bar{k}_0\bar{x}\right] \exp\left[-iE(\bar{k}_0)t/\hbar\right]/\Omega^{1/2}$ inside $r(\Omega)$, while $\Psi(\bar{x};t)$ vanishes approximately outside $r(\Omega)$. Then, the probability current density is, approximately

$$\frac{1}{\Omega} \frac{\hbar \bar{k}_0}{m_n}. \qquad (1.33)$$

Next, let a neutron beam C_n be represented by the statistical mixture in Eq. (1.31), with wave functions similar to those given in Eq. (1.32) (each with a complex function $a(\bar{k})_j$, as above). The beam C_n: (i) is contained in a similar large three-dimensional region $r(\Omega)$ with volume Ω and (ii) has, approximately, t-independent statistical properties. We shall suppose the following approximation in Eq. (1.31):

$$\sum_j a(\bar{k})_j a(\bar{k}')_j^* p_j \simeq \delta^{(3)}(\bar{k} - \bar{k}') \frac{\sigma(\bar{k})}{\Omega}, \qquad (1.34)$$

with some $\sigma(\bar{k}) \geq 0$ for any $\bar{k}$, describing the statistical mixture. Consequently, Eq. (1.31) can be replaced by

$$\rho(\bar{x},\bar{x}';t) \simeq \int \frac{d^3\bar{k}}{(2\pi)^3} \frac{\sigma(\bar{k})}{\Omega} \exp\left[i\bar{k}(\bar{x} - \bar{x}')\right] = \rho(\bar{x},\bar{x}'), \qquad (1.35)$$

which yields a t-independent statistical mixture, consistently with the above assumption (ii).

Upon integrating over the whole region $r(\Omega)$ with volume Ω:

$$\int d^3\bar{x}\rho(\bar{x},\bar{x}') = \int d^3\bar{x} \int \frac{d^3\bar{k}}{(2\pi)^3} \frac{\sigma(\bar{k})}{\Omega} = \int \frac{d^3\bar{k}}{(2\pi)^3} \frac{\sigma(\bar{k})}{\Omega}\Omega = \int \frac{d^3\bar{k}}{(2\pi)^3}\sigma(\bar{k}) = +1, \quad (1.36)$$

due to the explicit factor Ω^{-1} in Eq. (1.34). An ansatz for a statistical mixture yielding an example for Eq. (1.34), for a neutron beam in one spatial dimension, can be seen in Ref. [2]. Some useful forms for $\sigma(\bar{k})$ are given below. An alternative and physically equivalent description, if neutrons are represented by stationary statistical mixtures (i.e., with t-independent properties), is the following: (a) starting from $\Psi(\bar{x};t)\Psi(\bar{x}';t)^*$ ($\Psi(\bar{x};t)$ being given in Eq. (1.32)), one accepts that the amplitude $a(\bar{k})$ has random phase, (b) one performs statistical averaging $(<..>)$, assuming that $<a(\bar{k})>= 0$, and $<a(\bar{k})a(\bar{k}')^* >\simeq \delta^{(3)}(\bar{k} - \bar{k}')\left[\sigma(\bar{k})/\Omega\right]$, with the same $\sigma(\bar{k})$ as above. This alternative description leads to the same results as Eq. (1.35) and has been employed previously (see, for instance, Refs. [5, 37]).

The probability current density in the statistical mixture Eq. (1.35) is

$$\int \frac{d^3\bar{k}}{(2\pi)^3} \frac{\sigma(\bar{k})}{\Omega} \frac{\hbar\bar{k}}{m_n}. \tag{1.37}$$

An interesting example of statistical mixture for a neutron beam containing $N(\Omega)$ neutrons in the region $r(\Omega)$, with t-independent statistical properties, is the following: $\sigma(\bar{k})$ is nonvanishing and approximately constant if $\bar{k}$ lies within a small region about $\bar{k}_0$, while it practically vanishes outside that region, in such a way that Eq. (1.36) holds. A compact version of this example is

$$\sigma(\bar{k}) = (2\pi)^3\delta^{(3)}(\bar{k} - \bar{k}_0). \tag{1.38}$$

For the statistical mixture Eq. (1.38), the probability current density for one neutron is

$$\frac{1}{\Omega}\frac{\hbar\bar{k}_0}{m_n} \tag{1.39}$$

to be compared to Eq. (1.33), while for the $N(\Omega)$ neutrons in the beam is

$$\frac{N(\Omega)}{\Omega}\frac{\hbar\bar{k}_0}{m_n}. \tag{1.40}$$

Notice that $N(\Omega)/\Omega$ is the number of neutrons per unit volume, in the beam. Let us consider a detector receiving neutrons through a surface (with area S), located perpendicular to $\bar{k}_0$. Let n_1 neutrons penetrate into such a detector through S, during a time interval Δt. Then, if almost all neutrons penetrate, $(n_1 / S\Delta t)\left[\hbar\bar{k}_0 / m_n\right]$ equals Eq. (1.40), approximately. In this subsection, we have disregarded, so far, neutron spin. Statistical mixtures involving neutron spin states will be included in Subsection 1.7.1.

For a statistical mixture of neutrons in thermal equilibrium at absolute temperature T (in vacuum), the corresponding $\sigma(\bar{k})$ is given by the Maxwell-Boltzmann distribution:

$$\sigma(\bar{k}) = \left[\frac{\hbar^2}{2\pi m_n K_B T}\right]^{3/2} \exp\left[-\hbar^2 \bar{k}^2 / (2m_n K_B T)\right]. \tag{1.41}$$

Let a thermal neutron beam, extracted from a nuclear reactor and adequately colli-mated, be represented by a statistical mixture (like Eq. 1.41) at absolute temperature T. As a side remark, the latter is, typically, slightly higher than that of the moderator of the reactor. Let N_0 be the total number of neutrons extracted per second, with any velocity. The number of neutrons extracted per second, with de Broglie wavelengths between λ_{dB} and $\lambda_{dB} + d\lambda_{dB}$ is [7]

$$\frac{2N_0}{\lambda_{dB}} \left[\frac{E}{K_B T}\right]^2 \exp\left[-E / (K_B T)\right] d\lambda_{dB}, \tag{1.42}$$

where $E (= E_K)$ and λ_{dB} are related to each other through $E_K = (2\pi\hbar)^2 / (2m_n \lambda_{dB}^2)$.

1.5   REFLECTION AND REFRACTION OF SLOW NEUTRONS

1.5.1   REFLECTED AND REFRACTED STATES FOR AN IDEAL PLANAR SURFACE OF DISCONTINUITY: FRESNEL FORMULAS

Let us consider two different homogeneous media (medium 1 and medium 2), with densities ρ_j and amplitudes b_j, $j = 1, 2$, so that $\rho_1 b_1 \neq \rho_2 b_2$. We suppose that b_1 is real. Medium 1 and medium 2 occupy, respectively, the lower ($z < 0$) and the upper ($z > 0$) three-dimensional half-spaces. Then, the plane $z = 0$ is a surface of discontinuity separating both media. The z-axis points from medium 1 toward medium 2. Let a slow neutron, with total energy E, propagate initially in medium 1, with wavevec-tor $\bar{k}_{1,in}$, toward the surface of discontinuity $z = 0$. Then, there are two possibilities, namely, either the neutron can be reflected in $z = 0$ (propagating back in medium 1 with wavevector $\bar{k}_{1,re}$), or it can penetrate into medium 2, propagating (getting refracted) in it with wavevector $\bar{k}_2$. The incoming neutron pure state is represented by the plane wave:

$$\Phi(\bar{x})_{1,in} = \exp\left[i\bar{k}_{1,in} . \bar{x}\right]. \tag{1.43}$$

The reflected ($\Phi(\bar{x})_{1,re}$) and refracted ($\Phi(\bar{x})_2$) states of the neutron are:

$$\Phi(\bar{x})_{1,re} = A_{1,re} \exp\left[i\bar{k}_{1,re} \bar{x}\right], \Phi(\bar{x})_2 = A_2 \exp\left[i\bar{k}_2 \bar{x}\right]. \tag{1.44}$$

Notice that:

$$E = \frac{\hbar^2 \bar{k}_{1,in}^2}{2m_n} + \frac{2\pi\hbar^2}{m_n} b_1 \rho_1 = \frac{\hbar^2 \bar{k}_2^2}{2m_n} + \frac{2\pi\hbar^2}{m_n} b_2 \rho_2, \tag{1.45}$$

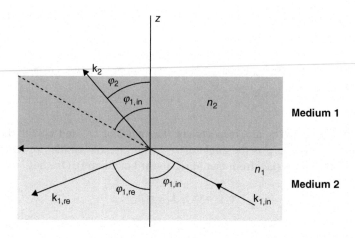

FIGURE 1.3 Two different homogeneous media (1 and 2) with refractive indices n_1 and n_2 ($n_1 < n_2$) are separated by an infinite planar surface. An incoming slow neutron, from medium 1, experiences reflection and refraction, due to the surface.

E being real and $\left|\overline{k}_{1,in}\right| \equiv k_{1,in} = k_{1,re} \equiv \left|\overline{k}_{1,re}\right|$. $A_{1,re}$ and A_2 are the reflection and refraction amplitudes, respectively. The total (stationary) wave functions in media 1 and 2 are, respectively, $\Phi(\overline{x})_1 = \Phi(\overline{x})_{1,in} + \Phi(\overline{x})_{1,re}$ ($z \leq 0$) and $\Phi(\overline{x})_2$ ($z \geq 0$). See Figure 1.3.

At the surface of discontinuity $z = 0$, the total wave function has to be continuous, so that for any point $\overline{x} = (x, y, 0)$ in the plane $z = 0$, one has:

$$\Phi(\overline{x})_1 = \Phi(\overline{x})_2. \tag{1.46}$$

Moreover, the first partial derivatives of $\Phi(\overline{x})_1$ and $\Phi(\overline{x})_2$ along z have to fulfill, for any point $\overline{x} = (x, y, 0)$ in the plane $z = 0$:

$$\frac{\partial \Phi(\overline{x})_1}{\partial z} = \frac{\partial \Phi(\overline{x})_2}{\partial z}. \tag{1.47}$$

Equations (1.46) and (1.47) warrant that both the probability density and the z-component of the probability current density vector are continuous across the surface of discontinuity $z = 0$. Let $J_z(\overline{x}, t)$ be the z-component of the probability current density vector given in Eq. (1.28). Let $J_z(\overline{x}, t) = J_{z,1}(\overline{x}, t)$ and $J_z(\overline{x}, t) = J_{z,2}(\overline{x}, t)$ for the total wave functions $\Phi(\overline{x})_1 \exp[-iEt / \hbar]$ and $\Phi(\overline{x})_2 \exp[-iEt / \hbar]$, respectively. Equations (1.46) and (1.47) yield:

$$J_{z,1}(\overline{x}, t) = J_{z,2}(\overline{x}, t). \tag{1.48}$$

So far, the analysis in this subsection holds for either real or complex b_2. Let us now suppose that b_2 is real, so that there is no absorption of neutrons in medium 2. Without loss of generality, it is always allowed to express the wavevectors as:

$$\bar{k}_{1,in} = (k_{1,in}\sin\varphi_{1,in},0,k_{1,in}\cos\varphi_{1,in}),\bar{k}_{1,re} = (k_{1,re}\sin\varphi_{1,re},0,-k_{1,re}\cos\varphi_{1,re}),$$
$$\bar{k}_2 = (k_2\sin\varphi_2,0,k_2\cos\varphi_2), \tag{1.49}$$

with $\left|\bar{k}_2\right| \equiv k_2$. $\varphi_{1,in}$ and φ_2 are, respectively, the angles of $\bar{k}_{1,in}$ and $\bar{k}_2$ with the z-axis. The reflection angle $\varphi_{1,re}$ is the angle of $\bar{k}_{1,re}$ with an axis parallel to, but pointing opposite to, the z-axis (i.e., from medium 2 toward medium 1). One has:

$$\varphi_{1,in} = \varphi_{1,re}, k_{1,in} = k_{1,re}, \tag{1.50}$$

$$n_1\sin\varphi_1 = n_2\sin\varphi_2, \tag{1.51}$$

where n_1 and n_2 are the indices of refraction for both media, given in Eqs. (1.26). Eqs (1.50), and (1.51) constitute the laws of reflection and refraction, respectively, for slow neutrons. Both laws enable to perform, in principle, computations for slow neutrons similar to those characterizing geometrical optics for light [30]. One gets:

$$A_2 = \frac{2n_1\cos\varphi_{1,in}}{n_1\cos\varphi_{1,in} + (n_2^2 - n_1^2(\sin\varphi_{1,in})^2)^{1/2}}, \tag{1.52}$$

$$A_{1,re} = \frac{n_1\cos\varphi_{1,in} - (n_2^2 - n_1^2(\sin\varphi_{1,in})^2)^{1/2}}{n_1\cos\varphi_{1,in} + (n_2^2 - n_1^2(\sin\varphi_{1,in})^2)^{1/2}}, \tag{1.53}$$

which are, for the reflection and refraction of slow neutrons, the counterparts of the Fresnel formulas for those of light [30]. The derivation of Eqs. (1.50) to (1.53), from Eqs. (1.46), (1.47), and (1.49) is given in exercise 1.1.

Let $J_{z,1,in}(\bar{x},t)$ and $J_{z,1,re}(\bar{x},t)$ be the z-components of the probability current density vectors for $\Phi(\bar{x})_{1,in}\exp[-iEt/\hbar]$ and $\Phi(\bar{x})_{1,re}(\bar{x},t)$, respectively. The coefficients of reflection (R) and transmission (T) are defined as:

$$R = \frac{\left|J_{z,1,re}(\bar{x},t)\right|}{J_{z,1,in}(\bar{x},t)}, T = \frac{J_{z,2}(\bar{x},t)}{J_{z,1,in}(\bar{x},t)}. \tag{1.54}$$

One gets easily:

$$R = \left|A_{1,re}\right|^2, T = \frac{(n_2^2 - n_1^2(\sin\varphi_{1,in})^2)^{1/2}}{n_1\cos\varphi_{1,in}}\left|A_2\right|^2. \tag{1.55}$$

Equation (1.48) yields, directly:

$$R + T = 1. \tag{1.56}$$

$J_{z,1,in}(\overline{x},t)$, $J_{z,1,re}(\overline{x},t)$, and $J_{z,2}(\overline{x},t)$ represent, respectively, the probabilities for the incoming, reflected and transmitted waves for the neutron go across a surface of unit area perpendicular to the z-axis, per unit time. Thus, Eq. (1.56) expresses the conservation of probability; the probability flux determined by the incoming wave $(\Phi(\overline{x})_{1,in} \exp[-iEt / \hbar])$ gives rise at the surface of discontinuity $(z = 0)$, necessarily, to the probability fluxes associated to the reflected $(\Phi(\overline{x})_{1,re} \exp[-iEt / \hbar])$ and transmitted $(\Phi(\overline{x})_2 \exp[-iEt / \hbar])$ waves, since there is no absorption of neutrons in medium 2 (as b_2 is real). A direct computation confirms easily, by using Eqs. (1.52), (1.53), and (1.55), that Eq. (1.56) holds.

The transformation of $\Phi(\overline{x})_{1,in}$ (medium 1) into $\Phi(\overline{x})_2$ (medium 2) constitutes a simple example of an important theorem, known as the extinction theorem (see Section 1.8).

The case of complex b_2 is also interesting and will be treated, with less detail, in exercise 1.2.

1.5.2 TOTAL INTERNAL REFLECTION OF SLOW NEUTRONS

If a slow neutron moves in vacuum with wavevector $\overline{k}$, then $E = E_K = \hbar^2 \overline{k}^2 / (2m_n)$ and $V(\overline{x}) \equiv 0$, so that the refractive index in Eqs. (1.25) and (1.26) is $n = 1$. Let such a slow neutron, having moved initially in vacuum, approach a material medium with $V(\overline{x}) \neq 0$ and real $b > 0$, which holds for most materials (nuclear absorption being neglected). The refractive index n is real and < 1 provided that E be such that $1 > (2\pi b\rho\hbar^2) / (m_n E)$. If E is such that $1 < (2\pi b\rho\hbar^2) / (m_n E)$, n is pure imaginary. From the standpoint of neutron optics (and upon comparing with ordinary optics of light [30]), a medium fulfilling the second condition could be regarded as less dense than vacuum. Let φ be the angle between $\overline{k}$ and the normal to the limiting surface separating vacuum from the medium. Then, by translating standard geometrical optics arguments [30], the neutron suffers total reflection back into vacuum and it does not penetrate into the medium if $\varphi \geq \varphi_{cr}$, where the critical angle φ_{cr} fulfills $\sin\varphi_{cr} = n(< 1)$. This total reflection into vacuum always occurs if E fulfills $1 < (2\pi b\rho\hbar^2) / (m_n E)$. See Figure 1.4. See also Subsection 1.5.3.

Upon going beyond the geometrical-optics approximation and employing the solution $\Phi(\overline{x})_2$ in Eq. (1.44), in such a case, the probability for the neutron to penetrate a distance $d > 0$ into the medium $(z > 0)$ decreases exponentially with d as $\exp[-(2d / \hbar)(2m_n((2\pi b\rho\hbar^2 / m_n) - E))^{1/2}]$. Let us introduce the glancing angle $\theta = \pi / 2 - \varphi$, and the critical glancing angle $\theta_{cr} = \pi / 2 - \varphi_{cr}$. Then, total reflection occurs for $\theta \leq \theta_{cr}$. with $b > 0$. For typical values of $b(> 0)$ and ρ, n and φ_{cr} are a bit smaller than $+1$ and $\pi / 2$, respectively. Then, one has $\theta_{cr} \simeq \lambda_{dB}[b\rho / \pi]^{1/2}$.

1.5.3 ULTRACOLD NEUTRONS

For ultracold neutrons, typically, E is of order 4×10^{-8} eV and λ_{dB} is about 1400 Å, which is still small compared to the distance d_{n-n} estimated in Subsection 1.4.2.1. Equivalently, E corresponds to an absolute temperature T about $\simeq 0.5 \times 10^{-3}$ K (kelvin degrees).

Let us consider the situation described in Subsection 1.5.1, with $\rho_1 b_1 = 0$ and b_2 real and positive, when the neutrons are ultracold, with energy $E(> 0)$ such that

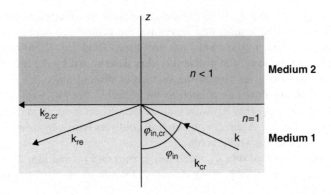

FIGURE 1.4 Medium 1 is vacuum. For medium 2: $n < 1$. Let an incoming slow neutron approach the planar surface of discontinuity from medium 1 with critical angle $\varphi_{in,cr}$ (wavevector $\bar{k}_{cr}$). Then, the wavevector $\bar{k}_{2,cr}$ of the refracted wave is parallel to the surface (while that of the reflected wave is not displayed in this specific case). If the incoming neutron has wavevector $\bar{k}$ with incidence angle $\varphi_{in} > \varphi_{in,cr}$, then there is only a reflected wave, with wavevector $\bar{k}_{re}$, but no refracted wave.

$E < (2\pi\hbar^2 b_2 \rho_2) / m_n$. In this case, there is no refracted wave in medium 2, whatever the incidence angle φ_{in} could be, and there is only a reflected wave (total or specular reflection) in medium 1. Stated equivalently, the refractive index for ultracold neutrons in medium 2 is pure imaginary. Notice that, even if no ultracold neutron could propagate a large distance in medium 2, it can always penetrate a small distance into the latter. There is always a nonvanishing probability for the ultracold neutron to enter a distance $d > 0$ into medium 2, but this probability decays exponentially with d: recall Subsection 1.5.2.

The difference of potential energies of a neutron under the field of gravity, for a difference of heights about 1 m, is about the energy E of an ultracold neutron.

Three-dimensional spatial domains suitably formed by the above media 1 (containing and confining ultracold neutrons) and 2 (into which those neutrons cannot penetrate) enable to handle ultracold neutrons in order to perform various experiments: they are named "neutron bottles." See, for instance, Ref. [2].

1.6   MAGNETIC INTERACTIONS OF A SLOW NEUTRON

Let a slow neutron propagate coherently through a material medium, being subject to both the effective optical potential given in Eq. (1.5) and to a constant and homogeneous magnetic field $\bar{B}$. Then, one has to take into account and include the effective interaction of the magnetic moment of the neutron with $\bar{B}$. Having in mind Subsection 1.3.1, notice that the inclusion of magnetic interactions gives rise to the more general $V(\bar{x})$ in Eq. (1.57). An approximate expression for the total effective potential acting on the slow neutron is

$$V(\bar{x}) = \frac{2\pi\hbar^2 \rho}{m_n} b I_2 - \mu_n \frac{2\bar{s}}{\hbar} \bar{B}, \qquad (1.57)$$

b being, like in Eq. (1.5), the scattering amplitude of the slow neutron by an atomic nucleus in the material. I_2 is the unit 2×2 matrix:

$$I_2 = \begin{pmatrix} 1 & 0 \\ 0 & 1 \end{pmatrix}. \tag{1.58}$$

μ_n was given in Subsection 1.2.1. $\bar{s} = (s_1, s_2, s_3)$ is the neutron spin operator, with $s_j = (\hbar / 2)\sigma_j$, $j = 1, 2, 3$. In turn, σ_j is the standard j-th 2×2 Pauli matrix [28, 31]:

$$\sigma_1 = \begin{pmatrix} 0 & 1 \\ 1 & 0 \end{pmatrix} \qquad \sigma_2 = \begin{pmatrix} 0 & -i \\ i & 0 \end{pmatrix} \qquad \sigma_3 = \begin{pmatrix} 1 & 0 \\ 0 & -1 \end{pmatrix}. \tag{1.59}$$

Notice that the spin state (Subsection 1.3.1) $|\alpha>$ is an eigenstate of s_3 with eigenvalue or spin projection α ($\alpha = +\hbar / 2$ and $\alpha = -\hbar / 2$): $s_3 |\alpha> = \alpha |\alpha>$. Then, the wave function has two components, namely,

$$\Psi(\bar{x}, t) = \begin{pmatrix} \Psi(\bar{x}, t)_+ \\ \Psi(\bar{x}, t)_- \end{pmatrix}. \tag{1.60}$$

Recall Subsection 1.3.1. See also Sections 5.3, 5.4 and 5.5 in Chapter 5. Then, with the above understanding for $\Psi(\bar{x}, t)$ and $V(\bar{x})$, Eq. (1.2) describes the propagation of the coherent wave for a slow neutron subject to the standard effective interaction potential due to a material medium and to a constant and homogeneous magnetic field.

We shall choose: $\bar{B} = (0, 0, B)$ (along the z-axis). Then, if Ψ reduces to $\Psi(\bar{x}, t)_+$, Eq. (1.2) becomes an equation for $\Psi(\bar{x}, t)_+$ alone, with

$$V(\bar{x}) = V(\bar{x})_+ = \frac{2\pi \hbar^2 \rho}{m_n} b - \mu_n B, \tag{1.61}$$

while if Ψ contains only $\Psi(\bar{x}, t)_-$, Eq. (1.2) becomes an equation for $\Psi(\bar{x}, t)_-$ alone, with

$$V(\bar{x}) = V(\bar{x})_- = \frac{2\pi \hbar^2 \rho}{m_n} b + \mu_n B. \tag{1.62}$$

Consequently, the two components of the wave function, associated to the two components of the neutron spin, are subject to different interactions. It is interesting to give the associated refractive indices $n_\pm$, by using Eq. (1.26) for a slow neutron with wavevector $\bar{k}$. For spin projections $\pm \hbar / 2$:

$$n_\pm^2 = 1 - \frac{1}{E}\left[\frac{2\pi b \rho \hbar^2}{m_n} - (\pm)\mu_n B \right] =$$

$$1 - \frac{4\pi}{k^2}\left[b\rho + (\pm)\frac{m_n}{2\pi \hbar^2}|\mu_n| B \right], \tag{1.63}$$

with $E = (\hbar^2 \bar{k}^2) / (2m_n)$. Then, as far as the propagation of a slow neutron also subject to the magnetic field is concerned, the medium behaves as a birefringent one, with two different refractive indices. This allows to obtain polarized beams of slow neutrons; after going through such a medium, most neutrons have their spins oriented toward one direction, while only a small fraction point along the opposite one.

Important examples of the magnetic interactions are commented shortly below.

We remind (Subsection 1.3.1.2) that the second contribution in Eq. (1.6) was the effective magnetic interaction of the neutron spin $\bar{s}$ with the spin $\bar{I}_j$ of the nucleus at location j.

Let the medium be ferromagnetic and be at thermal equilibrium, at suitably low temperature (below the associated Curie temperature). In this case, the magnetic moments of the atoms in the medium point, more or less, along the same direction and they produce an overall net magnetization which, in turn, gives rise to an overall magnetic field. Let us suppose that a slow neutron propagates inside such a ferromagnetic medium, being also subject to the strong interactions due to the atomic nuclei. We shall limit ourselves to a few comments. In principle, the neutron magnetic moment interacts with magnetic moments of the atoms in the medium. Then, one also has to take into account the contribution of the latter interaction. The magnetic moment of an atom receives its dominant contributions from the spins and the orbital angular momenta of the electrons bound in the former. Various estimates indicate that the magnitude of the interaction between the magnetic moment of the neutron and an atomic one has the same order as that of the strong interaction between the neutron and the corresponding nucleus. An approximate formula for the energy of the latter magnetic interaction between the neutron and an atom is

$$\mu_n \frac{2\bar{s}}{\hbar} \bar{B}, \tag{1.64}$$

where $(2\mu_n s) / \hbar$ refers to the neutron and has the same meaning as in Eq. (1.57). On the other hand, $\bar{B}$ is the global magnetic field due to the spins and the orbital angular momenta of the electrons bound in the atom. At a later stage, all magnetic interaction energies Eq. (1.64) associated to all atoms in the medium have to be added. A more detailed study lies outside our scope here. See, for instance, Refs. [2, 5, 7, 32]. See Chapter 5 for an updated and detailed study of neutron spin optics.

1.7 SOME APPLICATIONS OF THERMAL NEUTRONS

1.7.1 Neutron Scattering by a Medium

We consider an adequately collimated beam of slow neutrons (say, thermal, to fix the ideas), all of them approximately with the same wavevector $\bar{k}$, which approaches a material medium (macroscopically large, but finite). The treatment to be outlined in this subsection could also describe, through some minor modifications, the case in which the material medium be replaced by a microscopic system, for instance, a molecule. The beam is scattered, as a result of its interaction with the medium. The different neutrons in the beam are supposed to be adequately separated among themselves, from one another. Then, the scattering of the whole beam by the medium can be replaced, approximately, by the scattering of each individual neutron by the

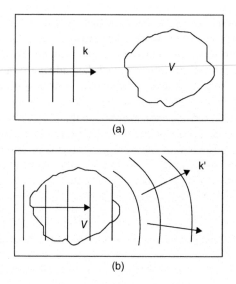

FIGURE 1.5 Scattering of a slow neutron by the effective potential $V(\bar{x})$, due to a material medium: (a) the incoming wave, with wavevector $\bar{k}$, approaches the medium, (b) due to the interaction, scattered waves (with possible wavevectors $\bar{k}'$) are generated for the outgoing neutron.

medium, repeated a number of times equal to the number of neutrons contained in the beam—the scatterings of the different neutrons are approximately independent on one another. In turn, the interaction of each neutron with the medium is described by Eq. (1.24), if the interaction of the spin of the neutron with those of nuclei in the medium is disregarded or by modifying Eq. (1.24) upon taking into account Eq. (1.6), if spin interactions are taken into account. See Figure 1.5.

Would a generic neutron in the incoming beam be represented by a pure state then, having in mind plane waves or Eq. (1.32) and under certain conditions, the incoming probability density flux would be given by Eq. (1.33). Alternatively, let the incoming beam be represented approximately by a stationary statistical mixture like Eq. (1.35), with $\sigma(\bar{k})$ given in Eq. (1.38). Then, the incoming probability density flux is given in Eq. (1.40).

The phenomenon of slow neutron scattering by a medium can be either relatively simple or rather complex depending, respectively, on whether the interactions of the incoming neutrons do not produce changes in the state of the material medium (elastic scattering) or they do induce such changes (inelastic scattering). We shall outline, from the outset, a short introduction to the more general case of inelastic scattering which also includes the case of elastic scattering. We shall represent the quantum-mechanical states of the material medium, which scatters neutrons, by "kets" ($|\Upsilon >$) and "bras" ($< \Upsilon|$), which provide compact notations, instead of by their corresponding wave functions. In particular, $|\Upsilon >$ accounts for the possible spin values and locations of the nuclei and the isotope distributions in the medium. The incoming and outgoing neutrons are represented by plane waves times spin states, namely, $\exp\left[i\bar{k}\bar{x}\right]/\Omega^{1/2}|\alpha >$ and $\exp\left[i\bar{k}'\bar{x}\right]/\Omega^{1/2}|\alpha' >$, respectively or, equivalently, by the kets $\left|\bar{k},\alpha\right>$ and $\left|\bar{k}',\alpha'\right>$. We remark that, typically, the incoming neutrons, even if approximately represented by plane waves (pure states regarding spatial dependences), after collimation and monochromatization may have not been subject to any process that make them to have a well-defined spin projection, that is, to be polarized. In such a case, they are represented by statistical mixtures of spin states $|\alpha>$, with probabilities $p(\alpha) \ge 0$ ($\Sigma_\alpha p(\alpha) = 1$).

Accordingly, in the initial state, the material medium with total energy $E_{mm,in}$ could be represented by the ket $|\Upsilon(in)>$, while $|\Upsilon(out)> (\ne |\Upsilon(in)>)$ is the ket accounting for one possible final state (with total energy $E_{mm,out}$) of the medium. We could alternatively suppose that the medium be in any of a possible set of initial states $|\Upsilon(in)>$ s, with probability $p(\Upsilon(in)) \ge 0$ ($\Sigma_{\Upsilon(in)} p(\Upsilon(in)) = 1$), that is, in a statistical mixture. The wavevectors and energies of the incoming and outgoing neutrons are, respectively: $\bar{k}$, $E = \left(\hbar^2 \left|\bar{k}\right|^2\right) / (2m_n)$ and $\bar{k}'$, $E' = \left(\hbar^2 \left|\bar{k}'\right|^2\right) / (2m_n)$. In the interaction leading from $|\Upsilon(in)>$ to $|\Upsilon(out)>$, which generates the scattering of the neutron, energy conservation reads

$$\frac{\hbar^2 \left|\bar{k}\right|^2}{2m_n} + E_{mm,in} = \frac{\hbar^2 \left|\bar{k}'\right|^2}{2m_n} + E_{mm,out}. \tag{1.65}$$

The change in the momentum of the neutron is $\bar{q} = \bar{k}' - \bar{k}$.

We shall now invoke the analysis of quantum-mechanical elastic scattering of a non-relativistic particle (say, the neutron) by a potential V. We shall not reproduce that analysis here, since it is well documented in various books on quantum mechanics [20, 31]. Actually, that analysis can be extended to the quantum-mechanical inelastic scattering of a neutron, subject to the potential V produced by a material medium constituting the target. In turn, the interaction with the neutron may induce that the state of the target changes. The net and practically useful result from those analyses is the resulting first Born approximation for the corresponding scattering amplitude and cross section associated to those transitions. In the actual situation, in which the state of the material medium can be modified, the amplitude for the (inelastic) scattering of the neutron by the medium, from the initial state to the possible final one in first Born approximation, is

$$\left\langle \alpha' \right| \left\langle \Upsilon(out) \right| \int d^3\bar{x} \left(\exp\left[-i\bar{k}'\bar{x}\right] / \Omega^{1/2}\right) V(\bar{x}) \left|\Upsilon(in)\right\rangle \left(\exp\left[i\bar{k}\bar{x}\right] / \Omega^{1/2}\right) \alpha \right\rangle$$
$$\equiv \Omega^{-1} \left\langle \bar{k}',\alpha' \right| \left\langle \Upsilon(out) |V| \Upsilon(in) \right\rangle \left| \bar{k},\alpha \right\rangle, \tag{1.66}$$

where one integrates over the region occupied by the medium, where V acts on the neutron. The probability per unit time for the scattering of slow neutrons with momenta lying within a small set with size $d^3\bar{k}'$ about $\bar{k}'$ (say, inside a solid angle $d\Omega(\bar{k}')$ with energy in a small interval dE' about E') is given by Fermi's golden rule. The latter leads to consider $(2\pi / \hbar) \left|\Omega^{-1} < \bar{k}'\right| < \Upsilon(out) V \left|\Upsilon(in) > \right| \bar{k} >\left|^2 \left[d^3\bar{k}'\Omega / (2\pi)^3\right]\right.$ (not writing here, in order to emphasize only spatial dependences, the neutron spin states, which are written correctly later in Eqs. (1.67) and (1.68)). See Figure 1.6. The above probability per unit time has to be divided by the probability current density for incoming neutrons, given

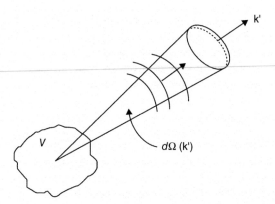

FIGURE 1.6 Outgoing neutron scattered per unit time, with momentum $\bar{k}'$ lying within a small solid angle $d\Omega(\bar{k}')$ about $\bar{k}'$.

in Eq. (1.33). Then, volume factors cancel out. This is not the end of the story, since we should also take into account all possible transitions to final states of the medium and energy conservation (namely, Eq. 1.65), in the possible transitions. All these yield the cross section for the above inelastic scattering, with definite $|\Upsilon(in)>, \alpha$ and α':

$$
\frac{|\bar{k}'|}{|\bar{k}|}\left[\frac{m_n}{2\pi\hbar^2}\right]^2 \sum_{\Upsilon(out)} \left|\langle \bar{k}',\alpha'|\langle \Upsilon(out)|V|\Upsilon(in)\rangle|\bar{k},\alpha\rangle\right|^2
$$

$$
\times \delta\left[\left(\frac{\hbar^2|\bar{k}|^2}{2m_n}+E_{mm,in}\right)-\left(E'+E_{mm,out}\right)\right]d\Omega(\bar{k}')dE',
\tag{1.67}
$$

with $E'=\left(\hbar^2|\bar{k}'|^2\right)/(2m_n)$. δ is the Dirac delta function and expresses energy conservation, namely, Eq. (1.65).

We shall also consider the cross section for any possible α'. We should also consider the possibility that the initial states $|\alpha>$ and $|\Upsilon(in)>$ of the medium constitute statistical mixtures with probabilities $p(\alpha)$ and $p(\Upsilon(in))$. Then, we have to average by summing over the possible initial states of the former (with $\Sigma_{\Upsilon(in)}p(\Upsilon(in))$ and so on for the initial spin states of the neutron). The result is

$$
\frac{|\bar{k}'|}{|\bar{k}|}\left[\frac{m_n}{2\pi\hbar^2}\right]^2 \sum_{\alpha}p(\alpha)\sum_{\Upsilon(in)}p(\Upsilon(in))\sum_{\alpha'}\sum_{\Upsilon(out)}\left|\langle \bar{k}',\alpha'|\langle \Upsilon(out)|V|\Upsilon(in)\rangle|\bar{k},\alpha\rangle\right|^2
$$

$$
\times \delta\left[\left(\frac{\hbar^2|\bar{k}|^2}{2m_n}+E_{mm,in}\right)-\left(E'+E_{mm,out}\right)\right]d\Omega(\bar{k}')dE'.
\tag{1.68}
$$

Equation (1.68) is the inelastic differential cross section for the scattering of slow neutrons by the medium (in lowest Born approximation). Equation (1.68) gives the number of outgoing neutrons scattered per unit time, with momenta lying within a small solid angle $d\Omega(\bar{k}')$ about $\bar{k}'$ and with energy in a small interval dE' about E', divided by the incoming neutron flux (Eq. 1.33). We shall not delve further into aspects of elastic scattering of slow neutrons by material media. For such a study, including in particular magnetic interactions, see Ref. [32].

1.7.2 Neutron Diffraction

1.7.2.1 Introduction to the Dynamical Theory of Neutron Diffraction

The study presented in this section is an extension to thermal neutrons of a related one for the diffraction of X-rays by crystals: see Ref. [38] for X-rays and Refs. [5, 21] for neutrons. Other possible formulations can be seen in Refs. [39, 40]. The name "dynamical theory of neutron diffraction" is intended to mean that multiple scattering effects of the slow neutron with the nuclei in the crystal are taken into account. That name distinguishes it from the so-called "kinematical theory of neutron diffraction" [5], in which multiple scattering is disregarded. In this chapter, we shall not treat specifically the kinematical theory of neutron diffraction, although it should be noticed that the latter also yields directly Bragg's law (Subsection 1.7.3.1).

We shall consider a nonmagnetized three-dimensional crystal, occupying the region $0 < z < z_0$ and being infinitely extended along the directions x and y. There is vacuum in $z < 0$ and $z > z_0$. An incoming slow neutron, represented by a plane wave $\exp\left[i\bar{k}\bar{x}\right]$ with wavevector $\bar{k} = (k_x, k_y, k_z)$ ($k_z > 0$) and energy $E = \hbar^2 \bar{k}^2 / 2m_n$, approaches the crystal from $z < 0$, penetrates into it, interacts with the atomic nuclei forming it and, finally, is either reflected back towards $z < 0$ or is transmitted through the crystal toward $z > z_0$ (each possibility with a given probability amplitude). We shall study those physical processes, with certain generality, leaving various specific cases of interest for the following subsections. The absence of magnetic interactions enables to describe the neutron by means of one unique stationary wave function $\Phi(\bar{x})$, disregarding the neutron spin. In $z < 0$ and in $z > z_0$, $\Phi(\bar{x})$ fulfills Eq. (1.4), with $V \equiv 0$. The solution of the latter equation in $z < 0$ is

$$\Phi(\bar{x})_{<0} = \exp\left[i\bar{k}\bar{x}\right] + \sum_{\bar{k}_{re}} A_{re}(\bar{k}_{re}) \exp\left[i\bar{k}_{re}\bar{x}\right], \tag{1.69}$$

$$E = \frac{\hbar^2 \bar{k}_{re}^2}{2m_n}, \tag{1.70}$$

with a set of possible reflected waves $A_{re}(\bar{k}_{re}) \exp\left[i\bar{k}_{re}\bar{x}\right]$, $A_{re}(\bar{k}_{re})$ being the reflection amplitude. Each possible reflected wavevector $\bar{k}_{re}$ has a negative z-component. The solution of Eq. (1.4), with $V \equiv 0$, in $z > z_0$ is

$$\Phi(\bar{x})_{>z0} = \sum_{\bar{k}_{tr}} A_{tr}(\bar{k}_{tr}) \exp\left[i\bar{k}_{tr}\bar{x}\right], \tag{1.71}$$

$$E = \frac{\hbar^2 \bar{k}_{tr}^2}{2m_n}, \tag{1.72}$$

with a set of possible transmitted waves $A_{tr}(\bar{k}_{tr})\exp\left[i\bar{k}_{tr}\bar{x}\right]$, $A_{tr}(\bar{k}_{tr})$ being transmission amplitude. Each possible transmitted wavevector $\bar{k}_{tr}$ has a positive z-component.

In the region $0 < z < z_0$, occupied by the crystal, $\Phi(\bar{x})$ fulfills Eq. (1.4), V being given by Eq. (1.24) and, by assumption, $b_a = b$ for any a (namely, the same coherent low energy scattering amplitude for any nuclei). We regard the crystal as infinitely extended and periodic along the x- and y-directions. Then, Eq. (1.24) defines a periodic function in both x and y. The latter does not hold strictly for the z-direction, since the crystal is contained in $0 < z < z_0$. We remind that the distances between neighboring atomic nuclei in the crystal lattice are of order a few Å and, so, much smaller (by several orders of magnitude) than z_0. Then, for certain computations below, we shall regard approximately $V(\bar{x})$ as if it were periodic along the whole z-axis as well, that is, we deal with a crystal lattice infinite along the three directions x, y, z, in which the positions of the atomic nuclei are $\bar{a} = n_1\bar{a}_1 + n_2\bar{a}_2 + n_3\bar{a}_3$, with $n_j = 0, \pm1, \pm2, \ldots$, $j = 1, 2, 3$, without limitations due to $0 < z < z_0$. The $\bar{a}_j$'s are three linearly independent vectors. We shall introduce the three basic "reciprocal lattice" vectors $\bar{b}_j$ through the following relationships: $\bar{a}_n.\bar{b}_j = 2\pi\delta_{n,j}, n, j = 1, 2, 3$, $\delta_{n,j}$ being the Kronecker delta ($\delta_{n,n} = 1$, $\delta_{n,j} = 0$ if $n \neq j$). As a consequence of the above physical periodicity (which is not exact, in strict mathematical terms, regarding the z-dependence), we shall accept that $V(\bar{x})$ is given approximately in $0 < z < z_0$ and any x and y as a threefold Fourier series:

$$V(\bar{x}) \simeq V_0 \sum_{\bar{b}} F(\bar{b})\exp\left[-i\bar{b}\bar{x}\right], \tag{1.73}$$

$$V_0 = \frac{2\pi\hbar^2}{m_n}\rho b. \tag{1.74}$$

The reciprocal lattice vectors $\bar{b}$ are given by

$$\bar{b} = l_1\bar{b}_1 + l_2\bar{b}_2 + l_3\bar{b}_3, \tag{1.75}$$

with $l_j = 0, \pm1, \pm2, \ldots$, $j = 1, 2, 3$. The physical meaning of the approximation is that $V(\bar{x})$ is given by the exactly periodic function in the right-hand side of Eq. (1.74) only for $0 < z < z_0$. All coefficients $F(\bar{b})$ are equal to unity, for any $\bar{b}$.

We shall keep open the possibility of general coefficients $F(\bar{b})$ in Eq. (1.74) (with the only restriction $F(\bar{b} = (0,0,0)) = +1$), in order to allow for the possibility of including physical situations wider or more general than the one corresponding to Eq. (1.24), for instance, effects of vibrations of the crystal, having thermal origin (giving rise to the Debye-Waller factor) [5]. As a practical recipe, we shall deal with Eq. (1.74) and arbitrary $F(\bar{b})$ (with the restriction $F(\bar{b} = (0,0,0)) = +1$).

We look for the solution $\Phi(\bar{x})$ of Eq. (1.4) in $0 < z < z_0$ with the following structure:

$$\Phi(\bar{x}) = \exp\left[i\bar{q}x\right]g(\bar{x}), \tag{1.76}$$

$$g(\bar{x}) = \sum_{\bar{b}} A(\bar{b})\exp\left[-i\bar{b}\bar{x}\right], \tag{1.77}$$

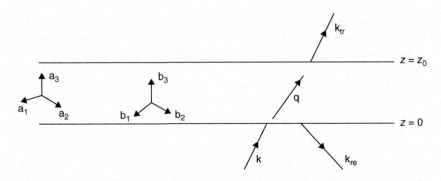

FIGURE 1.7 Diffraction of a slow neutron by a three-dimensional crystal. In general, for a given wavevector $\bar{k}$ of the incoming neutron, there are several possible reflected ($\bar{k}_{re}$) and transmitted ($\bar{k}_{tr}$) wavevectors. The figure also displays $\bar{q}$, the basic vectors $\bar{a}_1$, $\bar{a}_2$, and $\bar{a}_3$ of the crystal lattice and the basic vectors $\bar{b}_1$, $\bar{b}_2$, and $\bar{b}_3$ of the reciprocal lattice.

with certain unknown complex amplitudes $A(\bar{b})$. The wavevector $\bar{q}(=(q_x,q_y,q_z))$ is also unknown, so far. See Figure 1.7. Notice that the function $g(\bar{x})$ is periodic $g(\bar{x}+n_1\bar{a}_1+n_2\bar{a}_2+n_3\bar{a}_3)=g(\bar{x})$, for arbitrary integers n_j, $j=1,2,3$.

We replace in Eq. (1.4) for $0<z<z_0$, $\Phi(\bar{x})$ by Eqs. (1.76) and (1.77), and $V(\bar{x})$ by Eq. (1.73). Upon factoring out $\exp[i\bar{q}\bar{x}]$, we get

$$\sum_{\bar{b}}\left[\left(\frac{\hbar^2(\bar{q}-\bar{b})^2}{2m_n}-E\right)A(\bar{b})+V_0\sum_{\bar{b}'}F(\bar{b}-\bar{b}')A(\bar{b}')\right]\exp\left[-i\bar{b}\bar{x}\right]=0. \qquad (1.78)$$

Equation (1.78) holds in $0<z<z_0$, for arbitrary x, y, if and only if

$$\left(\frac{\hbar^2(\bar{q}-\bar{b})^2}{2m_n}-E\right)A(\bar{b})+V_0\sum_{\bar{b}'}F(\bar{b}-\bar{b}')A(\bar{b}')=0, \qquad (1.79)$$

for any $\bar{b}$ (given in Eq. 1.75). We introduce the following dimensionless parameter:

$$\xi=\frac{V_0}{E}=\frac{\rho b\lambda_{dB}^2}{\pi}, \qquad (1.80)$$

Use has been made of Eq. (1.70). ξ is useful and interesting for thermal neutrons because it is very small, which will enable to carry out approximations later. In fact, for typical values $\lambda_{dB}\approx1\text{Å}$, $b\approx10^{-5}\text{Å}$ and $\rho\approx10^{22}-10^{23}$ nuclei/cm^3, ξ ranges between 10^{-6} and 10^{-7}. ξ allows to recast Eq. (1.80) as

$$((\bar{q}-\bar{b})^2-\bar{k}^2)A(\bar{b})+\xi\bar{k}^2\sum_{\bar{b}'}F(\bar{b}-\bar{b}')A(\bar{b}')=0. \qquad (1.81)$$

Equation (1.81) is a homogeneous system of linear equations for the set of all amplitudes $A(\bar{b})$, for any $\bar{q}$ (thus far, unknown). Let Eq. (1.81) have one solution for all

$A(\bar{b})$'s, among which there is, at least, a nonvanishing one; then, the determinant formed by the coefficients of the linear system Eq. (1.81) has to vanish, necessarily. This yields one equation (independent on the $A(\bar{b})$'s) enabling to determine $\bar{q}$. The latter equation has, in general, a set of solutions for $\bar{q}$, which can be obtained at least approximately for thermal neutrons, due to the smallness of ξ. Each $\bar{q}$ in such a set gives rise, by solving Eq. (1.81) for the $A(\bar{b})$'s, to one $\Phi(\bar{x})$ with the physically correct structure in Eq. (1.76). The latter $\Phi(\bar{x})$ contains some $A(\bar{b})$ so far undetermined, since Eq. (1.81) is a homogeneous system of linear equations.

Let $\Phi(\bar{x})_{int}$ be a suitable linear superposition of $\Phi(\bar{x})$'s given in Eq. (1.76). The following boundary conditions have to be fulfilled for any x, y. At $z = 0$:

$$\Phi(\bar{x})_{<0} = \Phi(\bar{x})_{int}, \frac{\partial \Phi(\bar{x})_{<0}}{\partial z} = \frac{\partial \Phi(\bar{x})_{int}}{\partial z}, \tag{1.82}$$

and at $z = z_0$:

$$\Phi(\bar{x})_{int} = \Phi(\bar{x})_{>z_0}, \frac{\partial \Phi(\bar{x})_{int}}{\partial z} = \frac{\partial \Phi(\bar{x})_{>z_0}}{\partial z}. \tag{1.83}$$

Notice the subtlety of the above global approximation procedure. First, one has replaced the finite width of the crystal along the z-direction by an infinitely extended one along the latter, so as to arrive at Eqs. (1.76), (1.77), and (1.81). Later, one employs the structures in Eqs. (1.76) and (1.77) only for $0 < z < z_0$, so as to match $\Phi(\bar{x})_{int}$ to $\Phi(\bar{x})_{<0}$ and $\Phi(\bar{x})_{>z_0}$. It is required that $\Phi(\bar{x})_{int}$ be a linear superposition of suitable $\Phi(\bar{x})$'s in Eq. (1.76) in order to fulfill both Eqs. (1.82) and (1.83). The latter enable to obtain the amplitudes (so far undetermined, as commented above) in that linear superposition, as well as $A_{re}(\bar{k}_{re})$ and $A_{tr}(\bar{k}_{tr})$, as we shall exemplify below.

If $V \equiv 0$, there is also vacuum in $0 < z < z_0$, and hence $\xi = 0$. In this case, the solution is the following: all $A_{re}(\bar{k}_{re}) = 0$, $\bar{q} = \bar{k}$, all $A(\bar{b})$ vanish if $\bar{b} \neq 0$, except the one corresponding to $\bar{b} = (0,0,0)$ (with $A((0,0,0)) = 1$), and all $A_{tr}(\bar{k}_{tr})$ vanish if $\bar{k}_{tr} \neq \bar{k}$, except $A_{tr}(\bar{k}) = 1$.

Next, we shall deal with a crystal in $0 < z < z_0$ ($V \neq 0$) and thermal neutrons.

The following subsections, together with exercises 1.3, 1.4, and 1.5, will provide interesting examples.

1.7.2.2 Approximation with One Amplitude

Let us suppose that $\bar{k}$ fulfills $\bar{b}^2 - 2\bar{b}\bar{k} \neq 0$ and $\bar{b}^2 - 2\bar{b}.(k_x, k_y, -k_z) \neq 0$ for any $\bar{b} \neq (0,0,0)$. For such $\bar{b} \neq (0,0,0)$, $((\bar{q} - \bar{b})^2 - \bar{k}^2)$ is of order ξ^0 and the equation in Eq. (1.81) corresponding to that $\bar{b}$ is consistent with the corresponding $A(\bar{b})$ being of order ξ.

Let $\bar{b} = (0,0,0)$. Then, Eq. (1.81) implies that $(\bar{q} - \bar{b})^2 - \bar{k}^2 = \bar{q}^2 - \bar{k}^2$ is of order ξ and the equation corresponding to $\bar{b} = (0,0,0)$ enables to restrict consistently to the leading amplitude $A((0,0,0))$. Any other $A(\bar{b})$, with $\bar{b} \neq (0,0,0)$, is of order ξ. Equation (1.81) becomes, approximately:

$$((\bar{q})^2 - \bar{k}^2)A((0,0,0)) + \xi \bar{k}^2 A((0,0,0)) = 0, \tag{1.84}$$

since $F((0,0,0)) = 1$. We shall summarize the consequences of Eq. (1.84) and the determination of the reflected and transmitted wave functions. We use Eqs. (1.82) and (1.83). See exercise 1.3. There is one unique reflected wave $A_{re}(\overline{k}_{re})\exp\left[i\overline{k}_{re}\overline{x}\right]$ in $z < 0$, with $\overline{k}_{re} = (k_x, k_y, -k_z)$, and one unique transmitted one $A_{tr}(\overline{k}_{tr})\exp\left[i\overline{k}_{tr}\overline{x}\right]$ in $z > z_0$, with $\overline{k}_{tr} = \overline{k}$. The solutions, which are consistent with the incoming wave, are

$$\overline{q}_\pm = (k_x, k_y, q_{z,\pm}),$$

$$q_{z,\pm} = \pm k_z \left[1 - \xi \frac{\overline{k}^2}{k_z^2}\right]^{1/2}. \tag{1.85}$$

The existence of a double solution $q_{z,\pm}$ implies that of one amplitude $A((0,0,0)) = A((0,0,0))_+$ of order $1(= \xi^0)$ and of another one $A((0,0,0)) = A((0,0,0))_-$ of order ξ. See Figure 1.8. The wave function describing the propagation of the neutron inside the crystal $(0 < z < z_0)$ is the following linear superposition of $\exp\left[i\overline{q}_+\overline{x}\right]$ and $\exp\left[i\overline{q}_-\overline{x}\right]$:

$$\Phi(\overline{x})_{int} = A((0,0,0))_+ \exp\left[i\overline{q}_+\overline{x}\right] + A((0,0,0))_- \exp\left[i\overline{q}_-\overline{x}\right]. \tag{1.86}$$

Thus, the general linear superposition of suitable $\Phi(\overline{x})$'s in Eq. (1.76) in order to fulfill both Eqs. (1.82) and (1.83), now reduces to $\Phi(\overline{x})_{int}$ in Eq. (1.86).

The above solution coincides with the one which would have been obtained if we would have treated, from the outset, the medium as homogeneous and, so, have disregarded its crystal structure.

Based upon the analysis of the above solution (as ξ is small for thermal neutrons), one can justify that as the ratio $|k_z| / \left[k_x^2 + k_y^2\right]^{1/2}$ increases, the magnitudes of $A_{re}(\overline{k}_{re})$ and $A((0,0,0))_-$ decrease, so that the neglections of $A_{re}(\overline{k}_{re})\exp\left[i\overline{k}_{re}\overline{x}\right]$ and of $A((0,0,0))_- \exp\left[i\overline{q}_-\overline{x}\right]$ become increasingly justified. In such a case, if one keeps only the two amplitudes $A((0,0,0))_+$ and $A_{tr}(\overline{k}_{tr})$, one cannot fulfill the four equations in Eq. (1.123). The physical solution to the problem arising from such an overdetermined system is the following: In this case, we only impose the continuity

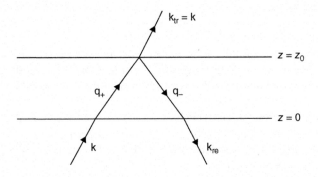

FIGURE 1.8 Diffraction of a slow neutron by a three-dimensional crystal, in the approximation neglecting all $A(\overline{b})$'s, with $\overline{b} \neq (0,0,0)$.

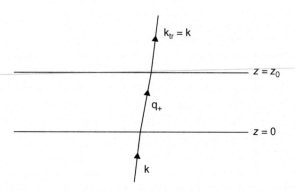

FIGURE 1.9 Diffraction of a slow neutron by a three-dimensional crystal, in the approximation neglecting all $A(\bar{b})$'s, with $\bar{b} \neq (0,0,0)$, when $|k_z| / \left[k_x^2 + k_y^2 \right]^{1/2}$ is not small. Then, the contributions from $\bar{q}_-$ and $\bar{k}_{re}$ are neglected.

of the $\Phi(\bar{x})$'s (but not that of $\partial \Phi(\bar{x}) / \partial z$) at $z = 0$ and $z = z_0$. Then, the first and third equations in Eq. (1.123) (the only ones to be considered now), by omitting $A_{re}(\bar{k}_{re})$ and $A((0,0,0))_-$, yield directly $A((0,0,0))_+ = +1$, and $A_{tr}(\bar{k}_{tr}) = \exp i\left[q_{z,+} - k_z \right] z_0$. See Figure 1.9.

1.7.2.3 Approximation with Two Amplitudes (Bragg and Laue)

Let us fix some $\bar{b} = \bar{b}_{Br} \neq (0,0,0)$ and suppose that, for the incoming $\bar{k}$, $\bar{b}_{Br}^2 - 2\bar{b}_{Br} \cdot \bar{k} \equiv \eta$ is close to 0. Plausibly, there exists, at least, one solution such that (i) $q_z - k_z$ is of order ξ, (ii) $((\bar{q} - \bar{b}_{Br})^2 - \bar{k}^2)$ is of order ξ, (iii) the equation in Eq. (1.81) corresponding to such $\bar{b}_{Br}$ is consistent with $A(\bar{b}_{Br})$ being of order $1(= \xi^0)$. All that suggests to restrict and simplify the present analysis of Eq. (1.81) by keeping only $\bar{b} = (0,0,0)$ and $\bar{b} = \bar{b}_{Br}$ and the corresponding amplitudes. Consequently, Eq. (1.81) is approximated through the following two equations ($F((0,0,0)) = 1$):

$$(\bar{q}^2 - \bar{k}^2)A((0,0,0)) + \xi \bar{k}^2 A((0,0,0)) + \xi \bar{k}^2 F(-\bar{b}_{Br})A(\bar{b}_{Br}) = 0 \qquad (1.87)$$

together with

$$((\bar{q} - \bar{b}_{Br})^2 - \bar{k}^2)A(\bar{b}_{Br}) + \xi \bar{k}^2 F(\bar{b}_{Br})A((0,0,0)) + \xi \bar{k}^2 A(\bar{b}_{Br}) = 0. \qquad (1.88)$$

Alternatively, one could study directly the model characterized by Eqs. (1.87) and (1.88) (omitting the remaining amplitudes $A(\bar{b})$ and equations, with $\bar{b} \neq \bar{b}_{Br}$), without accepting, from the outset, the above plausible (i), (ii), and (iii). The approximate validity of (i), (ii), and (iii) would follow, a posteriori, from the analysis of Eqs. (1.87) and (1.88). Equations (1.87) and (1.88) constitute a homogeneous linear system for the two amplitudes $A((0,0,0))$ and $A(\bar{b}_{Br})$. That system has, at least, one nonidentically vanishing solution for those amplitudes if the determinant formed by the coefficients vanishes. In turn, the last condition yields:

$$\left[\bar{q}^2 - \bar{k}^2 + \xi \bar{k}^2 \right]\left[(\bar{q} - \bar{b}_{Br})^2 - \bar{k}^2 + \xi \bar{k}^2 \right] - (\xi \bar{k}^2)^2 F(\bar{b}_{Br})F(-\bar{b}_{Br}) = 0. \qquad (1.89)$$

A general study of Eq. (1.89) has been carried out in Ref. [41]. Consistently with the latter, we set $\bar{q} = \bar{k} + \varepsilon |\bar{k}| (0,0,1)$ and replace it in Eq. (1.89), which becomes an algebraic equation of fourth degree for the unknown ε. A numerical solution of that algebraic equation has been made in Ref. [41]. There is a physically interesting simplification and approximate solution (the more reliable the larger the ratio $|k_z| / [k_x^2 + k_y^2]^{1/2}$), to which we shall now proceed. We accept the existence of, at least, one solution such that in $\bar{q} = \bar{k} + \varepsilon |\bar{k}| (0,0,1)$, ε being of order ξ. Then, Eq. (1.89) can be reduced, approximately, to one algebraic equation of second degree for ε. See exercise 1.4. One finds, for $0 < z < z_0$:

$$\Phi(\bar{x})_{int} \simeq A((0,0,0))_+ \left[\exp[i\bar{q}_+\bar{x}] + a_+ \exp\left[i\left(\bar{q}_+ - \bar{b}_{Br}\right)\bar{x}\right] \right]$$
$$+ A((0,0,0))_- \left[\exp[i\bar{q}_-\bar{x}] + a_+ \exp\left[i\left(\bar{q}_- - \bar{b}_{Br}\right)\bar{x}\right] \right], \tag{1.90}$$

$$\bar{q}_\pm = \bar{k} + \varepsilon_\pm |\bar{k}| (0,0,1), \tag{1.91}$$

$\varepsilon_\pm$ and $a_\pm$ are given in Eqs. (1.125) and (1.126), respectively. In principle, $\Phi(\bar{x})_{int}$ has to be a linear superposition of suitable $\Phi(\bar{x})$'s with amplitudes $A((0,0,0))_\pm$, in order to fulfill the boundary conditions. Notice that the corresponding $A(\bar{b}_{Br})$ can be regarded to be included into $A((0,0,0))_\pm$. The boundary conditions enable to determine both $A((0,0,0))_\pm$ together with the amplitudes $A_{re}(\bar{k}_{re})$ and $A_{tr}(\bar{k}_{tr})$ and the vectors $\bar{k}_{re}$ and $\bar{k}_{tr}$, in general.

We remind that ξ is small for thermal neutrons and that the ratio $|k_z| / [k_x^2 + k_y^2]^{1/2}$ is not small so that, in the present case, certain amplitudes (in particular, those corresponding to some reflected waves) are negligible. As a prescription regarding the amplitudes to be retained and their determination, their total number should be equal to that of equations fulfilled by them at a latter stage, following from the boundary conditions. Such a recipe, holding necessarily in the general analysis without approximation, should also be valid in approximate treatments. Consequently, if certain amplitudes have been neglected, the number of relevant equations (implied by the boundary conditions) for the amplitudes retained should also decrease, in order to avoid overdeterminations. Of course, for $z > z_0$ there is always one transmitted wave with $\bar{k}_{tr} = \bar{k}$, that is, $A_{tr,1}(\bar{k}) \exp[i\bar{k}\bar{x}]$. As valid approximations, it is natural: (i) to make use of the continuity of the various wave functions $\Phi(\bar{x})$ in $z = 0$ and $z = z_0$ only (but not that of $\partial \Phi(\bar{x}) / \partial z$), (ii) that the leading contributions would correspond to either a reflected wave with $\bar{k}_{re} = \bar{k} - \bar{b}_{Br} - \varepsilon_2 |\bar{k}| (0,0,1)$ with no further transmitted wave (Bragg), or to an additional transmitted wave with $\bar{k}_{tr} = \bar{k} - \bar{b}_{Br} + \varepsilon_3 |\bar{k}| (0,0,1)$, without any appreciable reflected wave (Laue). See Figures 1.10 (Bragg) and 1.11 (Laue).

We shall cover both mutually excluding possibilities (Bragg and Laue) compactly as follows. We approximate the wave function in $z < 0$ as

$$\Phi(\bar{x})_{<0} = \exp[i\bar{k}\bar{x}] + A_{re}(\bar{k}_{re}) \exp[i\bar{k}_{re}\bar{x}] \tag{1.92}$$

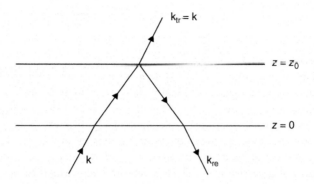

FIGURE 1.10 Bragg diffraction of a slow neutron by a three-dimensional crystal. Then, $\bar{k}_{re} = \bar{k} - \bar{b}_{Br} - \varepsilon_2 |\bar{k}| (0,0,1)$. Inside the crystal, there are four wavevectors: $\bar{q}_+$, $\bar{q}_+ - \bar{b}_{Br}$, $\bar{q}_-$ and $\bar{q}_- - \bar{b}_{Br}$ (although only two are displayed).

and the wave function in $z > z_0$ as

$$\Phi(\bar{x})_{>z_0} = A_{tr,1}(\bar{k}) \exp\left[i\bar{k}\bar{x} \right] + A_{tr,2}(\bar{k}_{tr}) \exp\left[i\bar{k}_{tr}\bar{x} \right].\qquad(1.93)$$

In order to determine $\varepsilon_2(>0)$, $\varepsilon_3(>0)$ and the corresponding amplitudes, one obtains (exercise 1.5) an inhomogeneous linear system of four equations containing formally five unknowns, namely, $A((0,0,0))_+$, $A((0,0,0))_-$, $A_{re}(\bar{k}_{re})$, $A_{tr,1}(\bar{k})$, and $A_{tr,2}(\bar{k}_{tr})$. Such a formal system is physically meaningful and has unique solutions in the following two different situations: (i) $A_{tr,2} = 0$ with $A_{re}(\bar{k}_{re}) \neq 0$ (Bragg diffraction), and (b) $A_{tr,2} \neq 0$ with $A_{re}(\bar{k}_{re}) = 0$ (Laue diffraction). The two corresponding solutions for $A((0,0,0))_+$, $A((0,0,0))_-$, $A_{re}(\bar{k}_{re})$ and $A_{tr,1}(\bar{k})$ (Bragg) and for $A((0,0,0))_+$, $A((0,0,0))_-$, $A_{tr,1}(\bar{k})$, and $A_{tr,2}(\bar{k}_{tr})$ (Laue) are presented rather briefly in exercise 1.5. A more detailed discussion of both kinds of diffraction phenomena can be seen in Ref. [5].

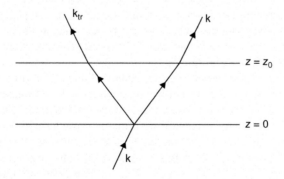

FIGURE 1.11 Laue diffraction of a slow neutron by a three-dimensional crystal. Then, $\bar{k}_{tr} = \bar{k} - \bar{b}_{Br} + \varepsilon_3 |\bar{k}| (0,0,1)$. Inside the crystal, there are four wavevectors: $\bar{q}_+$, $\bar{q}_+ - \bar{b}_{Br}$, $\bar{q}_-$ and $\bar{q}_- - \bar{b}_{Br}$ (although only two are displayed).

1.7.3   NEUTRON DIFFRACTION BY CRYSTALS: SOME APPLICATIONS

1.7.3.1   Application of Bragg Diffraction to Produce Monochromatic Beams of Slow Neutrons

Let us consider an incoming beam of slow neutrons, already appreciably or rather well collimated (the momenta of practically all neutrons in the beam having approximately the same directions, within some small solid angle). However, the energies of the neutrons in the beam are, in principle, appreciably different among themselves, constituting a continuous spectrum (wide, in various usual cases). The phenomenon of Bragg diffraction by a thin crystal lattice (Subsection 1.7.2.3) allows for that incoming beam to give rise to a reflected neutron beam which be not only rather well collimated but also approximately monochromatic (quasi-monochromatic, say, with energies within a narrow interval). The reflected beam just corresponds to $\bar{k}_{re} = \bar{k} - \bar{b}_{Br} - \varepsilon_2 |\bar{k}| (0,0,1)$ and to $A_{re}(\bar{k}_{re})$ (Subsection 1.7.2.3).

In a thin crystal lattice, let a set of atomic nuclei lie on a (diffracting) plane, π_0, determined by the expression $n_1\bar{a}_1 + n_2\bar{a}_2 + n_3\bar{a}_3$, as two of the three integers (n_1, n_2, n_3) vary independently (the third one being determined by the former two). Let us consider an additional series of (diffracting) planes, $\pi_1, \pi_2,...$, formed by similarly arranged atomic nuclei, so that (i) each plane is similar and parallel to one another and to π_0, (ii) all planes in that series and π_0 are neighbors of one another, there being no atomic nuclei lying between any pair of neighboring planes and any plane being separated from its two neighbors by the same constant crystal spacing d. d is of order a few Angstroms. Let a slow neutron, with de Broglie wavelength λ_{dB}, enters into the thin crystal lattice, and let θ be the angle between its momentum and any of the parallel planes in the series π_0, π_1, π_2. The neutron experiences Bragg diffraction of order $n(= 1,2,...)$ by those parallel planes if

$$2d \sin\theta = n\lambda_{dB} \qquad (1.94)$$

(Bragg's law, Subsection 1.2.5). Its justification is easy, starting from $\bar{k}_{re} = \bar{k} - \bar{b}_{Br} - \varepsilon_2 |\bar{k}| (0,0,1)$ (Subsection 1.7.2.3) provided that one omits the contribution in ε_2, approximates $\bar{k}_{re}^2 \simeq (\bar{k} - \bar{b}_{Br})^2$ and supposes, for simplicity, that the diffracting planes in the series $\pi_0, \pi_1, \pi_2, ...$ correspond to $n_1\bar{a}_1$ (as the integer n_1 varies). Then, one has $\bar{b}_{Br} = n\bar{b}_1$, $d = |\bar{a}_1|$ and $|\bar{b}_{Br}| = (2\pi n)/|\bar{a}_1|$. The angle between $\bar{k}_{re}$ and $\bar{k}$ is, approximately, 2θ. See Figure 1.12.

If the incoming beam is already appreciably collimated and if the possible $\bar{k}_{re}$ is adequately limited (i.e., the reflected beam be appreciably collimated as well), the thin crystal lattice selects only those neutrons with λ_{dB}'s such that Eq. (1.94) be fulfilled. The thin crystal lattices employed for that purpose have dimensions of order a few centimeters and they are constituted, typically, by Cu, Al, Pb, Fe, Si, Be, etc.

Another curious application of Bragg's law is the following. Let us consider a slow neutron with $\lambda_{dB} > 2d$. Then, as $n = 1,2,...$, there exists no angle θ such that Eq. (1.94) could be fulfilled. Then, that neutron can propagate inside the thin crystal lattice without experiencing Bragg diffraction. Conversely, a neutron with $\lambda_{dB} < 2d$ will certainly be subject to Bragg diffraction, with large probability, inside that thin crystal lattice. The facts just discussed provide a basis to separate neutrons having

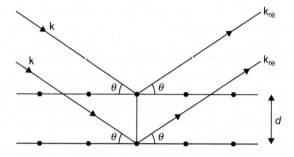

FIGURE 1.12 Bragg's law. The approximation $\bar{k}_{re} = \bar{k} - \bar{b}_{Br}$ is assumed.

different energy ranges from a beam. As a neutron beam enters into an adequate thin crystal lattice (Be, graphite and Bi), the neutrons which are more energetic suffer Bragg diffraction and emerge along directions which are quite different from those of lower energy neutrons. The latter propagate without experiencing Bragg diffraction.

1.7.3.2 Application of Laue Diffraction to Neutron Interferometry

We proceed to describe here neutron interferometers in a way less schematic than in Subsection 1.4.1. See also Chapter 2. Bonse and Hart [42] constructed firstly an X-ray interferometer for X-rays, based upon the diffraction of the latter by thin crystal lattices. Diffraction phenomena for neutrons and X-rays by such crystals turn out to be analogous to each other. The corresponding interferometers for neutrons, following a similar pattern, were developed later [43].

A typical interferometer (of the Bonse-Hart class), based upon a (Si) monocrystal, is displayed schematically in Figure 1.13. The monocrystal is cut partially, so as to form three parallel plates, all with the same width and at equal spacings between

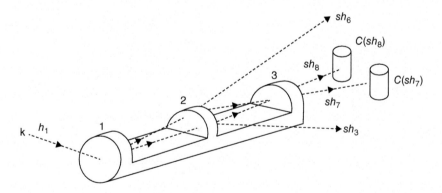

FIGURE 1.13 Neutron interferometer. The incoming beam h_1 experiences Laue diffraction in the three plates 1, 2, and 3, successively. The partial beams sh_3 and sh_6, after their generation in the plate 2, leave the interferometer. The final partial beams sh_7 and sh_8 are detected in the counters $C(sh_7)$ and $C(sh_8)$, respectively. The notations for the partial beams sh_1, sh_2, sh_4, and sh_5 are not displayed. No obstacle is located in the path of any partial beam. Compare with Figure 1.14.

two neighboring ones. The three plates continue to constitute part of the monocrystal and stand out over the remainder of it, which provides mechanical support. The cuts made in the monocrystal are such that the diffraction phenomena generated by the three parallel plates, when the neutrons reach them successively, are mostly of the Laue type (Subsection 1.7.2.3). In what follows, we shall use the words "beam" and "partial beams" referring to slow neutrons traveling inside the interferometer, with the following specific understanding: During certain time interval there is, with very high probability, only one neutron in the whole interferometer before the next neutron enters into the latter (Subsection 1.4.1).

The (adequately collimated) incoming beam h_1, with wavevector $\bar{k}$, experiences Laue diffraction in the first plate and gives rise to two partial beams, sh_1 and sh_2. In turn, each of the later partial beams is transmitted through and suffers Laue diffraction in the second plate, each of them giving rise to other two partial beams: Then, the four partial beams sh_3, sh_4, sh_5, and sh_6 are generated after the action of the second plate. Two of them (sh_3 and sh_6) leave the interferometer without having traveled across the third plate. The remaining two partial beams (sh_4 and sh_5) travel across the third plate and, in so doing, interfere with each other, experience Laue diffraction and give rise to the last two partial beams sh_7 and sh_8, which subsequently leave the interferometer and are detected in certain counters, $C(sh_7)$ and $C(sh_8)$. One introduces an additional body or obstacle, having adequately limited size, in the path of one (and only one) of the two partial beams leaving the second plate, before reaching the third plate, for instance, sh_5. In this way, an additional complex amplitude γ (and, so, an additional phase) are generated in the partial beam sh_5, in comparison with the other partial beam (sh_4) also reaching the third plate. The amplitude γ could, in case that the additional obstacle be a plate having width d, just correspond to $A_{tr}(\bar{k}_{tr}) = \exp i\left[q_{z,+} - k_z \right] z_0$ (see Subsection 1.7.2.2, now with $z_0 = d$). See Figure 1.14.

The analysis of Laue diffraction and the corresponding amplitudes ($A_{tr,1}(\bar{k})$ and $A_{tr,2}(\bar{k}_{tr})$) (see Subsection 1.7.2 and exercise 1.5) enable to deal directly with the

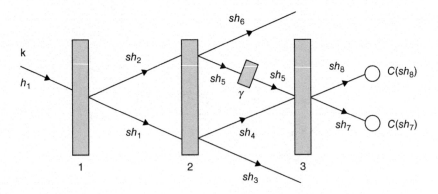

FIGURE 1.14 Neutron interferometer (seen from above). This figure displays the same process as in Figure 1.13, except that one obstacle is located in the path of the partial beam sh_5. All partial beams are displayed. The obstacle located in the path of sh_5 gives rise to the additional amplitude γ.

successive Laue diffractions in each of the three plates. We shall treat briefly the wave functions associated to the partial beams sh_7 and sh_8 based, in turn, in the amplitudes in Eq. (1.93) (not displaying explicitly, for simplicity, dependences on wavevectors).

$$\Phi(\bar{x};sh_7) = A(sh_7)\exp\left[i\bar{k}\bar{x}\right],$$
$$A(sh_7) = A_{tr,2}A_{tr,2}A_{tr,1} + A_{tr,1}\gamma A_{tr,2}A_{tr,2}$$

$$(1.95)$$

$$\Phi(\bar{x};sh_8) = A(sh_8)\exp\left[i\bar{k}_{tr}\bar{x}\right],$$
$$A(sh_8) = A_{tr,1}A_{tr,2}A_{tr,1} + A_{tr,2}\gamma A_{tr,2}A_{tr,2}$$

$$(1.96)$$

$\Phi(\bar{x};sh_7)$ and $\Phi(\bar{x};sh_8)$ are the counterparts of the wave functions including $A_{tr,1}(\bar{k})$ and $A_{tr,2}(\bar{k}_{tr})$, respectively, in Eq. (1.93). Those wave functions determine probability current densities, to be detected in the counters $C(sh_7)$ and $C(sh_8)$, respectively. In turn, those probability current densities are proportional, respectively, to (i) $|A(sh_7)|^2$ and $|A(sh_8)|^2$, (ii) the probabilities that the waves associated to $\Phi(\bar{x};sh_7)$ and $\Phi(\bar{x};sh_8)$ propagate across surfaces of unit area in the counters $C(sh_7)$ and $C(sh_8)$, per unit time, (iii) the numbers of neutrons traveling across surfaces of unit area in the counters $C(sh_7)$ and $C(sh_8)$, per unit time, when the whole process and the corresponding detections are repeated, by using a large number of identical copies. Recall Subsections 1.4.1 and 1.5.1. Upon varying the amplitude γ associated to the additional obstacle in the path of the partial beam sh_5 (i.e., d, if the obstacle is a plate), the numbers of neutrons detected in the counters $C(sh_7)$ and $C(sh_8)$ vary. A study of how those numbers of neutrons detected vary, as γ varies, displays the existence of maxima and minima. In turn, the latter establish experimentally the interferences produced between the corresponding partial beams. The generation of those interferences between partial beams can be recast into the statement that a neutron traveling along the interferometer has interfered with itself. We shall omit further details; see Ref. [5].

Neutron interferometers have been employed to test experimentally the corresponding theoretical foundations outlined above. The results of the experiments for the numbers of detected neutrons show a very good agreement with the theoretical predictions (based upon Eqs. (1.95), (1.96) and in various variants thereof, omitted here for brevity). More generally, those interferometers, with various modifications, have been used to perform different experimental tests of quantum mechanics: The results of the various experiments are consistent with the quantum-mechanical predictions, in the different experiments. See, for instance, Refs. [5, 33–35]. See also Chapter 2.

1.7.4 COMPARISON BETWEEN SLOW NEUTRON DIFFRACTION AND X-RAY DIFFRACTION

X-rays and thermal neutrons have been and continue to be extremely useful and valuable for the investigation of matter (biological matter being included). The informations provided by both of them turn out to be complementary of each other, in several important aspects. The (de Broglie) wavelength λ_{dB} for thermal neutrons is about three orders of magnitude smaller than that of ordinary or visible light (which ranges between 4×10^3 to 7×10^3 Å). For X-rays, the wavelength is about or smaller than 5×10^2 Å (1Å $= 10^{-10}$ m).

The energies of quanta of vibration (phonons) in crystals are of order 0.04 eV, that is, of the same order as the energies (E_K) of thermal neutrons (Subsection 1.2.1). The energy of one photon in X-rays is, typically, several orders of magnitude larger (say, a factor 10^5 larger).

The interaction mainly responsible for X-ray scattering and diffraction by matter is the electromagnetic one. More specifically, that interaction is determined (and dominated) by the distribution of the electric charges of the atoms in the medium and, so, the magnetic properties of the latter do not give rise, typically, to appreciable additional effects. The corresponding X-ray amplitudes for scattering: (i) are not isotropic, but display angular dependences, (ii) the associated probabilities increase in a regular way, essentially, with the atomic number of the atom (or atoms) constituting the material medium, (iii) those probabilities do not change if the atoms are respectively replaced by possible isotopes.

On the contrary, the interaction responsible for scattering and diffraction of thermal neutrons by matter is, primarily, the nuclear one, that is, the strong neutron-nucleus interaction. The latter gives rise to scattering amplitudes which: (i) are isotropic (without angular dependences), (ii) change, in an irregular way, with atomic number, (iii) the associated probabilities do change if the atoms are respectively replaced by possible isotopes, (iv) the associated probabilities do change with the spin of the nuclei. The magnetic interaction of the thermal neutrons with the magnetic moments of the atoms (Section 1.6) could be rather appreciable or important.

For X-rays propagating through a material medium, the associated index of refraction n_{rX} has a real part fulfilling $Re n_{rX} < 1$. For comparison, we remind that for ordinary or visible light through typical material media with index of refraction n_{vl} (relative to vacuum), one has $n_{lv} > 1$, usually. Absorption effects are, usually, quite important for X-rays. Thus, the counterpart of the linear coefficient μ for X-rays in a material medium is, typically, of order 10^2 to 10^3 cm^{-1}. X-rays absorption increases with the atomic number of the atomic number of the atoms in the medium.

The absorption of thermal neutrons by the atomic nuclei of the material medium is, typically, small or very small. Then, $Im b$ can be neglected in a large number of cases (but not always). In such a case, Eq. (1.26) implies $n < 1$ for $b > 0$, which holds for almost all atomic nuclei. The exceptional case $n > 1$ occurs for the few atomic nuclei with $b < 0$, mentioned in Table 1.1.

1.7.5 Neutron Holography

1.7.5.1 Introduction: General Overview

The chronology in the contributions to atomic-resolution holography started in 1986 [44]. It included new concepts as photoelectron and X-ray fluorescence holography. These new technologies generated important applications in material sciences like the study of adsorbates, dopants, and disordered systems, as examples.

Holography is an optical technique discovered by Dennis Gabor in 1947. For his pioneering work, he received the Nobel Prize in Physics in 1971. This optical technique was initially devoted to the optimization of imaging systems. Gabor's research focused on electron microscopy, which led him to the invention of holography. The basic purpose was to obtain a perfect optical imaging. All information (i.e., the totality thereof) has to be used—not only the amplitude, as in usual optical imaging, but also the phase of the waves. In this way, a complete holo-spatial picture can be obtained.

Essentially, a hologram is an interferogram containing in its structure the complete information, in amplitude and phase, of the light wave diffracted by the object and whose image we want to reconstruct. Therefore, the holographic technique involves interferometry. The encoded information can be retrieved by the light wave diffracted by the recorded hologram (reading the hologram). Indeed, the setup to record a hologram is an interferometer. Figure 1.15 displays the pioneering experimental setup developed by Gabor to record a hologram [45]. Then, it is known nowadays as Gabor's hologram or on-axis hologram.

In the top scheme of Figure 1.15, the recording is performed by illuminating with a light source a two-dimensional object (say, a slide). The divergent beam strikes the slide, under a certain angle, producing a diffracted wave, named object wave. A photographic plate is located at a certain distance so as to record the diffracted pattern, and the hologram is obtained. Once processed, the photographic plate is then located in the reconstruction setup (bottom). A light source is illuminating the recorded plate with a divergent beam formed by the optical system preceding the plane where the photographic plate is located. A second optical system is located close to the hologram and then the diffracted beam is recorded in a new photographic plate (or a photodetector). In this plate, the data of the slide structure (the object in the actual case) is codified, and the image reproduces it. We notice that, indeed, in the plate there are two superimposed images, the one corresponding to the virtual image, located previous to the hologram, and a second one corresponding to the real image, located at the photographic plate plane. These twin images cannot be differentiated, and, so, a certain noise is recorded, as well, along with the expected reconstructed image. This setup, being the first one used by Gabor, is also the one employed for the purpose of neutron holography, as we shall see later on. Figure 1.16 displays Gabor's hologram of a slide with some text. We notice that the reconstructed image is a noisy one, due to the superposition of the twin images (real and virtual, respectively), as it is formed by using an on-axis holographic technique.

Since the launching of laser sources earlier in 1960, holography has enjoyed a great development. Moreover, the quality of holograms has dramatically increased. One of the main reasons is the use of illumination with laser sources, presenting quasi-monochromaticity, high convergence, high spatial coherence, hereby providing sources with a high degree of spatio-temporal coherence. This fact implies that the recorded interferogram has high visibility factor or image contrast, and, therefore, the image presents high quality. We notice that, in the pioneering work of Gabor, he actually used a mercury arc lamp for the setup, which has low coherence. In 1962, Emmett Leith and Juris Upatnieks read Gabor's paper and, "simply out of curiosity," decided to duplicate Gabor's technique using

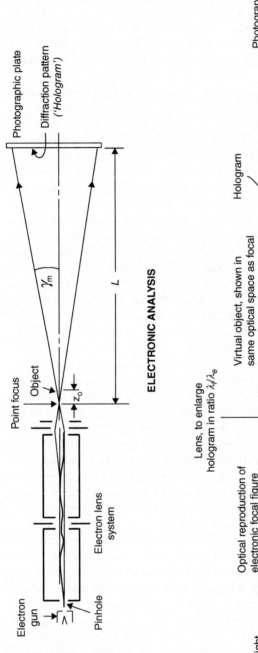

ELECTRONIC ANALYSIS

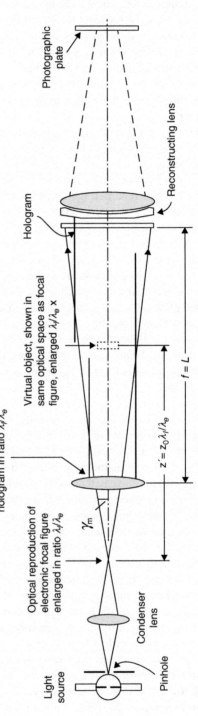

FIGURE 1.15   The two schemes published by Dennis Gabor in 1949 illustrating the setup for recording (top) and reading (bottom) of an on-axis optical hologram (Figure 3 in Ref. [45]).

FIGURE 1.16 Reconstructed image from Gabor's hologram.

Source: Department of Optics, Complutense University of Madrid, Spain. The hologram was recorded with a He-Ne laser ($\lambda = 632.8$ nm) and a spectroscopic plate was used as recording photomaterial.

a He-Ne laser and an off-axis technique borrowed from their work in the development of side-reading radar. The result was the first laser transmission hologram of 3D objects. Since those pioneering techniques, holography has become a powerful nondestructive optical technique with a large and wide range of applications involving material sciences, metrology, biomedicine, microscopy, as some of many other examples. At present, analogic holography has been substituted by digital techniques overcoming the problems arising from the high required sensibility, needed by the use of holographic photomaterials [47].

We have emphasized the first Gabor's hologram, because the neutron holography technique is based upon this type of on-axis holography.

In optical holography, the image resolution is restricted by the wavelength of the light to several hundred nanometers. However, from the development of other nonluminous sources, originating from elementary particles collisions and alternative nonvisible particle sources, a number of holographic techniques are being applied for investigations on material science and crystalline structures, because the wavelength emitted by the source is the order of the interatomic dimensions of the crystal (on the order of 1 Å). Two examples are the γ-ray and the X-ray holographies, for which the reconstructed hologram is giving information on the atomic and molecular structure of a precise crystal by a real image [48]. In the particular case of X-ray holography, the main problem is the low contrast of the image. Opposite, for γ-rays, nuclei (as those used for Mössbauer spectroscopy) can emit γ-rays with very well-defined energy providing a much better contrast. The main challenge for this technique is that only few isotopes have the required resonance to be used as reference sources [49].

Neutron measurements can eventually fill the gap produced by the limitation of X-ray energy range. This fact provides the implementation of an alternative technique—the neutron holography, feasible because of the already mentioned wave-particle dual nature of neutron beams.

The mathematical description of the recording and reconstruction in a holographic process is quite simple. It only needs to introduce the condition to obtain an interference pattern. For definiteness, we shall describe it for the case of light. The extension to neutron holography would follow a similar pattern (provided that quantum-mechanical issues be duly taken care of).

Let us consider two classical scalar complex quasi-monochromatic waves denoted by U_1 and U_2. These two complex waves superimpose, under certain partial coherence conditions, and the total complex amplitude U is

$$U = U_1 + U_2. \tag{1.97}$$

Then, the total intensity is

$$I = \langle UU^* \rangle = \langle U_1 U_1^* \rangle + \langle U_2 U_2^* \rangle + \langle U_1 U_2^* \rangle + \langle U_2 U_1^* \rangle, \tag{1.98}$$

where <> brackets denote time average. We recast Eq. (1.98), with obvious notations, as ($< U_1^* U_2 >=< U_2 U_1^* >$):

$$I = I_1 + I_2 + \langle U_1 U_2^* \rangle + \langle U_1^* U_2 \rangle. \tag{1.99}$$

In the application to holography, the two complex amplitudes, U_1 and U_2 correspond to the object and the reference waves, respectively. Equation (1.99) gives the total intensity distribution recorded in the holographic photomaterial (or in a photodetector. We may consider that both U_1 and U_2 are coming from the same light source and, then, the superposition of both of them has a certain degree of coherence, defined by the mutual coherence function [30, 46]:

$$\langle U_1(\bar{x}_1;t)^* U_2(\bar{x}_2;t+\tau) \rangle = \Gamma(\bar{x}_1,\bar{x}_2;\tau). \tag{1.100}$$

Spatial and temporal dependences will be displayed explicitly, as it suits. In Eq. (1.100), Γ depends on the spatial variables and on τ which is the time delay between the two beams. The visibility of the interference pattern is ($\bar{x} = \bar{x}_1 - \bar{x}_2$) [46]:

$$\begin{aligned} V(\bar{x},\tau) &= \frac{I_1 + I_2 + 2(I_1 I_2)^{1/2}\gamma_{12} - (I_1 + I_2 - 2(I_1 I_2)^{1/2}\gamma_{12})}{I_1 + I_2 + 2(I_1 I_2)^{1/2}\gamma_{12} + (I_1 + I_2 - 2(I_1 I_2)^{1/2}\gamma_{12})} \\ &= \gamma_{12} \frac{2(I_1 I_2)^{1/2}}{I_1 + I_2}. \end{aligned} \tag{1.101}$$

In Eq. (1.101), γ_{12} is the normalized mutual coherence function. The visibility or contrast factor is proportional to the square modulus of γ_{12}, to I_1 and to I_2. Let us assume that the transmittance $t_r = t_r(x, y)$ in amplitude [46] of the hologram is proportional to the recorded intensity (say, under a linear regime). Then,

$$t_r(x,y) \propto |U_1 + U_2|^2 = |U_O + U_R|^2 = |U_O|^2 + |U_R|^2 + U_O U_R^* + U_R U_O^*, \quad (1.102)$$

where $U_1 \equiv U_O$ refers to the object wave and $U_2 \equiv U_R$ to the reference wave. The slight change of notation could make the explanation more understandable. Then, if we now illuminate the holographic plate for the reconstruction with the reference beam U_R the diffracted complex (transmittance in) amplitude is

$$t_r U_R \propto I_R U_R + I_O U_R + I_R U_O + U_R^2 U_O^*. \quad (1.103)$$

We observe that both, the amplitude and phase of the object, are codified in the reconstruction and two twin images are formed. Both images are phase shifted by π. Figure 1.17 displays the two processes: the one showing the hologram (recording) and the one displaying the image of the object (reading).

We notice that in Figure 1.17, for the particular case of Gabor's hologram, the angle θ between the reference and the object waves is zero, so that the two images overlap.

1.7.5.2 Atomic-Resolution Slow Neutron Holography

In the previous subsection, we have analyzed a general description of the optical holographic technique. Now, we shall focus on the specific neutron holographic technique.

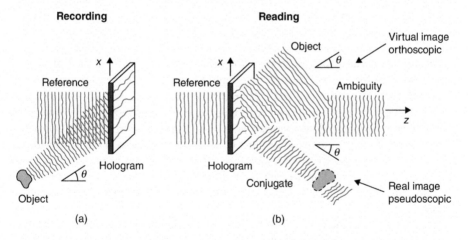

FIGURE 1.17 Two schemes for explaining the whole holographic technique process. Left: The recording of the interferogram in the holographic plate (or photomaterial, or photodetector). Right: The reading of the hologram. It is then illuminated by the reference wave. Two conjugate waves are formed, giving rise to two conjugate images. Both are carrying information on the object. The observer can be located, for example, at the direction of the virtual image, and will see the object with a perspective. If it is a 3D object, the real image gives a distorted depth perception (pseudoscopic).

In 2001, L. Cser et al. [50] demonstrated the feasibility of neutron holography with the use of a slow neutron point source and for particular crystal structures. They showed that atomic-resolution neutron holography can be achieved, under certain conditions for the neutron beam source. In a first step, a point-like monochromatic source of slow neutrons is produced inside a certain crystal under study; such a source generates the reference wave. The complex amplitude and geometry of the wave fit with the one of a spherical wave. A second wave (object wave) is created by elastic scattering of the latter radiation from an object (single atoms, in the present case). The intensity distribution of the radiation generated from the point-like source interacting with the hologram, for a given direction (say, some given $\bar{k}$) and at a distance R, reads (similarly to the X-ray holography case)

$$I(\bar{k}) \propto \frac{I_0}{R^2}\left[1 + 2Re\left(\sum_j a_j(\bar{k})\right) + \left|\sum_j a_j(\bar{k})\right|^2\right]. \qquad (1.104)$$

In Eq. (1.104), $a_j(\bar{k})$ denotes the complex scattering amplitude associated to a single nucleus at the j-th location. According to the description for pure states and statistical mixtures (see Subsection 1.4.3), we recall here that a neutron beam is represented by a statistical mixture (Eq. (1.30)) and has time-independent statistical properties (Eqs. (1.34) and (1.35)). Since for slow neutron holography we require a monochromatic beam, we consider that in the actual case the corresponding probability p_j that matters is maximal (i.e., very close to unity). Moreover, we may notice that the term: $2Re(\sum a_j(\bar{k}))$ in Eq. (1.104) contains the information on the phase of the object wave and, therefore, contains the holographic information. Equation (1.104) is a representation of the general law of interference, as in the classical formulation for the monochromatic case. The first term denotes the reference wave and the third term represents the object wave. Then, according to Eq. (1.103), this case corresponds to $|U_R|^2 = I_0 / R^2$. Moreover, one has $a_j(\bar{k}) = (1/r_j)b_j \exp\left[\left[i(r_j k - \bar{r}_j\bar{k})\right]\right]$, where b_j is the scattering length (say, the coherent amplitude considered in Section 1.3) which, for slow neutrons, has no angular dependence. $\bar{r}_j$ is the position vector of the scattered wave with respect to the source and for a particular nucleus at the j-th position. If the slow neutron beam is not a pulsed one, then, according to Eq. (1.30) in this chapter, there is a maximum quantum-mechanical probability flux (per unit area and unit time interval) for the statistical mixture and the holographic experiment is realizable.

Under the experimental conditions, several restrictions should be imposed, as it is for the case of X-ray holography. The main conditions concern the monochromaticity of the source and the wavelength: $\delta\lambda / \lambda \ll \lambda / r$. We notice that, for slow neutrons, λ is the de Broglie wavelength (λ_{dB}): for thermal neutrons, the latter is the order of 1.8 Å, and the corresponding condition may be fulfilled. However, the main challenge is the weakness of the monochromatic neutron beam, whose flux is typically six orders of magnitude smaller than, for example, that of an X-ray beam from a third-generation synchrotron radiation source. In addition, the fluctuation of an atom in the crystal structure may contribute to a decrease of the intensity. Also, the amplitude of the hologram is affected by the absorption of neutrons and gamma rays by the sample.

As for other additional requirements, it is advantageous to cast the crystal under spherical shape in order to avoid distortions in the scattered wave. Bragg reflections have to be corrected for, as in any other standard transmission holographic technique.

The recorded hologram is usually digitized, and an additional digital image treatment is needed. Due to the fact that the holographic term $(2Re(\sum a_j(\bar{k}))$ carries low intensity, there are requirements for the digitized recorded image to contain an appropriate high number of counts in each pixel (i.e., larger than the order of 106).

Despite all the mentioned stringent requirements, slow neutron holography has reached certain level of development. In 2002, L. Cser et al. [51], using the technique of the internal detector, obtained a hologram of a spherically shaped single crystal of $Pb_{0.9974}Cd_{0.0026}$ with a diameter of about 7 mm, used as the sample. In the holographic experiment the Cd atoms behaved as highly efficient detectors. The lead nuclei played the role of the object, while the cadmium nuclei served as point-like detectors inside the sample. Figure 1.18 displays the schematic diagram of the 12 Pb atoms surrounded by the Cd atom at the center. The actual disposition of the phase-centered-cubic (fcc) lattice was recorded at the hologram. One of the main characteristics of this result was the achievement, for the first time, of the determination of the structure, with atomic resolution, of a quasi-crystal by means of the neutron holographic technique.

Following with the optimization of neutron holographic techniques, in 2017 Hayashi et al. [52] have developed new experiments by using a pulsed slow

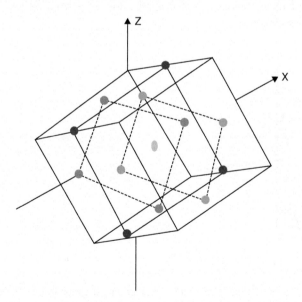

FIGURE 1.18 Schematic diagram of the spatial disposition of Pb atoms, with Cd atom centered, in a $Pb_{0.9974}Cd_{0.0026}$ quasi-crystal, as obtained from slow neutron holographic technique (see Ref. [50]).

neutron source. The beam was issued from the spallation source facility in Japan (J-Parc). Those authors have studied a Eu-doped CaF_2 single crystal and obtained a clear three-dimensional atomic image around trivalent Eu substituted for divalent Ca, revealing an interesting feature of the local structure that allows it to maintain charge neutrality. In this work, the pulsed neutron source emitted at $\lambda_{dB} = 5.6$ Angstrom. The 3D real-space image was reconstructed from the theoretical holograms normalized by the 2D incident monitor pattern and by applying a convenient algorithm. Figure 1.19 displays the neutron holograms of environmental structure around Eu in CaF_2. The pulsed neutron source technique seems to overcome the problem of the low neutron intensity associated to neutron continuous flux sources. Notice that, in this particular work, the holographic experimental setup did not exactly fit with Gabor's hologram scheme: rather, the photodetector was located at $90°$ from the neutron beam source, then, enhancing the scattering of the neutron beam.

The quite active field of material science studies for quasi-crystal structures is, at present, triggering and enhancing further developments in neutron holography. These new contributions are now devoted to macroscopic objects, for which neutron holography provides information on the internal structure of a particular object and reveals neutron beam structures, as well. As an example, in 2016, D. Sarenac et al. [53] have obtained a spiral phase diffractive plate by means of neutron interferometry in a Mach-Zender type interferometer. The resulting hologram is a standard fork dislocation image, which could be used, eventually, to reconstruct neutron beams with various orbital angular momenta [53].

From all the above examples, it is clear that slow neutron holography is an active field in holographic techniques and has applications in material science and neutron beams characterization.

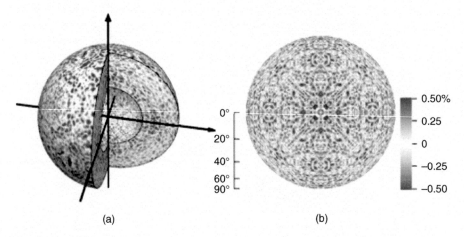

(a) (b)

FIGURE 1.19 (a) Structure of the volume hologram corresponding to the environmental structure around Eu in CaF_2. (b) The corresponding 2D hologram at $|\vec{k}| = 4.05$ Å^{-1}.
Source: see Ref. [52].

1.7.6 EFFECTS OF GRAVITY ON SLOW NEUTRONS: THE PRINCIPLE OF EQUIVALENCE AT THE QUANTUM LEVEL

We shall now turn to the influence of gravity on slow neutrons, briefly. The difference of potential energies of a neutron, subject to gravity, between two different positions with heights z_1 and $z_0 (< z_1)$, vertically on the surface of the Earth, is $V_{grav,0} = m_n.g.(z_1 - z_0)$ (g being the acceleration of gravity, $g = 9.8$ m·s^{-2}, m denoting 1 m). For $z_1 - z_0$ about 1 m, $V_{grav,0}$ is about 10^{-7} eV, which is comparable to typical values for Eq. (1.5). We shall regard z_0 and z_1 as reference heights. If the slow neutron is at the arbitrary height z, the difference of potential energies between z and z_0, also subject to gravity, is $V_{grav}(z) = V_{grav,0}[1 + (z - z_1)/(z_1 - z_0)]$. For $|z - z_1|$ not exceeding 1 cm. which, in practice, will be the case for (essentially all or almost) horizontal motions of neutrons to be considered here, $V_{grav}(z)$ will be approximately equal to $V_{grav,0}$. In turn, a constant $V_{grav,0}$ cancels out in various physically interesting quantities. Then, for a slow neutron moving on the surface of the Earth with differences of height not exceeding 1 cm, and as a zeroth-order approximation, the effects of gravity on it will be neglected and so shall we proceed safely in our study. In case that more accuracy in the treatment of gravity effects were required, one could start with Eq. (1.2), $V(\bar{x})$ being replaced by $V(\bar{x}) + V_{grav}(z)$ and $V(\bar{x})$ being now given through Eq. (1.5) in usual cases. In search for better accuracy, for differences of height not exceeding 1 cm), one could regard $V_{grav}(z)$ as a small perturbation of $V(\bar{x})$. Quantum states of neutrons in the Earth's gravitational field have been observed experimentally, as another demonstration of the universality of the quantum properties of matter [54].

There are interference phenomena of a neutron with itself due to Earth's gravity. This phenomenon has been shown experimentally by using neutron interferometry. The experiment made use of an interferometer of the Bonse-Hart type. The two relevant partial beams traveling inside the interferometer followed paths at different heights. This introduced an additional amplitude (a phase factor), due to the influence of gravity (say, of V_{grav}), in one of the partial beams. As the two partial beams were recombined at the detector, there was a phase difference between them, giving rise to an interference pattern. See Ref. [55]. The experimental results agreed with the quantum-mechanical prediction. See also Refs. [33, 34]. The results of those experiments were consistent with the validity of the Schrödinger equation when gravity is included (i.e., when $V(\bar{x})$ is replaced by $V(\bar{x}) + V_{grav}(z)$), in the scales of distances and energies actually explored. That is, the experiment in Ref. [55] shows that, gravity acts microscopically like any other force placed into the Schrödinger equation, as emphasized in Ref. [34].

There is a subtle issue involving gravity, nonrelativistic classical physics, the nonrelativistic Schrödinger equation and the (weak) equivalence principle. The equality between the inertial and gravitational masses for any given body or particle is supported by successive experiments [56]. Throughout this chapter, masses are always understood to be the inertial ones. That fundamental equality has given rise to the (weak) equivalence principle in classical physics, which played a key role in the foundations of classical general relativity [56]. The latter principle has been posed into various similar forms. A simplified one, which suffices for our purposes, is: the trajectory of a classical point mass subject to a gravitational field depends only on its initial position and velocity, and is independent of its composition and structure. That is, a neutron and a proton (with mass m_p), in classical mechanics and subject only to the gravitational field (in vacuum), follow the same motion if they had the same initial positions and velocities. Let us consider an inertial reference frame S_i (namely, one in which any free classical particle, not subject to external forces, moves along a straight line, with constant velocity), assume that Earth is, approximately, an inertial reference frame (disregarding its motion around the Sun and its own rotation) and let S_{n-i} be a non-inertial reference frame which falls vertically toward Earth with a constant acceleration equal to the acceleration of gravity (say, S_{n-i} is an elevator in free fall toward Earth). By assumption, S_{n-i} occupies only a limited spatial region, in which the gravitational field due to Earth can be well approximated through the standard constant acceleration of gravity. The motions of the above classical neutron and proton (in vacuum) with respect to S_{n-i} follow entirely similar straight paths, with constant velocities. Let us now deal with the (nonrelativistic) quantum-mechanical separate evolutions of a microscopic neutron and proton (also in vacuum) in some limited vacuum region (no material medium), in both S_i and S_{n-i}. We denote spatial coordinates measured with respect to S_i and S_{n-i} as $\bar{x}_i$ and $\bar{x}_{n-i}$, respectively (the time t being the same in both reference frames). In S_i, the Schrödinger equation for the microscopic neutron wave function $\Psi_{i,n}(\bar{x}_i;t)$ is given in Eq. (1.2) with $V(\bar{x}_i) = V_{grav}(z_i)$, and so on for the microscopic proton wave function $\Psi_{i,p}(\bar{x}_i;t)$ (with mass m_p). In S_{n-i}, since the acceleration of Earth's gravity is exactly cancelled for both particles by the constant acceleration of the latter accelerated reference frame, the Schrödinger equations for the microscopic neutron and proton wave functions $\Psi_{n-i,n}(\bar{x}_{n-i};t)$ and $\Psi_{n-i,p}(\bar{x}_{n-i};t)$ are the interaction-free ones:

$$\left[-\frac{\hbar^2}{2m_n}\Delta_{n-i} \right]\Psi_{n-i,n}(\bar{x}_{n-i};t) = i\hbar\frac{\partial\Psi_{n-i,n}(\bar{x}_{n-i};t)}{\partial t}, \tag{1.105}$$

$$\left[-\frac{\hbar^2}{2m_p}\Delta_{n-i} \right]\Psi_{n-i,p}(\bar{x}_{n-i};t) = i\hbar\frac{\partial\Psi_{n-i,p}(\bar{x}_{n-i};t)}{\partial t}. \tag{1.106}$$

Partial derivatives in Δ_{n-i} are understood to be performed with respect to $\bar{x}_{n-i}$. See Ref. [33] for the relationships between wave functions in S_i and S_{n-i}. We stress that Eqs. (1.105) and (1.106) refer to the same S_{n-i}. After the preceding development, we now arrive at the announced subtle issue: as Eqs. (1.105) and (1.106) depend on the (different) masses m_n and m_p, the quantum motions of neutron and proton are not the same in S_{n-i}, but they depend on the type of particle, contrary to the classical macroscopic case. This is due to the presence of $\hbar$, the dimensionful constant genuine of microscopic scales and phenomena. This constitutes a qualification of the (weak) equivalence principle at the quantum level and, at least, at the nonrelativistic regime. Alternatively, the standard form of the (weak) equivalence principle, as reminded above, appears to apply exactly only in the classical limit. See Ref. [34] for a wide and lucid discussion.

1.8 THE EXTINCTION THEOREM: A NEW APPROACH

1.8.1 A QUALITATIVE DISCUSSION

The phenomenon of the refraction of the neutron wave, discussed in Subsection 1.5.1, provides a good example of the following physical fact: the incoming wave $\Phi(\overline{x})_{1,in}$, propagating initially in medium 1 (the half-space $z < 0$), becomes completely "extinguished" and the transmitted wave, propagating in medium 2 (the half-space $z > 0$), is $\Phi(\overline{x})_2$. Let us discuss it in more detail. The neutron is subject to different interactions in the two half-spaces. The neutron wave has a physically possible form and propagation in medium 2 which differ from that in medium 1. The incoming neutron wave is genuine of medium 1 ($z < 0$). That wave disappears beyond $z = 0$ and it is replaced by that genuine of medium 2 ($z > 0$). This extinction of one wave with the subsequent emergence of another wave is an example of the so-called "extinction theorem." The important physical phenomenon expressed by the latter was first analyzed for classical electromagnetic waves by Oseen [58] and Ewald [57], and extended later to neutron wave propagation. See Refs. [2, 5, 19, 30].

Moreover, the extinction theorem has a far more general character, holding for more general surfaces S separating two different media (1 and 2). In fact, its validity requires, essentially, only that S be (i) either a closed finite surface (enclosing medium 1, with medium 2 being outside S, or conversely) [5], (ii) or an infinite one (like, for instance, in the case discussed in Subsection 1.5.1). The extinction theorem holds independently on the shape of S. It also holds for more general media (in particular, periodic ones) as discussed below.

We shall now discuss two interpretations of the extinction theorem, concentrating on slow neutron waves. Let an incoming wave propagate in medium 1 and approach the surface S separating the former from medium 2.

A first interpretation of the theorem states that a_1) new waves are generated in the surface S, b_1) the extinction of the incoming wave (genuine of medium 1) in medium 2 is due to a completely destructive interference between the former incoming wave with part of the new waves generated in S, c_1) the wave genuine of and propagating in medium 2 is the net result of the remainder of the waves generated in S.

A second interpretation of the theorem states that: a_2) upon arrival of the incoming wave from medium 1 in medium 2, new waves are generated in the latter (due to the different interaction suffered by the neutron), b_2) the extinction of the incoming wave (genuine of medium 1) in medium 2 is due to a completely destructive interference between the former incoming wave with part of the new waves generated in medium 2, c_2) the wave genuine of and propagating in medium 2 is the net result of the remainder of the waves generated in medium 2.

Both interpretations turn out to be complementary of and compatible with each other, as emphasized in Ref. [5].

Within the framework of the second interpretation, further approximate justifications of the extinction theorem were given in Ref. [59], for a generic microscopic particle, specifically based upon multiple scattering theory, Ref. [60], for classical electromagnetic waves (taking into account, approximately, their vector character), also based upon multiple scattering theory, and Refs. [61] and [62], for a quantum-mechanical particle, when medium 1 was vacuum and a periodic potential occupied medium 2 (say, a perfect semi-infinite three-dimensional crystal). The treatment of dynamical diffraction in Subsections 1.7.2.1 to 1.7.2.3 employed mostly the second interpretation of the extinction theorem.

We shall outline below a new formulation of the extinction theorem, in the framework of the second interpretation and employing Green's functions and scattering integral equations. Our treatment below is essentially different from those in Refs. [2] and [5].

The new approach to be introduced here has the following methodological motivation. Upon studying wave propagation (generated by some incoming wave) in two different media separated by a surface of discontinuity S, the formulation of the standard boundary conditions (the counterparts of Eqs. (1.46) and (1.47)) on S lead, in general, to inhomogeneous linear equations of first kind, namely, of the type $K_1 A = A_{1,in}$, for an (in principle) infinite set of unknown amplitudes A. The kernel K_1 is some known linear operator and $A_{1,in}$ is some known set of amplitudes, determined by the incoming wave. Such inhomogeneous linear equations, although can be solved easily in simple cases (like that in Subsection 1.5.1), are rather hard to solve in general and, moreover, cannot be analyzed by successive iterations. Then, it is potentially very interesting to recast, if possible, such a problem into equivalent inhomogeneous linear equations of second kind, namely, of the type $A = A_{2,in} + K_2 A$, the kernel K_2 being another known linear operator and $A_{2,in}$ being another known set of amplitudes, also determined by the incoming wave. The solution A of the second linear equations (namely, those of second kind) can be analyzed by successive iterations and one could try several techniques available (for instance and if possible, Fredholm's theory of linear integral equations of second kind). Upon adopting the second policy, to which we shall adhere, the pending problem would be to get "good" inhomogeneous linear equations of second kind which be well behaved. Our treatment below will follow and provide an example of that policy. First, we shall find an inhomogeneous linear equation of second kind (Eq. 1.108) posing certain problems, but providing the basis for further inhomogeneous linear equations of second kind (Eqs. (1.114) and (1.116)), which will be adequate.

1.8.2 The New Approach Illustrated in a Simple Case

We shall consider a simplification of the case treated in Subsection 1.5.1, where the derivation of the Fresnel formulas for slow neutrons omitted any reference to the extinction theorem. Now, medium 1 (vacuum, with $\rho_1 = 0$ and $b_1 = 0$) and medium 2 (a homogeneous one, with constant $V = V_2 \neq 0$, $\rho_2 b_2 \neq 0$, given in Eq. (1.5)) occupy, respectively, the lower ($z < 0$) and the upper ($z > 0$) three-dimensional half-spaces and S is the plane $z = 0$. Our treatment below will also lead to the Fresnel formulas and will be somewhat lengthier; the present justification will be aimed to display the generation of new waves by effective interactions in the extended medium 2 (and not by S) and, in so doing, to illustrate the power of the technique.

We shall write: $\bar{x} = (x, y, z) = (\mathbf{x}, z)$, $\mathbf{x}$ being a two-dimensional vector. Similarly, $(\mathbf{k}_{1,in}, k_{1,in,z}) = \bar{k}_{1,in}$. Let

$$\Phi(\bar{x})_{1,in} = \exp\left[i\bar{k}_{1,in}\bar{x}\right] = \Phi(\mathbf{x}, z)_{1,in} = \exp\left[i(\mathbf{k}_{1,in}\mathbf{x} + k_{1,in,z}z)\right] \qquad (1.107)$$

be the incoming plane wave, given in Eqs. (1.43), (1.49), and let $\Phi(\overline{x}) = \Phi(\mathbf{x}, z)$ be the t-independent total wave function for any $\overline{x}$, in both media ($z < 0$ and $z > 0$). It fulfills Eq. (1.2), with $V(\overline{x}) = 0$ for $z < 0$ and $V(\overline{x}) = V_2 \neq 0$ for $z > 0$. In turn, Eq. (1.2) can be recast into the well-known (t-independent) linear integral equation for scattering in quantum mechanics, for any $(\mathbf{x}, z)$ [20, 59]:

$$\Phi(\mathbf{x}, z) = \Phi(\mathbf{x}, z)_{1, in} + \int_0^{+\infty} dz' \int d^2 \mathbf{x}' G_1(\mathbf{x} - \mathbf{x}', z - z') V_2 \Phi(\mathbf{x}', z'), \qquad (1.108)$$

$$G_1(\mathbf{x} - \mathbf{x}', z - z') = \int_{-\infty}^{+\infty} \frac{dl_z'}{2\pi} \int \frac{d^2 \mathbf{l}'}{(2\pi)^2} \frac{\exp[i(\mathbf{l}'(\mathbf{x} - \mathbf{x}') + l_z'(z - z'))]}{E + i\varepsilon - (\hbar^2 / 2m_n)(\mathbf{l}'^2 + l_z'^2)}, \qquad (1.109)$$

G_1 is the t-independent three-dimensional Green's function in vacuum (medium 1), also given in Eq. (1.12). Equation (1.108) expresses the total wave function $\Phi(\mathbf{x}, z)$ for any $\overline{x}$ in terms of the incoming $\Phi(\mathbf{x}, z)_{1, in}$ in medium 1 and the effect of the interaction in medium 2, through G_1 and $V(\overline{x}) = V_2 \neq 0$ for $z > 0$. We emphasize that Eq. (1.108) is a representation for $\Phi(\mathbf{x}, z)$ in $z < 0$ (in terms of $\Phi(\mathbf{x}', z')$ in $z' > 0$) and a strict integral equation for $\Phi(\mathbf{x}, z)$ in $z > 0$ (in terms of $\Phi(\mathbf{x}', z')$ in $z' > 0$). Equation (1.108) gives rise to conceptual difficulties, arising from the fact that, as $V(\overline{x}) = V_2 \neq 0$, the product $G_1(\mathbf{x} - \mathbf{x}', z - z') V_2$ does not tend to vanish as $z' \to +\infty$. The successive iterations of Eq. (1.108) for $\Phi(\mathbf{x}, z)$ in $z > 0$ display what may be called "secular terms": in fact, the first iteration (upon direct substitutions and integrations) displays terms linear in z, and higher iterations lead to polynomials of higher order in z. Such "secular terms" would imply an unphysical (divergent) increase of $\Phi(\mathbf{x}, z)$ as $z \to +\infty$ and, so, are necessarily spurious, say, artifacts of the direct successive iterations of Eq. (1.108) for $\Phi(\mathbf{x}, z)$ in $z > 0$. Those "secular terms" can indeed be eliminated, by reabsorbing them into a new plane wave containing the index of refraction for medium 2 (like in Refs. [59, 60–62]), thereby generating the wave genuine of medium 2 and, hence, implementing the extinction theorem. The successive eliminations of the "secular terms," order by order, as one iterates Eq. (1.108) for $\Phi(\mathbf{x}, z)$ in $z > 0$, turns out to be a laborious and cumbersome procedure. One key remark: the n-th iteration of the integral equation for $\Phi(\mathbf{x}, z)$ in $z > 0$ contains integrals like $\int_0^{+\infty} dz_1' \ldots \int_0^{+\infty} dz_n'$, that is, only the structure $\int_0^{+\infty} dz'$ appears. Those integrals generate precisely the "secular terms." Thus, Eq. (1.108) is not adequate, as it stands.

In order to bypass the generation of "secular terms" (and their subsequent cumbersome eliminations), we shall now outline a new procedure. It will be based upon a new integral equation the iterations of which do not give rise to the occurrence of $\int_0^{+\infty} dz_1' \ldots \int_0^{+\infty} dz_n'$. For that purpose, we recast Eq. (1.4) as

$$\left[-\frac{\hbar^2}{2m_n} \Delta + V(\overline{x})_2' \right] \Phi(\overline{x}) = (E - V_2) \Phi(\overline{x}), \qquad (1.110)$$

$$V(\overline{x})_2' \equiv V(\overline{x}) - V_2. \qquad (1.111)$$

Notice that $V(\overline{x})_2' = -V_2, z < 0$ while $V(\overline{x})_2' = 0, z > 0$. In turn, Eq. (1.110) leads to the new (t-independent) linear integral equation for the same $\Phi(\mathbf{x}, z)$, for any $(\mathbf{x}, z)$:

$$\Phi(\mathbf{x}, z) = \int_{-\infty}^{0} dz' \int d^2\mathbf{x}' G_2(\mathbf{x} - \mathbf{x}', z - z')[-V_2]\Phi(\mathbf{x}', z'), \tag{1.112}$$

$$G_2(\mathbf{x} - \mathbf{x}', z - z') = \int_{-\infty}^{+\infty} \frac{dl_z'}{2\pi} \int \frac{d^2\mathbf{l}'}{(2\pi)^2}$$

$$\times \frac{\exp\left[i(\mathbf{l}'(\mathbf{x} - \mathbf{x}') + l_z'(z - z'))\right]}{E - V_2 + i\varepsilon - (\hbar^2 / 2m_n)(\mathbf{l}'^2 + l_z'^2)}, \tag{1.113}$$

G_2 is the t-independent three-dimensional Green's function for medium 2, supposing that it occupies the whole three-dimensional space: G_2 is obtained from G_1, just by replacing in the latter E by $E - V_2$. We emphasize a crucial point: There can be no inhomogeneous term in Eq. (1.112), for the actual case in which the incoming plane wave is, precisely, $\Phi(\mathbf{x}, z)_{1,in}$ in medium 1 (coming from $z \rightarrow -\infty$). There would have been an inhomogeneous term in Eq. (1.112) if the incoming neutron would have come from $z \rightarrow +\infty$, which is not the case here! Next, we replace $\Phi(\mathbf{x}', z')$, with $z' < 0$, in the right-hand side of Eq. (1.112) by the right-hand side of Eq. (1.108). We get

$$\Phi(\mathbf{x}, z) = Inh + \int_{-\infty}^{0} dz' \int d^2\mathbf{x}' G_2(\mathbf{x} - \mathbf{x}', z - z')[-V_2]$$

$$\times \int_{0}^{+\infty} dz'' \int d^2\mathbf{x}'' G_1(\mathbf{x}' - \mathbf{x}'', z' - z'')V_2\Phi(\mathbf{x}'', z''), \tag{1.114}$$

$$Inh \equiv \int_{-\infty}^{0} dz' \int d^2\mathbf{x}' G_2(\mathbf{x} - \mathbf{x}', z - z')[-V_2]\Phi(\mathbf{x}', z')_{1,in}. \tag{1.115}$$

Equation (1.114), with the inhomogeneous term *Inh*, is a linear integral equation for $\Phi(\mathbf{x}, z)$ in $z > 0$ (in terms of $\Phi(\mathbf{x}', z')$ in $z' > 0$). The key point (and the crucial advantage!) is that the n-th iteration of the integral Eq. (1.114) for $\Phi(\mathbf{x}, z)$ in $z > 0$ contains always alternating integrals $\int_{-\infty}^{0} dz_1'$ and $\int_{0}^{+\infty} dz_2'$ (the successive occurrence of the unique structure $\int_{0}^{+\infty} dz_1$ being avoided completely). Accordingly, the successive iterations of Eq. (1.114) for $\Phi(\mathbf{x}, z)$ in $z > 0$ contain no secular terms. A direct, albeit lengthy, computation enables to solve Eq. (1.114) for $\Phi(\mathbf{x}, z)$ in $z > 0$, without approximation. The result is $\Phi(\mathbf{x}, z) = \Phi(\overline{x})_2 = A_2 \exp\left[i\overline{k}_2\overline{x}\right]$ (i.e., Eq. (1.44), with A_2 given in Eq. (1.52)), which clearly implement the extinction theorem. See exercise 1.6. The key $\int_{0}^{+\infty} dz'' \int d^2x''$ in the second term on the right-hand side of Eq. (1.114) and, upon iterations, the subsequent alternating integrals $\int_{-\infty}^{0} dz_1'$ and $\int_{0}^{+\infty} dz_2'$ have given

rise consistently to the cancellation of waves genuine of medium 1 and the generation of waves genuine of medium 2 (say, in the framework of the second interpretation of the extinction theorem).

Similarly, a linear integral equation for $\Phi(\mathbf{x}, z)$ in $z < 0$ is immediately obtained
by replacing $\Phi(\mathbf{x}', z')$, with $z' > 0$, in the right-hand side of Eq. (1.108) by the right-
hand side of Eq. (1.114). We get

$$
\Phi(\mathbf{x}, z) = \Phi(\mathbf{x}, z)_{1,in} + \int_0^{+\infty} dz' \int d^2\mathbf{x}' G_1(\mathbf{x} - \mathbf{x}', z - z') V_2
$$

$$
\times \int_{-\infty}^0 dz'' \int d^2\mathbf{x}'' G_2(\mathbf{x}' - \mathbf{x}'', z' - z'')[-V_2]\Phi(\mathbf{x}'', z'')
$$

(1.116)

Equation (1.116) for $\Phi(\mathbf{x}, z)$ in $z < 0$ has the same good properties as Eq. (1.114), namely,
its n-th iteration contains always alternative integrals $\int_{-\infty}^0 dz_1'$ and $\int_0^{+\infty} dz_2'$. Another
direct, also lengthy, computation enables to solve Eq. (1.116) for $\Phi(\mathbf{x}, z)$ in $z < 0$, with-
out approximation. The result is, consistently: $\Phi(\mathbf{x}, z) = \Phi(\overline{x})_1 = \Phi(\overline{x})_{1,in} + \Phi(\overline{x})_{1,re}$
with $A_{1,re}$ given in Eq. (1.53).

Equations (1.114) and (1.116) are adequate inhomogeneous linear equation of second kind.

One could proceed to other problems of slow neutron wave propagation in two
different extended media separated by an infinite surface S, with shape different
from that of an infinite plane. In a naive (scattering-theory-like) formulation with
one Green's function (the counterpart of Eqs. (1.108) and (1.109), one would eventually face "secular terms": then, the problem would be to absorb them so that genuine waves in each media appear and the corresponding extinction theorem become
explicit. For certain S, the above new formulation could be extended in principle (say,
the counterparts of Eqs. (1.114) and (1.116)), so as to bypass the "secular terms" and
give rise to, at least, some approximation method. A necessary condition would be
that some adequate integral representations for the corresponding Green's functions
(more than one!) for the corresponding media be available.

1.9 CONCLUSION AND FINAL COMMENTS

We have concentrated on introduced various basic subjects of neutrons optics. We
shall enumerate them:

General aspects of (mostly, slow) neutron physics have been summarized, in an
updated way as much as possible (Section 1.2).

Quantum mechanics of slow neutrons has been treated, with a focus on a simplified justification of the basic approximate optical potential (Section 1.3).

The probabilistic interpretation of the slow neutron wave function (pure states
and statistical mixtures) has been analyzed, in connection to neutron interferometry
experiments (Section 1.4).

Reflection and refraction of slow neutrons by an ideal planar surface and the Fresnel formulas have been studied (Section 1.5), paving the way for a new fresh analysis of the basic extinction theorem later (Section 1.8).

Neutron spin and its magnetic interactions have been introduced rather briefly (Section 1.6), an updated and detailed study of the field of neutron spin optics being presented in Chapter 5.

Several important developments for thermal neutrons have been dealt with, namely, scattering by a medium, dynamical diffraction by crystals and some applications (Bragg and Laue diffractions), neutron diffraction versus X-rays diffraction, neutron holography and effects of gravity (Section 1.7).

The important extinction theorem for slow neutrons and a new approach to it in a simple context (in connection to Section 1.5) have been considered (Section 1.8).

Our selection and presentation aim to provide suitable introductory connections with the following chapters, which will present updated accounts of very interesting, active, and more advanced fields of research and applications.

Besides other omissions, and for brevity, we have not treated other important experiments with thermal neutrons in interferometers. We shall quote two of them: (i) the influence of Earth's rotation (with a suitable arrangement, a phase factor being introduced in one of the two partial neutron beams traveling through the interferometer) [63], (ii) the overall change of sign of the neutron wave function due to a complete rotation of an angle 2π about an axis [64]. The results of those experiments were consistent, in each case, with the predictions of quantum mechanics. Other aspects of neutron optics lie definitely outside our scope in this chapter as they are dealt with elsewhere along this book. One of them includes diffraction phenomena by apertures in plane opaque (very absorbing) screens. Neutrons emitted by a source are diffracted by a diffracting opening and subsequently detected: Fraunhofer diffraction occurs if the incoming wave on the aperture and the diffracted wave received at the detector are approximately plane waves. If the latter conditions are not met, one has Fresnel diffraction. See Chapter 2 in this book (Subsections 2.2.1, 2.2.2, 2.2.3, and 2.2.3.1). For other presentations of neutron optics, dealing with subjects omitted here, see Refs. [2, 5, 65, 66].

1.10   ACKNOWLEDGMENTS

R. F. Álvarez-Estrada acknowledges Mineco (Ministry of Economy and Competitiveness, Spain) for Projects FIS2015-65078-C2-1-P and PGC2018-094684-B-C21 (partially funded by FEDER) for financial support. We are grateful to Prof. Ángel S. Sanz and to Mr. David Fernandez for kind help with the files.

1.11   EXERCISES

1.1. Justify Eqs. (1.50), (1.51), (1.52), and (1.53).

1.2. Study the reflection and the refraction of a slow neutron by a planar surface separating two different media, when b_2 is complex ($Imb_2 < 0$) and derive the generalizations of Eqs. (1.52) and (1.53).

1.3. Consider the diffraction of a slow neutron by a thin crystal lattice, treated in Subsection 1.7.2.2, when the incoming wavevector $\overline{k}$ fulfills $\overline{b}^2 - 2\overline{b}\overline{k} \neq 0$ and

$\bar{b}^2 - 2\bar{b}.(k_x, k_y, -k_z) \neq 0$ for any $\bar{b} \neq (0,0,0)$. Derive the conditions character-izing the waves propagating inside the crystal and those corresponding to the reflected and transmitted waves, outside the crystal.

1.4 We consider the diffraction of a neutron by a thin crystal lattice (see Subsection 1.7.2.3), when the incoming wavevector $\bar{k}$ and $\bar{b}_{Br} \neq (0,0,0)$ are such that $\bar{b}_{Br}^2 - 2\bar{b}\bar{b}_{Br}\bar{k} \equiv \eta$ is small. We accept the existence of, at least, one solution such that $\bar{q} = \bar{k} + \varepsilon|\bar{k}|(0,0,1)$, ε being of order ξ. Reduce Eq. (1.89), approximately, to an algebraic equation algebraic equation of second degree for ε. In that case, characterize the amplitudes yielding the wave function inside the crystal.

1.5. We continue dealing with the same physical situation as in exercise 1.4. We impose only the continuity of the various wave function $\Phi(\bar{x})$ in $z = 0$ and in $z = z_0$. Formulate the conditions yielding ε_2 and ε_3 and the amplitudes $A((0,0,0))_\pm$, $A_{tr,1}(\bar{k})$, $A_{tr,2}(\bar{k}_{tr})$, and $A_{re}(\bar{k}_{re})$.

1.6. Derive Eq. (1.50) from Eq. (1.114), as an example of the extinction theorem.

1.12 SOLUTIONS TO EXERCISES

1.1. Equations (1.49), (1.46), and (1.47) imply, for any x:

$$\exp ik_{1,in} \sin \varphi_{1,in} x + A_{1,re} \exp ik_{1,re} \sin \varphi_{1,re} x = A_2 \exp ik_2 \sin \varphi_2 x,$$

$$k_{1,in} \cos \varphi_{1,in} \exp ik_{1,in} \sin \varphi_{1,in} x - k_{1,re} \cos \varphi_{1,re} A_{1,re} \exp ik_{1,re} \sin \varphi_{1,re} x \qquad (1.117)$$

$$= k_2 \cos \varphi_2 A_2 \exp ik_2 \sin \varphi_2 x.$$

Equation (1.117) hold only if all exponentials are equal to one another for any x. The latter, since $k_{1,in} = k_{1,re}$ and by recalling Eq. (1.26), imply the validity of Eqs. (1.50) and (1.51). Then, upon factorizing exponentials, one gets

$$1 + A_{1,re} = A_2,$$
$$k_{1,in} \cos \varphi_{1,in} (1 - A_{1,re}) = k_2 \cos \varphi_2 A_2. \qquad (1.118)$$

The solution of the system formed by both Eq. (1.118) for $A_{1,re}$ and A_2 leads directly to Eqs. (1.52) and (1.53).

1.2. As b_2 is complex, so should be at least one component of $\bar{k}_2$. Equations (1.44), (1.45), (1.46), and (1.47) continue to hold when b_2 is complex. The same is true for $\bar{k}_{1,in}$ and $\bar{k}_{1,re}$, as given in Eq. (1.49), in which, without loss of generality, we are including the validity of Eq. (1.50). On the other hand, the expression for $\bar{k}_2$ in Eq. (1.49) should be replaced by

$$\bar{k}_2 = (k_{1,in} \sin \varphi_{1,in}, 0, n_2' k_{1,in} \cos \varphi_{1,in}). \qquad (1.119)$$

In turn, n_2' is determined through Eq. (1.45). One gets directly

$$n_2' = \left[1 - \frac{4\pi\rho_2 b_2}{k_{1,in}^2 \cos^2 \varphi_{1,in}} \right]^{1/2}. \qquad (1.120)$$

Notice that n_2' is complex and, then, so is the z-component of $\bar{k}_2$. After imposing Eqs. (1.46) and (1.47) in the present case, one sees that, with the actual choice for the components (x, y) of $\bar{k}_{1,in}$, $\bar{k}_{1,re}$, and $\bar{k}_2$, exponentials factor out, like in the case with real b_2, treated in exercise 1.1. Equation (1.51) is no longer valid for complex b_2.

Instead of Eq. (1.118), we now get

$$1 + A_{1,re} = A_2,$$
$$k_{1,in} \cos\varphi_{1,in}(1 - A_{1,re}) = A_2 n_2' k_{1,in} \cos\varphi_{1,in}. \tag{1.121}$$

The solution of the system formed by Eq. (1.121) is

$$A_{1,re} = \frac{1 - n_2'}{1 + n_2'},$$
$$A_2 = \frac{2}{1 + n_2'}. \tag{1.122}$$

1.3. We start from Eq. (1.84), and factor out $A((0,0,0))$. We impose the boundary conditions at $z = 0$ (corresponding to the incoming wave $\exp\left[i k \bar{x}\right]$) and at $z = z_0$: recall Eqs. (1.82) and (1.83). The validity of those equations for any x, y implies that there is a unique reflected wave with $\bar{k}_{re} = (k_x, k_y, -k_z)$, and a unique transmitted wave with $\bar{k}_{tr} = \bar{k}$. The solutions for Eq. (1.84), which are compatible with those boundary conditions, are given in Eq. (1.85) and lead to Eq. (1.86). After factoring out exponential dependences in x and y, Eqs. (1.82) and (1.83) become an inhomogeneous linear system formed by four equations for the four amplitudes $A_{re}(\bar{k}_{re})$, $A_{tr}(\bar{k}_{tr})$, $A((0,0,0))_+$, and $A((0,0,0))_-$:

$$1 + A_{re}(\bar{k}_{re}) = A((0,0,0))_+ + A((0,0,0))_-,$$

$$k_z\left[1 - A_{re}(\bar{k}_{re})\right] = q_{z,+} A((0,0,0))_+ + q_{z,-} A((0,0,0))_-,$$

$$A((0,0,0))_+ \exp iq_{z,+} z_0 + A((0,0,0))_- \exp iq_{z,-} z_0 = A_{tr}(\bar{k}_{tr}) \exp ik_z z_0, \tag{1.123}$$

$$q_{z,+} A((0,0,0))_+ \exp iq_{z,+} z_0 + q_{z,-} A((0,0,0))_- \exp iq_{z,-} z_0$$

$$= k_z A_{tr}(\bar{k}_{tr}) \exp ik_z z_0.$$

We shall omit the explicit solution of the above system.

1.4. One has $\bar{q}^2 - \bar{k}^2 \simeq 2\varepsilon|\bar{k}|k_z$, $\left[(\bar{q} - \bar{b}_{Br})^2 - \bar{k}^2 \simeq 2\varepsilon|\bar{k}|(k_z - (\bar{b}_{Br})_z) + \eta$, $(\bar{b}_{Br})_z$ being the z-component of $\bar{b}_{Br}$: notice that we have retained the small contribution of η. Then, Eq. (1.89) can be approximated by the following equation of second degree in $\varepsilon_1 \equiv \varepsilon / \xi$ (in principle, of order ξ^0).

$$\left[2\frac{k_z}{|\bar{k}|}\varepsilon_1 + 1\right] \cdot \left[2\frac{k_z - (\bar{b}_{Br})_z}{|\bar{k}|}\varepsilon_1 + \frac{\eta}{|\bar{k}|^2\xi} + 1\right] \tag{1.124}$$

$$- F(\bar{b}_{Br})F(-\bar{b}_{Br}) = 0.$$

The solution of Eq. (1.124) is, expressed directly through ε, is

$$\varepsilon_{\pm} = \xi \frac{|\bar{k}|}{2k_z} \left[-\frac{1 + \alpha_1 - \alpha_2}{2\alpha_1} + (\pm) \left(\frac{F(\bar{b}_{Br})F(-\bar{b}_{Br})}{\alpha_1} + \frac{(-1 + \alpha_1 + \alpha_2)^2}{4\alpha_1^2} \right)^{1/2} \right], \quad (1.125)$$

with $\alpha_1 = (k_z - (\bar{b}_{Br})_z)/k_z$ and $\alpha_2 = -\eta/(|\bar{k}|^2 \xi)$. Upon keeping only one equation in Eq. (1.88) (for instance, the first one), one gets the following expression for the ratio $A(\bar{b}_{Br})/A((0,0,0))$ for both cases with $\varepsilon = \varepsilon_{\pm}$:

$$a_{\pm} = -\frac{\bar{q}^2 - \bar{k}^2 + \xi \bar{k}^2}{\xi \bar{k}^2 F(-\bar{b}_{Br})} \simeq -\frac{2\varepsilon_{\pm} |\bar{k}k_z}{\xi \bar{k}^2 F(-\bar{b}_{Br})}. \quad (1.126)$$

1.5. One has $\bar{k}_{tr}^2 = (\bar{k} - \bar{b}_{Br} + \varepsilon_3 |\bar{k}|(0,0,1))^2 = \bar{k}^2$ and $\bar{k}_{re}^2 = (\bar{k} - \bar{b}_{Br} - \varepsilon_2 |\bar{k}|(0,0,1))^2$ $= \bar{k}^2$. These equations determine ε_3 and ε_2, respectively. Equations (1.90), (1.91), and (1.92), together with the validity of the first Eq. (1.82) (for $z = 0$ and any x, y), imply:

$$1 = A((0,0,0))_+ + A((0,0,0))_- ,$$
$$A_{re}(\bar{k}_{re}) = A((0,0,0))_+ a_+ + A((0,0,0))_- a_-. \quad (1.127)$$

On the other hand, Eqs. (1.92) and (1.93), combined with the first Eq. (1.83) (for $z = z_0$ and arbitrary x, y), yield:

$$A((0,0,0))_+ \exp\left[i\left(k_z + \varepsilon_+ |\bar{k}|\right) z_0 \right] + A((0,0,0))_-$$
$$\exp\left[i\left(k_z + \varepsilon_- |\bar{k}|\right) z_0 \right] = A_{tr,1}(\bar{k}) \exp\left[ik_z z_0 \right],$$
$$A((0,0,0))_+ a_+ \exp\left[i\left(k_z + \varepsilon_+ |\bar{k}|\right) z_0 \right] + A((0,0,0))_-$$
$$a_- \exp\left[i\left(k_z + \varepsilon_- |\bar{k}|\right) z_0 \right] = A_{tr,2}(\bar{k}_{tr}) \exp\left[i\left(k_z + \varepsilon_3 |\bar{k}|\right) z_0 \right]. \quad (1.128)$$

Equations (1.127) and (1.128) constitute an inhomogeneous linear system of four equations with five unknowns: $A((0,0,0))_+$, $A((0,0,0))_-$, $A_{re}(\bar{k}_{re})$, $A_{tr,1}(\bar{k})$, and $A_{tr,2}(\bar{k}_{tr})$.

The so-called Bragg diffraction corresponds to the case $A_{tr,2}(\bar{k}_{tr}) = 0$. Then, Eqs. (1.127) and (1.128) do constitute an inhomogeneous linear system of four equations with four unknowns, and one gets readily

$$A((0,0,0))_+ = -\frac{a_- \exp\left[i\varepsilon_- |\bar{k}| z_0 \right]}{a_+ \exp\left[i\varepsilon_+ |\bar{k}| z_0 \right] - a_- \exp\left[i\varepsilon_- |\bar{k}| z_0 \right]},$$
$$A((0,0,0))_- = \frac{a_+ \exp\left[i\varepsilon_+ |\bar{k}| z_0 \right]}{a_+ \exp\left[i\varepsilon_+ |\bar{k}| z_0 \right] - a_- \exp\left[i\varepsilon_- |\bar{k}| z_0 \right]}. \quad (1.129)$$

$A_{re}(\overline{k}_{re})$ and $A_{tr,1}(\overline{k})$ can be obtained directly in terms of $A((0,0,0))_\pm$, given in Eq. (1.129), by using the second Eq. (1.127) and first Eq. (1.128), respectively.

Laue diffraction occurs when $A_{re}(\overline{k}_{re}) = 0$. Then, Eqs. (1.127) and (1.128) also constitute an inhomogeneous linear system of four equations with four unknowns. One gets easily:

$$A((0,0,0))_+ = -\frac{a_-}{a_+ - a_-},$$

$$A((0,0,0))_- = \frac{a_+}{a_+ - a_-}. \tag{1.130}$$

$A_{tr,1}(\overline{k})$ and $A_{tr,2}(\overline{k}_{tr})$ follow in terms of $A((0,0,0))_\pm$ given in Eq. (1.130), by using the two equations in Eq. (1.128).

1.6. We shall outline the derivation of Eq. (1.52), as an example of the extinction theorem.

We shall cast G_1 and G_2 into more adequate forms, upon integrating over l'_z in both Eqs. (1.109) and (1.113). The integrations are carried out through residue integration in the complex l'_z-plane. Here, $E' = E$ for G_1 and $E' = E - V_2$ for G_2. We use

$$\int_{-\infty}^{+\infty} \frac{dl'_z}{2\pi} \frac{\exp[il'_z(z - z')]}{E' + i\varepsilon - (\hbar^2 / 2m_n)(\mathbf{l}'^2 + l'^2_z)}$$

$$= -\frac{im_n \exp[il_z(E', \mathbf{l}')|z - z'|]}{\hbar^2 l_z(E', l')}, \tag{1.131}$$

$$l_z(E', \mathbf{l}') = + \left[E' - (\hbar^2 / 2m_n)\mathbf{l}'^2 \right]^{1/2}. \tag{1.132}$$

Consequently,

$$G_1(\mathbf{x} - \mathbf{x}', z - z')$$

$$= \int \frac{d^2\mathbf{l}'}{(2\pi)^2} \exp[i\mathbf{l}'(\mathbf{x} - \mathbf{x}')] \frac{(-i)m_n \exp[il_z(E, \mathbf{l}')|z - z'|]}{\hbar^2 l_z(E, \mathbf{l}')}, \tag{1.133}$$

$$G_2(\mathbf{x} - \mathbf{x}', z - z')$$

$$= \int \frac{d^2\mathbf{l}'}{(2\pi)^2} \exp[i\mathbf{l}'(\mathbf{x} - \mathbf{x}')] \frac{(-i)m_n \exp[il_z(E - V_2, \mathbf{l}')|z - z'|]}{\hbar^2 l_z(E - V_2, \mathbf{l}')}. \tag{1.134}$$

We shall make systematic use of

$$\int \frac{d^2\mathbf{x}'}{(2\pi)^2} \exp[i\mathbf{x}'(\mathbf{l} - \mathbf{l}')] = \delta^{(2)}(\mathbf{l} - \mathbf{l}'), \tag{1.135}$$

$\delta^{(2)}$ being the two-dimensional Dirac delta function. First, we shall deal with the inhomogeneous term Inh (Eq. (1.115)) in Eq. (1.114). We replace G_2 in it by Eq. (1.134), and integrate successively over $\mathbf{x}'$ (by using Eq. (1.135)), over $\mathbf{l}'$ (by using the two-dimensional delta function) and over z'. Then,

$$Inh = \frac{(-V_2)(-i)m_n \exp\left[i(\mathbf{k}_{1,in}\mathbf{x} + l_z(E - V_2, \mathbf{k}_{1,in})z)\right]}{\hbar^2 l_z(E - V_2, \mathbf{k}_{1,in})}$$

$$\times \frac{1}{i\left[k_{1,in,z} - l_z(E - V_2, \mathbf{k}_{1,in})\right]}. \tag{1.136}$$

We now turn to the second term on the right-hand side of Eq. (1.114). We replace G_1 and G_2 by using Eqs. (1.133) and (1.134). We integrate successively over two different vectors $\mathbf{l}'$ twice and over $\mathbf{x}'$ and z'. Then, Eq. (1.114) becomes, for $z > 0$:

$$\Phi(\mathbf{x}, z) = Inh + \int d^2\mathbf{x}'' \int_0^{+\infty} dz'' \Phi(\mathbf{x}'', z'')(-V_2)V_2$$

$$\times \int \frac{d^2\mathbf{l}'}{(2\pi)^2} \exp\left[i(\mathbf{l}'(\mathbf{x} - \mathbf{x}''))\right] \exp\left[i(l_z(E - V_2, \mathbf{l}')z + l_z(E, \mathbf{l}')z'')\right] \tag{1.137}$$

$$\times \frac{1}{i\left[k_{1,in,z} - l_z(E - V_2, \mathbf{k}_{1,in})\right]} \frac{(-i)m_n}{\hbar^2 l_z(E, \mathbf{l}')} \frac{(-i)m_n}{\hbar^2 l_z(E - V_2, \mathbf{l}')}.$$

Next, we consider $g(\mathbf{q}) \equiv \int d^2\mathbf{x} \int_0^{+\infty} dz \exp[-i\mathbf{q}\mathbf{x}] \exp[il_z(E, \mathbf{q})z] \Phi(\mathbf{x}, z)$. Then, upon applying $\int d^2\mathbf{x} \int_0^{+\infty} dz \exp[-i\mathbf{q}\mathbf{x}] \exp[il_z(E, \mathbf{q})z]$, Eq. (1.137) yields

$$g(\mathbf{q}) = \frac{1}{i\left[k_{1,in,z} - l_z(E - V_2, \mathbf{k}_{1,in})\right]} \frac{(-i)m_n(2\pi)^2 \delta^{(2)}(\mathbf{k}_{1,in} - \mathbf{q})}{\hbar^2 l_z(E - V_2, \mathbf{k}_{1,in})}$$

$$\frac{1}{il_z(E, \mathbf{q}) + l_z(E - V_2, \mathbf{k}_{1,in})} + g(\mathbf{q})B, \tag{1.138}$$

$$B = (-V_2)V_2 \frac{1}{i\left[l_z(E, \mathbf{q}) + l_z(E - V_2, \mathbf{q})\right]} \frac{(-i)m_n}{\hbar^2 l_z(E, \mathbf{q})} \frac{(-i)m_n}{\hbar^2 l_z(E - V_2, \mathbf{q})}$$

$$\times \frac{1}{i\left[l_z(E, \mathbf{q}) + l_z(E - V_2, \mathbf{q})\right]}. \tag{1.139}$$

Equation (1.138) can be solved compactly, and yields $g(\mathbf{q})$. By plugging $g(\mathbf{q})$ into the right-hand side of Eq. (1.137), we get the following exact structure for $z > 0$:

$$\Phi(\mathbf{x}, z) = \exp\left[i(\mathbf{k}_{1,in}\mathbf{x} + l_z(E - V_2, \mathbf{k}_{1,in})z))A_2'\right], \tag{1.140}$$

with some explicit constant A_2', omitted for simplicity. After some direct algebraic computation, one finds: $A_2' = A_2$, with the same A_2 given in Eq. (1.52), that is, the Fresnel amplitude for refraction, as it should be. In conclusion, the use of Eq. (1.114) implements the extinction theorem.

REFERENCES

1. J. Chadwick, Possible existence of a neutron, *Nature*, 129, 312 (1932).
2. J. Byrne, *Neutrons, Nuclei and Matter: An Exploration of the Physics of Slow Neutrons* (Institute of Physics Publishing, Bristol, 1994).
3. T. E. Mason, T. J. Gawne, S. E. Nagler, M. B. Nestor, and J. M. Carpenter, The early development of neutron diffraction: science in the wings of the Manhattan Project, *Acta Cryst. A*, 69, 3744 (2013).
4. E. Fermi, *Collected Papers*, edited by E. Segre (University of Chicago Press, Chicago, 1962).
5. V. F. Sears, *Neutron Optics: An Introduction to the Theory of Neutron Optical Phenomena and Their Applications* (Oxford University Press, Oxford, 1989).
6. V. F. Sears, Fundamental aspects of neutron physics, *Phys. Rep.*, 82, 1–29 (1982).
7. G. E. Bacon, *Neutron Diffraction*, 3rd ed. (Clarendon Press, Oxford, 1975).
8. D. F. R. Mildner, "Neutron Optics," in *Handbook of Optics*, Volume III, 2nd ed., The Optical Society of America, edited by M. Bass, J. M. Enoch, E. W. Van Stryland, and W. L. Wolfe (McGraw-Hill, New York, 2001).
9. K. S. Krane, *Introductory Nuclear Physics* (John Wiley and Sons, New York, 1988).
10. J. M. Pendlebury et al., Revised experimental upper limit on the electric dipole moment of the neutron, *Phys. Rev. D*, 92, 092003 (2015).
11. Y. Kasesaz et al., BNCT project at Tehran Research Reactor: current and prospective plans, *Prog. Nucl. Energy*, 91, 107–115 (2016).
12. J. Finney and U. Steigenberger, Neutrons for the future, *Physics World*, 27–32, December (1997).
13. W. S. Kiger III, S. Sakamoto, and O. K. Harling, Neutronic design of a fission converter-based epithermal neutron beam for neutron capture therapy, *Nuclear Science and Engineering*, 131, 1–22 (1999).
14. G. Nebbia and J. Gerl, Detection of buried landmines and hidden explosives using neutron, X-ray and gamma-ray probes, *Europhysics News*, 36, 119–123 (2005).
15. W. Kockelmann et al., Neutrons in cultural heritage research, *Journal of Neutron Research*, 14, 37–42 (2006).
16. D. J. Hughes and M. T. Burgy, Reflection of neutrons from magnetized mirrors, *Phys. Rev.*, 81, 498–506 (1951).
17. L. L. Foldy, The multiple scattering of waves. I. General theory of isotropic scattering by randomly distributed scatterers, *Phys. Rev.*, 67, 107–119 (1945).
18. M. Lax, Multiple scattering of waves, *Rev. Mod. Phys.*, 23, 287–310 (1951).
19. M. Lax, Multiple scattering of waves. II. The effective field in dense systems, *Phys. Rev.*, 85, 621–629 (1952).
20. A. Messiah, *Quantum Mechanics*, Vol. II (North-Holland, Amsterdam, 1962).
21. M. L. Goldberger and F. Seitz, Theory of the refraction and the diffraction of neutrons by crystals, *Phys. Rev.*, 71, 294–310 (1947).
22. R. Weinstock, Inelastic scattering of slow neutrons, *Phys. Rev.*, 65, 1–20 (1944).
23. O. Halpern, M. Hamermesh, and M. H. Johnson, The passage of neutrons through crystals and polycrystals, *Phys. Rev.*, 59, 981–996 (1941).
24. C. G. Shull and E. O. Wollan, Coherent scattering amplitudes as determined by neutron diffraction, *Phys. Rev.*, 81, 527–535 (1951).

25. E. Fermi, Sul moto dei neutroni nelle sostanze idrogenati [On the movement of neutrons in hydrogenated substances] *Ric. Sci.*, 7, 13 (1936).

26. H. Ekstein, Multiple elastic scattering and radiation damping. I, *Phys. Rev.*, 83, 721–729(1951).

27. H. Ekstein, Multiple scattering and radiation damping. II, *Phys. Rev.*, 89, 490–501 (1953).

28. L. E. Ballentine, *Quantum Mechanics. A Modern Development* (World Scientific, Singapore, 1998).

29. D. I. Blokhintsev, *Mecanique Quantique* (Masson et Cie., Paris, 1967).

30. M. Born and E. Wolf, *Principles of Optics* (Pergamon Press, Oxford, 1975).

31. L. Landau and E. Lifchitz, *Mecanique Quantique*, (Editions Mir, Moscow, 1967).

32. W. Marshall and S. W. Lovesey, *Theory of Thermal Neutron Scattering* (Clarendon Press, Oxford, 1971).

33. D. M. Greenberger and A. W. Overhauser, Coherence effects in neutron diffraction and gravity experiments, *Rev. Mod. Phys.*, 51, 43–78 (1979).

34. D. M. Greenberger, The neutron interferometer as a device for illustrating the strange behavior of quantum systems, *Rev. Mod. Phys.*, 55, 875–905 (1983).

35. A. Steyerl, "Neutron Wave Optics with Ultracold Neutrons," in *The Wave Particle Dualism*, edited by S. Diner et al. (D. Reidel Publishing Company, Dordrecht, 1984).

36. H. Chen, R. G. Downing, D. F. R. Mildner, W. M. Gibson, M. A. Kumakhov, I. Yu. Ponomarev, and M. V. Gubarev, Guiding and focusing neutron beams using capillary optics, *Nature*, 357, 391–393 (1992).

37. M. L. Calvo, Neutron fibers: a three-dimensional analysis of bending losses, *J. Phys. D: Appl. Phys.*, 33, 1666–1673 (2000).

38. B. W. Batterman and H. Cole, Dynamical diffraction of X rays by perfect crystals, *Rev. Mod. Phys.*, 36, 681–717 (1964).

39. R. F. Alvarez-Estrada, Particle diffraction by a thin rigid crystal lattice: a formal approach, *Ann. Phys.*, 85, 56–114 (1974).

40. R. F. Alvarez-Estrada and M. L. Calvo, Diffraction of a classical electromagnetic wave by a thin periodic slab: a rigorous approach, *J. Phys. A: Math. Gen.*, 11, 1855–1864 (1978).

41. A. Zeilinger and T. J. Beatty, Bragg diffraction and surface reflection of neutrons from perfect crystals at grazing incidence, *Phys. Rev. B*, 27, 7239–7250 (1983).

42. U. Bonse and M. Hart, An X-ray interferometer, *Appl. Phys. Lett.*, 6, 155–156 (1965).

43. H. Rauch, W. Treimer, and U. Bonse, Test of a single crystal neutron interferometer, *Phys. Lett. A*, 47, 369–371 (1974).

44. A. Szke, in Short Wavelength Coherent Radiation: Generation and Applications, edited by D. T. Attwood and J. Boker, AIP Conference Proc. 147, American Institute of Physics, 361 (1986).

45. D. Gabor, Microscopy by reconstructed wavefront, *Proc. Royal Society A*, 197, 454–487 (1949).

46. L. Mandel and E. Wolf, *Optical Coherence and Quantum Optics* (Cambridge University Press, Cambridge, U.K., 1999).

47. M. L. Calvo and P. Cheben, "Fundamentals and Advances in Holographic Materials for Optical Data Storage," in *Advances in Information Optics and Photonics*, edited by A. T. Friberg and R. Dndliker, International Commission for Optics, Vol. VI, (SPIE Press, Bellingham, 2008).

48. R. Fitzgerald, Physics today, X-rays and γ-rays holography improve views of atoms in solids, *Phys. Today*, 54(4), 2123 (2001).

49. P. Korecki, G. Materlik, and J. Korecki, Complex gamma-ray hologram: solution to twin images problem in atomic resolution imaging, *Phys. Rev. Lett.*, 86, 1534–1537 (2001).

50. L. Cser, G. Krexner, and Gy. Trk, Atomic-resolution neutron holography, *Europhys. Lett.*, 54(6), 747–752 (2001).

51. L. Cser, Gy. Trk, G. Krexner, I. Sharkov, and B. Farag, Holographic imaging of atoms using thermal neutrons, *Phys. Rev. Lett.*, 89(7), 175504-1 to 175504-4 (2002).

52. K. Hayashi et al., Multiple-wavelength neutron holography with pulsed neutrons, *Sci Adv.*, 3(8), e1700294 (2018).

53. https://physicsworld.com/a/neutron-holograms-image-the-interiors-of-objects/ [checked May 2019]. See also: D. Sarenac et al., Holography with a neutron interferometer, *Optics Express*, 24(20), 22528–22535 (2016).

54. V. V. Nesvizhevsky et al., Quantum states of neutrons in the earth's gravitational field, *Nature*, 415, 297–299 (2002).

55. R. Colella, A. W. Overhauser, and S. A. Werner, Observation of gravitationally induced quantum interference, *Phys. Rev. Lett.*, 34, 1472–1474 (1975).

56. S. Weinberg, *Gravitation and Cosmology: Principles and Applications of the General Theory of Relativity* (John Wilkey and Sons, Inc., New York, 1972).

57. P. P. Ewald, Zur begrundung der kristalloptik.teil i-iii., *Ann. Phys. Leipzig*, 49, 1–38, 117–143 (1916); 54, 519–97 (1917).

58. C. W. Oseen, Über die wechselwirkung zwischen zwei elektrischen dipolen und die drehung der polarisationsebene in kristallen und flssigkeiten, *Ann. d. Phys.*, 353(17), 1–53 (1915).

59. M. L. Golberger and K. M. Watson, *Collision Theory* (John Wiley and Sons, New York, 1965).

60. M. L. Calvo, "Contribution to the study of electromagnetic wave scattering by fixed defects in an isotropic medium" (Ph.D. Thesis, in Spanish), Universidad Complutense de Madrid (1977).

61. M. E. Villalón, "Theoretical study of scattering by periodic potentials: infinite and semiinfinite media" (Ph.D. Thesis, in Spanish), Universidad Complutense de Madrid (1976).

62. R. F. Alvarez-Estrada and M. E. Villalón, On the applicability of the Lippmann-Schwinger approach to the scattering of particles by semi-infinite crystals, *Ann. Phys.*, 111, 239–250 (1978).

63. S. A. Werner, J.-L. Staudenmann, and R. Colella, Effect of earth's rotation on the quantum mechanical phase of the neutron, *Phys. Rev. Lett.*, 42, 1103–1106 (1979).

64. S. A. Werner, R. Colella, A. W. Overhauser, and C. F. Eagen, Observation of the phase shift of a neutron due to precession in a magnetic field, *Phys. Rev. Lett.*, 35, 1053–1055 (1975).

65. G. L. Squires, *Introduction to the Theory of Thermal Neutron Scattering*, 3rd ed. (Cambridge University Press, Cambridge, 2012).

66. M. Utsuro and V. K. Ignatovich, *Handbook of Neutron Optics* (Wiley VCH Verlag, Weinheim, 2010).

2 Neutron Matter-Wave Diffraction: A Computational Perspective

Ángel S. Sanz

CONTENTS

2.1 INTRODUCTION

"A range of phenomena similar or analogous to those of classical optics is exhibited by slow neutrons." With this challenging sentence, Klein and Werner started a review work [1], where they established a one-by-one correspondence about the analogies between slow neutrons and light optics at three levels: geometrical optics, wave optics, and crystal optics. In many instances, the motivation behind such experiments has been of a purely fundamental nature, just to test and challenge important questions in quantum mechanics. This is the case, for example, of the experiments performed in the 1980s by Zeilinger and coworkers on cold neutron diffraction [2]. This interest for fundamental neutron physics can also be perceived after having a look at the two editions of the classic *Neutron Interferometry*, by Rauch and Werner, the first subtitled *Lessons in Experimental Quantum Mechanics* [3], while for the second the authors have chosen *Lessons in Experimental Quantum Mechanics, Wave-Particle Duality, and Entanglement* [4]. Furthermore, the issue of neutron entanglement has also been considered from a experimental viewpoint in recent times [5], as well as the measure of weak values [6]. Certainly, these topics are of much interest, although they go out of the scope of the chapter. Nonetheless, both the physical approach here described and the numerical tools discussed should suffice to tackle them, at least, at a phenomenological level.

For a theoretician, this breeding ground is quite challenging and deserves much attention for the problems involved. There is much theory developed to tackle these problems, much of it from general scattering theory, and also, of course, from wave optics. The purpose of this chapter, though, is to provide some alternative tools to explore new theoretical routes, particularly at a computationally level. To that end, the chapter has been prepared in way that it might serve the reader to acquire some skills and put them at work on the myriad of problems within the realm of neutron optics. As it also happens with the theoretical techniques so far developed and used in neutron optics, the computational tools here proposed are not new as such, because they have been in use for a while in other fields of physics. However, they can be of much interest in the field for the different viewpoint they may provide on the problems investigated.

More specifically, the main goal of the chapter is to provide the reader with simple and convenient tools to numerically simulate the time evolution of neutron diffraction and interference phenomena (and, by extension, also entanglement) at a phenomenological level. These phenomena can easily be tackled by standard quantum mechanics, i.e., a direct application of Schrödinger's equation plus the notion of optical potential accounting for the sudden-type interaction of neutron matter waves with the boundaries of the corresponding diffracting element. This enables two interesting routes or pathways for a phenomenological study and analysis of the experimental observations, a traditional-like wave-based approach and a hydrodynamic (Bohmian) one, as well as different numerical approaches to obtain the corresponding quantities of interest.

Accordingly, the work has been structured into three different parts or sections. Section 2.2 is essentially devoted to the analysis of the system and to introduce some considerations in order to perform the modeling. This is essential in order to get an accurate description; even if it is not at a first-principle level, at least the model has to capture the main features of the phenomenon under study. Hence, in a first part, within Section 2.2, there is a description and analysis of the experiment chosen to serve of working example, namely the experiment performed by Zeilinger and cowork-ers [2]. The second and third parts are devoted to the development of the analytical grounds, which starting from standard quantum mechanics, eventually reach the elementary tools of Fourier optics. More specifically, the general tools are described in the second part, while a more particular analysis of diffraction in terms of Gaussian wave packets is accounted for in the third part of the section. Finally, in a fourth part of the section, the Bohmian or hydrodynamic formulation of quantum mechanics is introduced and discussed. This formulation allows us to understand quantum processes in terms of well-defined trajectories or probability streamlines. Unlike other calculations based on classical trajectories, it should be noticed that Bohmian trajectories do not arise from any sort of approximation or semiclassical approach to quantum problems, but they are in compliance with the evolution displayed by the wave function. It is interesting to note that this formulation was precisely used formerly to describe different fundamental aspects in neutron interferometry back in the mid-1980s by Dewdney and coworkers [7–13]. Section 2.3 introduces some methods to produce simulations, from relatively simple and computationally cheap ones, to more sophisticated algorithms, which require some level of expertise. Section 2.4 is devoted to the analysis of some numerical simulations describing neutron diffraction by both one and two slits. These simulations take as a basis or working system the experiments performed by Zeilinger and coworkers. Finally, some concluding remarks are summarized in Section 2.5.

2.2 THEORETICAL PHENOMENOLOGICAL MODELING

Because neutrons are electrically neutral, their penetrating power is higher than that of charged particles with similar incident energies. Physically, this means that, while the interaction between charged particles and the electron cloud surrounding nuclei is relevant, neutrons essentially interact with nuclei themselves. From the viewpoint of theoretical modeling, this entails an important difference. Typically, the interaction of charged particles with matter is described by means of smooth (or relatively smooth) potential energy functions. This actually goes beyond the charged particles themselves and also applies to atomic and molecular diffraction, where short-range interactions, related to the electronic structure of both projectile and target, play a fundamental role. This is why, in order to tackle these problems, different methodologies have been developed in the literature. Now, the case of neutrons is very particular, because they undergo strong, sudden interaction with atomic nuclei, which is described by a delta-type effective pseudo-potential, namely the Fermi pseudo-potential [3, 14].

Yet, at low energies, neutrons are characterized by relatively large wavelengths spanning tenths of Bohr radii (large coherent lengths), which allows us to simplify the study of their diffraction dynamics (something similar also happens, for instance, in the case of channeling along neutron wave guides), substituting the effect of single collisions by effective potential functions. To some extent, this treatment is analogous to the one performed with light when it passes through a dielectric material: the complex behavior inside the material is substituted by an effective quantity, the refractive index, which describes "macroscopically" how the material behaves when light goes across it.

2.2.1 The Working System

As it has been mentioned above, a number of diffraction experiments have been reported in the literature with different types of massive particles in order to prove the generality of the wave behavior of matter and, with it, of quantum mechanics itself. Among them, the diffraction experiments with cold atoms from single- and double-slit setups reported by Zeilinger and coworkers in 1988 [2] deserve special attention, since they can be considered one of the former most beautiful and elegant proofs of the wave nature of matter, where the diffraction patterns obtained cannot be attributed to any artifact (e.g., electrical interactions, as it could happen with former experiments performed with electrons) other than the action of boundaries on the incoming (neutron) matter wave.

In these experiments, the neutron wavelength was controlled to range between 15 Å and 30 Å by means of a quartz prism, which plays the same monochromating role as an optical prism when it is used to select a given visible light wavelength. From the de Broglie relation, we find that the energy range corresponding to these wavelengths is, approximately, 360 μeV down to 90 μeV (i.e., the range of cold neutrons). On the other hand, given the relationship between thermal de Broglie wavelength and temperature, in terms of the relation

$$\lambda_{th} = \sqrt{\frac{2\pi\hbar^2}{mk_BT}}, \tag{2.1}$$

such a wavelength (energy) range corresponds to temperatures that go from about 1.3 K down to 0.3 K, approximately. A sketch of the experimental setup can be seen either in the original work [2], or in related monographs [3, 14], which allows us to get an idea of the dimensions of the device compared with an analogous light-diffraction experiment performed on an optical bench, which typically includes the use of lenses to importantly reduce the setup dimensions. As it can be seen, the system consists of a number of slits accommodated along the pathway followed by the neutrons. The first two slits, S_1 and S_2, which were part of the monochromating system, plus the entrance slit, S_3, determine the wavelength spread, although only the latter defines the width of the impinging neutron beam on the target slit, S_5, at 5 m from S_3. Diffraction features are then measured by collecting the neutron flux traversing a last (scanning) slit, S_4, at another 5 m from S_5.

While S_2 was kept with a relatively large width (100 μm), for both S_3 and S_4 a fifth of such a width was considered (20 μm), which can easily be understood from simple estimates based on single-slit diffraction considerations. In the case of S_3, a simple estimate based on the well-known single-slit diffraction formula for a slit of width 20 μm shows that, at 5 m from it, the width of the central maximum will span a distance of about 1 mm, which is much larger than the opening dimensions considered for S_5 (≤100 μm). In other words, the wavefronts impinging on S_5 can be considered to be, for all practical purposes, nearly plane. Likewise, the maxima associated with the diffracted waves at S_5 are going to have large dimensions when they reach S_4, at 5 m away, as it can readily be noticed by a quick inspection at the diffraction patterns reported in Ref. [2].

The experimental data collected were then analyzed by numerically (best) fitting them to an integral expression of the intensity, which takes into account the subsequent diffractions undergone by the neutron matter wave at slits S_3 and S_4, and the convolving effect due to the finite size of S_5, the wave angular spread, and its wavelength spread. As it is shown here, such data can also be analyzed by means of an effective model that allows us to also investigate the time evolution of the wave from S_5 to S_4 (an incident plane wave is assumed to reach S_5).

2.2.2 STANDARD QUANTUM ANALYSIS WITH PARAXIAL CONDITIONS

The correct theoretical description of the passage of the neutron matter wave across the experimental setup sketched in Ref. [2] is provided by Schrödinger's equation,

$$i\hbar \frac{\partial \Psi(\mathbf{r},t)}{\partial t} = -\frac{\hbar^2}{2m}\nabla^2\Psi(\mathbf{r},t) + V(\mathbf{r})\Psi(\mathbf{r},t), \tag{2.2}$$

where $V(\mathbf{r})$ represents a relatively complex potential function that describes the action of any element (slits, prism, etc.) encountered by the neutron wave as it progresses through the setup. This would provide us with an "exact" description of the transit of the electron beam through the full experimental setup (although by "exact" it should be understood "macroscopic," i.e., without taking into account the many components of such elements, but just their averaged or combined action in terms of relatively smooth potential functions). However, if the main interest relies on the diffraction pattern observed, it is only necessary to consider what happens between S_5 and S_4 (subsequently different levels of refinement can be assumed in order to include effects related to former elements). Notice that, as it was pointed out before, the wave impinging on S_5 is nearly flat, and hence one can just focus on what is going on from S_5 onward. In other words, this means that, in a first approximation, our potential model can be reduced to just the shape of the diffracting object, i.e., the channel-type structure that makes S_5, without or with the boron wire depending on whether we are considering the single- or the double-slit diffraction case.

Now, notice that despite the small amount of kinetic energy carried by the neutrons, it is still enough to make them relatively fast regarding diffractive effects. In other words, their longitudinal or axial propagation is going to be faster than along the transverse or radial direction, since they move at an average speed of about 180 m/s (for an average wavelength of 22.5 Å). This means that in about 28 ms, the neutron wave diffracted by S_3 has covered a distance of 5 m, while its main distinctive feature, the central diffraction maximum, has spanned a distance of about 2 mm. This has two important consequences. First, the average time spent by the wave inside the channel of S_5, with a length of about 500 μm, is negligible (about 2.8 μs). Hence, the passage can be assumed to be instantaneous. Accordingly, the action of the potential function at S_5 can be substituted by setting an initial condition (initial neutron wave function) that is in compliance with the features that characterize the slit. Regarding the perpendicular coordinate, the slit has a height of about $400\ \mu$m and the beam size is about $60\ \mu$m, which means that we can neglect broadening along such a direction and just assume longitudinal symmetry. This thus allows us to neglect diffraction along the y-axis, which reduces Eq. (2.2) to the free-evolving particle equation,

$$i\hbar\frac{\partial\Psi(x,z,t)}{\partial t} = -\frac{\hbar^2}{2m}\left(\frac{\partial^2}{\partial x^2}+\frac{\partial^2}{\partial z^2}\right)\Psi(x,z,t), \tag{2.3}$$

The second consequence is related to the paraxial conditions entailed by the fast propagation along the z-direction. Accordingly, if the wave vector satisfies the relation $k^2 = k_x^2 + k_z^2$, where $k = 2\pi/\lambda$, it can be assumed that $k_z \gg k_x$ and $k \approx k_z$. The neutron wave function can then be recast at any time as a product state,

$$\Psi(x,z,t) \approx \psi(x,t)e^{ik_z z - iE_z t/\hbar}, \tag{2.4}$$

The longitudinal component is a plane wave with average momentum $p_z = \hbar k_z$ and energy $E_z = p_z^2/2m = \hbar^2 k_z^2/2m$. The transverse component, $\psi(x,t)$, is a solution of the time-dependent, free-particle Schrödinger equation

$$i\hbar\frac{\partial\psi(x,t)}{\partial t} = -\frac{\hbar^2}{2m}\frac{\partial^2\psi(x,t)}{\partial x^2}, \tag{2.5}$$

with initial condition $\psi(x,0)$.

Solutions of Eq. (2.5) can be readily found by computing the Fourier transform of $\psi(x,t)$,

$$\psi(x,t) = \frac{1}{\sqrt{2\pi}}\int \tilde{\psi}(k_x,t)e^{ik_x x}\,dk_x, \tag{2.6}$$

Physically, this is just a way to recast $\psi(x,t)$ as a linear combination or superposition of plane waves, $e^{ik_x x}$, each one contributing with a weight and phase specified by $\tilde{\psi}(k_x,t)$. Notice that $\tilde{\psi}(k_x,t)$ is just the representation of the wave function in the momentum (reciprocal) space, which depends on the reciprocal variable or momentum k_x. Substituting Eq. (2.6) into Eq. (2.5) leads to

$$i\hbar\frac{\partial\tilde{\psi}}{\partial t} = \left(\frac{\hbar^2 k_x^2}{2m}\right)\tilde{\psi}, \tag{2.7}$$

which, after integration in time, yields

$$\tilde{\psi}(k_x,t) = \tilde{\psi}(k_x,0)e^{-i\hbar k_x^2/2m}, \tag{2.8}$$

The initial condition $\tilde{\psi}(k_x,0)$ corresponds to the representation of the initial wave function, $\psi(x,0)$, in the momentum space,

$$\tilde{\psi}(k_x,0) = \frac{1}{\sqrt{2\pi}}\int \psi(x,0)e^{-ik_x x}\,dx, \tag{2.9}$$

The solution Eq. (2.8), with initial condition Eq. (2.9), allows us to rearrange Eq. (2.6) as

$$\psi(x,t) = \frac{1}{\sqrt{2\pi}} \int \tilde{\psi}(k_x,0)e^{ik_x x - i\hbar k_x^2/2m} \, dk_x$$

$$= \frac{1}{2\pi} \iint \psi(x',0)e^{ik_x(x-x') - i\hbar k_x^2/2m} \, dk_x \, dx', \tag{2.10}$$

This expression can be simplified by integrating in k_x, which renders

$$\psi(x,t) = \sqrt{\frac{m}{2\pi i\hbar t}} \int \psi(x',0)e^{im(x-x')^2/2\hbar t} \, dx', \tag{2.11}$$

after integrating in k_x. Notice in the latter expression that the quantity

$$\mathcal{K}(x,x') \equiv \sqrt{\frac{m}{2\pi i\hbar t}} e^{im(x-x')^2/2\hbar t}, \tag{2.12}$$

corresponds to the free-particle kernel or propagator [15].

Given that the longitudinal component of the wave function is a plane wave that propagates with velocity $v = \hbar k_z/m \approx \hbar k/m$, the problem can be reparameterized in terms of the longitudinal coordinate, z. That is, the wave function, Eq. (2.11), describing the wave function evolution along the transverse direction (accounted for by the x coordinate) can be recast in terms of this coordinate,

$$\psi(x,z) \approx \sqrt{\frac{k}{2\pi i z}} \int \psi(x',0)e^{ik(x-x')^2/2z} \, dx', \tag{2.13}$$

with $z = vt \approx (\hbar k / m)t$. As it can be noticed, this relation is quite convenient, because it is mass independent. Hence, it can be used either for massive particles, such as neutrons, or by light, which makes more apparent the connection between matter and radiation (specifically, the fact that we talk about neutron optics).

From the above discussion, it is clear that the difference between different diffraction problems relies on the initial value $\psi(x,z_0)$, with z_0 being the specific position along the z-axis from which the (transverse) wave function starts its propagation (this fact is illustrated in the numerical simulations presented in Section 2.4). If the diffracting element is assumed to be infinitesimally thin along the longitudinal direction, as it is the case here, a reasonable working hypothesis for the initial ansatz is that it is a modulated function of the incident plane wave. For instance, if the borders of the diffracting element are sharp, a fair assumption is that the transverse wave function just after crossing it, $\psi(x,z_0)$, is a wave with constant amplitude, same momentum as the incident wave, and limited transverse size. On the other hand, there could also be some interaction with the borders, for instance, if absorption is relevant. In this case, a nearly Gaussian-shaped modulating factor could be used to describe such an initial ansatz. The properties of these two types of ansatz, square (top-hat) and Gaussian, will be analyzed in more detail in Section 2.4, because they represent cases of particular interest.

In general, the functional form, Eq. (2.13), cannot be solved analytically except for simple transmission amplitudes. Nonetheless, in the far field (*ff*) or Fraunhofer regime it acquires a simpler form and admits some additional analytical solutions. In this regime, we have $x \gg x'$, which allows us to approximate the phase factor in Eq. (2.13) by

$$\frac{k(x-x')^2}{2z} \approx \frac{kx^2}{2z} - \frac{kxx'}{z}. \tag{2.14}$$

This allows us to recast Eq. (2.13) as

$$\psi_{ff}(x,z) \approx \sqrt{\frac{k}{2\pi iz}}\, e^{ikx^2/2z} \int \psi(x',0) e^{-ik_x x'}\, dx', \tag{2.15}$$

where we introduce the definition $k_x = kx/z$, with x/z being the direction cosine with respect to the coordinate origin at S_5. Additionally, this quantity can also be related to the transverse momentum, as will be seen below. Comparing this integral with Eq. (2.9), we find that, except for a phase factor, the wave function in the far field is proportional to the representation of the initial wave function in the momentum space:

$$\psi_{ff}(x,z) \approx \sqrt{\frac{k}{iz}}\, \tilde{\psi}(k_x) e^{ikx^2/2z}. \tag{2.16}$$

The Fraunhofer diffraction can then be understood as the Fourier image of the initial wave function. Physically, this means that in the far field (asymptotically), the global shape of the probability density is independent of the distance from the grating, mimicking the transverse momentum distribution: $|\psi_{ff}(x,z)|^2 \propto |\tilde{\psi}(k_x)|^2$. In other words, the aspect ratio of the wave function remains invariant with z or, equivalently, with time. Furthermore, because $\tilde{\psi}(k_x)$ is related to the grating transmission, the far-field wave function is just a manifestation of the grating structure, thus providing us with information about it and not only about properties of the diffracted atom. Note again the analogy with optics [16], where the wave amplitude in the far field or Fraunhofer regime is just the Fourier transform of the aperture function evaluated at a spatial frequency precisely given by k_x.

Regarding the phase factor that appears in Eq. (2.16), it has not been recast in terms of k_x on purpose, because of its quadratic dependence on x. As it can be noticed, if we apply the usual (transverse) momentum operator, $\hat{p}_x = -i\hbar \partial/\partial x$ to Eq. (2.16), we find

$$\hat{p}_x \psi_{ff} \propto \left(\frac{\hbar k x}{z}\, \tilde{\psi} - \frac{i\hbar k}{z}\, \frac{\partial \tilde{\psi}}{\partial k_x} \right) e^{ikx^2/2z}$$

$$= \frac{\hbar k}{z} \left[\int (x-x') \psi(x',0) e^{-ik_x x'}\, dx' \right] e^{ikx^2/2z} \propto \frac{\hbar k x}{z}\, \psi_{ff} = \hbar k_x \psi_{ff}, \tag{2.17}$$

where we have considered the identity

$$\frac{\partial}{\partial x} = \frac{\partial k_x}{\partial x}\frac{\partial}{\partial k_x} = \frac{k}{z}\frac{\partial}{\partial k_x} \tag{2.18}$$

in the first line plus the approximation $x \gg x'$ in the last step. Hence, to some extent, it can be said that the far-field wave function evolves locally (at each point) as an effective plane wave with (effective) momentum $p_x = \hbar k x / z = \hbar k_x$. Taking this fact into account, Eq. (2.16) can be recast in terms of plane waves, as

$$\psi_{ff}(x,z) \propto \tilde{\psi}(k_x)e^{ik_x x}. \tag{2.19}$$

From the above discussion, it is thus interesting to introduce another quantity of interest, namely the *local transverse momentum* [17],

$$K_x(x,z) = \mathrm{Im}\left[\frac{1}{\psi(x,z)}\frac{\partial\psi(x,z)}{\partial x}\right], \tag{2.20}$$

which allows us to inquire questions specifically related to the momentum flux along the transverse direction. Specifically, we can quantify the local value of the transverse momentum K_x as a function of the x coordinate at a given distance z. Notice that this momentum is connected to the usual quantum flux,

$$J_x = \frac{\hbar}{m}\mathrm{Im}\left[\frac{1}{\psi}\frac{\partial\psi}{\partial x}\right] = \frac{\hbar}{m}K_x|\psi|^2, \tag{2.21}$$

commonly used in the Bohmian formulation of quantum mechanics [18]. In the far field, e.g., it typically coincides with the transverse momentum value associated with the different diffraction orders, as it can readily be seen by substituting the ansatz Eq. (2.26) into Eq. (2.20) and then considering the corresponding limits [17, 18]. This renders

$$K_x(x,z) \approx k_{x,\ell}. \tag{2.22}$$

Analogously, if we substitute the far-field expression Eq. (2.16) into Eq. (2.20), we find that

$$K_x(x,z) = \frac{kx}{z} = k_x, \tag{2.23}$$

which justifies both our choice of k_x and the rewriting of Eq. (2.16) in the form of Eq. (2.19).

2.2.3 GAUSSIAN-SLIT MODELS

In order to better understand the nature of the model here accounted for to analyze neutron diffraction at a phenomenological level, let us consider the simplest case of Gaussian slits. The use of Gaussian basis sets is quite common in atomic and molecular physics; in the context of slit diffraction, they were formerly considered by Feynman and Hibbs [15], because of the analyticity of the problem. For a general application to grating diffraction, the interested reader may consult Refs. [17, 19]. Physically, this kind of slits represent a situation where crossing through the center of slit is maximum and decreases gradually (Gaussian-like) toward the borders of the opening, something that can be related to an absorption process in the case of neutrons. By a simple inspection to experimental diffraction patterns, like those reported in Ref. [2], it can readily be noticed that this hypothesis is quite reasonable, because of the decay displayed by the envelope of the diffraction maxima.

A broad analysis of this type of slits and different applications can be found in Ref. [20]. In this section, we are going to restrict ourselves to the case of Gaussian single and double slits, just the cases to compare with the experimental data shown in Ref. [2]. Thus, let us consider that, in a first approximation, the neutron beam incident onto S_5 is described by a nearly plane wave,

$$\psi_{\text{inc}}(x,0) \sim e^{ik_{x0}x}, \tag{2.24}$$

where $k_{x0} = k_0 \sin\theta$, with θ being the incident angle (later on we shall particularize to the case of normal incidence, i.e., $k_{x0} = 0$). As seen above, this is a reliable guess, since the diffracted wave produced by S_3 has a nearly constant amplitude when it is incident onto the opening of S_5. The wave amplitude describing the neutron state just after crossing the slits, according to the Gaussian model, will be

$$\psi(x,0) \approx \frac{1}{\sqrt{N}} \sum_{j=1}^{N} \left(\frac{1}{2\pi\sigma_0^2} \right)^{1/4} e^{-(x-x_j)^2/4\sigma_0^2 + ik_{x0}x}, \tag{2.25}$$

where x_j and σ_0 denote, respectively, the centroid position and the width of the jth Gaussian wave packet. As it can be noticed, Eq. (2.25) represents the general case of a coherent superposition of N Gaussian amplitudes, with particular interest in the description of single or two Gaussian-slit diffraction ($N = 1$ or $N = 2$, respectively), but also applicable in the case of nearly flat (single or double) diffracted amplitudes, as it will be seen in this chapter. In this superposition, the symbol "$\approx$" comes from the fact that all the overlapping terms in the normalization condition are assumed to be negligible, i.e., if there are two slits, they are far apart. The prefactors $1/\sqrt{N}$ and $(2\pi\sigma_0^2)^{-1/4}$ are introduced in order to keep ψ normalized. However, there could be a bias in the flux crossing each slit, which would imply that the coherent superposition would arise from an asymmetric contribution from each diffracted wave amplitude. This is easily accounted for by assuming different values for the initial width, σ_0, associated with each Gaussian.

A nice feature about free Gaussian states is that their time evolution is fully analytical. A simple method to determine it is provided in Section 2.3.1, although direct substitution of Eq. (2.25) into Eq. (2.11) also leads quickly to the analytical solution

$$\psi(x,z) \approx \frac{1}{\sqrt{N}} \left(\frac{1}{2\pi\tilde{\sigma}_z^2} \right)^{1/4} \sum_{j=1}^{N} e^{-(x-x_{j,z})^2/4\sigma_0\tilde{\sigma}_z^2 + ik_{x0}(x-x_{j,z}) + ik_{x0}x_j + ik_{x0}^2 z/2k}, \tag{2.26}$$

where the quantities

$$x_{j,z} = x_j + \left(\frac{\hbar k_{x0}}{mv}\right)z = x_j + \left(\frac{k_{x0}\lambda}{2\pi}\right)z, \tag{2.27a}$$

$$\tilde{\sigma}_z = \sigma_0\left(1 + \frac{iz}{2k\sigma_0^2}\right) \tag{2.27b}$$

provide us with the centroid position and width of the jth Gaussian after it has traveled a distance z along the longitudinal direction. In Eq. (2.26) all slits are assumed to be identical; this is not a limitation of the model, but just a convenient feature here. Nonetheless, further refinements can be introduced in the model in order to describe in a simple way aspects such as the inhomogeneity in the illumination of each slit by means of an opacity factor [17], although to leave the analysis at the simplest level here they are kept out of the scope of the chapter.

Because of the approximations considered, the quantum state Eq. (2.26) does not describe the time evolution itself of the neutron wave function, but the transverse diffusion of the neutron beam at given distances from the grating (at $z = z_0$). Of course, the centroid of each wave packet evolves according to a classical-type trajectory $z_t = z(t)$ (see Section 2.3.1). Accordingly, each transverse cut or "slice" of the wave packet can be uniquely mapped onto a precise slice at any subsequent time (i.e., if $z_0 \to z_t$, then $z_0 + \delta \to z_t + \delta_t$ for any value of δ). This is an important property that is usually missed in standard quantum mechanical courses, but that can be better appreciated through the formalism of the Bohmian formulation of quantum mechanics. To some extent, it can be understood as a direct analog in configuration space of the symplectic structure describing classical Hamiltonian flows, which are volume preserving in the phase space.

2.2.3.1 Application: Single-Slit Diffraction

Although later on we are going to work with the paraxial solution, in order to get a wider perspective and therefore a better understanding of the method, let us consider the general solution to the two-dimensional Schrödinger's equation Eq. (2.3), which can easily be determined following the method described in Section 2.3.1. Thus, if the incoming neutron wave is incident onto the plane of S_5 with a certain transverse component $k_{x0} = k_z$, the initial ansatz ($t = 0$) can be represented as a Gaussian amplitude (i.e., a Gaussian wave packet) given by

$$\Psi(x,z,0) = \left(\frac{1}{2\pi\sigma_x\sigma_z}\right)^{1/2} e^{-x^2/4\sigma_x^2 + ip_0x/\hbar}e^{-z^2/4\sigma_z^2 + ip_zz/\hbar}, \tag{2.28}$$

where σ_x and σ_z describe the neutron beam width at the origin of the coordinate system, centered at the center of the slit, and $p_0 = \hbar k_{x0}$. Physically, this means that the slit has an effective width of the order of σ_x; that the beam is localized along the longitudinal direction, having a width of the order of σ_z (with the limit $\sigma_z \to \infty$ denoting the case of an incident plane wave with infinite extension); and that the slit is relatively long along the y-direction ($\sigma_y \gg \sigma_x, \sigma_z$).

The evolution in time of Eq. (2.28) is described by the wave packet

$$\Psi(x,z,t) = \left(\frac{1}{2\pi\tilde{\sigma}_{x,t}\tilde{\sigma}_{z,t}}\right)^{1/2} e^{-x^2/4\sigma_x\tilde{\sigma}_{x,t}+ip_0x/\hbar+iE_0t/\hbar} \; e^{-(z-z_t)^2/4\sigma_z\tilde{\sigma}_{z,t}+ip_z(z-z_t)/\hbar+iE_zt/\hbar}, \quad (2.29)$$

where $z_t = v_z t$ is the instantaneous position of the Gaussian centroid (a classical trajectory describing a rectilinear uniform motion), and $E_z = p_z^2/2m = 2\pi^2\hbar^2/m\lambda^2$ is the energy associated with the translational (longitudinal) motion, and $E_0 = p_0^2/2m$. Moreover, we also have the quantum mechanical quantity

$$\tilde{\sigma}_{i,t} = \sigma_i\left(1 + \frac{i\hbar t}{2m\sigma_i^2}\right), \quad (2.30)$$

with $i = x,z$, from which we obtain the instantaneous width of the wave packet along each direction

$$\sigma_{i,t} = |\tilde{\sigma}_{i,t}| = \sigma_i\sqrt{1 + \frac{\hbar^2 t^2}{4m^2\sigma_i^4}} \quad (2.31)$$

gives the instantaneous width along each direction. From Eq. (2.31), a timescale $\tau_i = 2m\sigma_i^2/\hbar$ separating two regimes with different "spreading rates" can be defined (see below for further details). If $t \ll \tau_i$, the Gaussian widths almost remain the same ($\sigma_{i,t} \approx \sigma_i$). On the other hand, if $t \ll \tau_i$, the Gaussian widths undergo a linear increase with time ($\sigma_{i,t} \approx \hbar t/2m\sigma_i$).

According to the previous discussion, by choosing $\sigma_z \gg \sigma_x \sim w$, with w being the slit width, the spreading of the wave packet Eq. (2.30) along the longitudinal coordinate z will be negligible, and only along the transverse direction x it will be noticeable. Hence, Eq. (2.29) can be recast in the approximate form

$$\Psi(x,z,t) = \left(\frac{1}{2\pi\tilde{\sigma}_{x,t}\sigma_z}\right)^{1/2} e^{-x^2/4\sigma_x\tilde{\sigma}_{x,t}+ip_0x/\hbar+iE_0t/\hbar} \; e^{-(z-z_t)^2/4\sigma_z^2+ip_z(z-z_t)/\hbar+iE_zt/\hbar}, \quad (2.32)$$

which shows that spreading essentially takes place along the transverse direction (first exponential term), while along the longitudinal direction there is only a translation of a constant Gaussian envelop (second exponential term) at a constant speed ($v_z = 2\pi h/m\lambda$) plus a translational phase factor (third exponential term). Notice that the first and third terms are just the two separate contributions that appeared in Eq. (2.4), with $\psi(x,t)$ being the first one, and where the constant Gaussian envelope arises from the fact that the initial state is a Gaussian. So, from a practical viewpoint, the analysis of just the first (exponential) contribution in Eq. (2.32),

$$\psi(x,z,t) = \left(\frac{1}{2\pi\tilde{\sigma}_{x,t}^2}\right)^{1/4} e^{-x^2/4\sigma_x\tilde{\sigma}_{x,t}+ip_0x/\hbar+iE_0t/\hbar}, \quad (2.33)$$

to get a full understanding of the diffraction process that takes place along the transverse direction as time proceeds, or, analogously, in terms of the z distance, due to the linear relation between this quantity and time.

As can readily be noticed, the (transverse) density at a time, t, or, equivalently, a distance, z, from S_5 is going to display a Gaussian shape,

$$\rho(x,t) = \left(\frac{1}{2\pi\sigma_t}\right)^{1/2} e^{-x^2/2\sigma_t}. \tag{2.34}$$

So far, all of this is analogous to Gaussian beam optics, where given a beam with Gaussian waist, it propagates forward and backward with a Gaussian profile while develops a phase factor, as it is apparent from Eq. (2.33). In order to get a different perspective on this spreading, it is insightful to analyze the average or expectation value of the energy, which reads as

$$\bar{E} = \langle \hat{H} \rangle = \frac{p_0^2}{2m} + \frac{p_s^2}{2m}. \tag{2.35}$$

Apart from the classical-type kinetic term, which depends on the transverse momentum, p_0, there is a second contribution, which comes from what in Bohmian mechanics (see below) is called the *quantum potential*, directly connected with the spreading of the wave packet. Hence, we can define an effective *spreading momentum* [21], as

$$p_s = \frac{\hbar}{2\sigma_0}. \tag{2.36}$$

This momentum also appears when we compute the wave-packet energy dispersion (variance),

$$\Delta E \equiv \sqrt{\langle \hat{H}^2 \rangle - \langle \hat{H} \rangle^2} = \sqrt{\frac{2p_s^2}{m}} \sqrt{\frac{p_0^2}{2m} + \frac{p_s^2}{4m}}. \tag{2.37}$$

The relationship between p_s and the wave-packet spreading becomes more apparent when Eq. (2.36) is analyzed in the light of Heisenberg's uncertainty principle: a spreading of the size of σ_0 gives rise to a spreading in momenta of the order of p_s. In the case of nondispersive wave functions, p_s will vanish, as happens when dealing with a plane wave or with Airy wave packets. Actually, one could be tempted to think that classical-like regimes correspond to situations or conditions where spreading momenta vanish (in accordance with Ehrenfest's theorem).

Given the two types of motions governing the time evolution of a wave packet, it is clear that the ratio between the corresponding velocities, v_0 ($= p_0/m$) and v_s ($= \hbar/2m\sigma_0$), is going to play an important role in diffractive processes. To better understand this point, let us consider Eq. (2.31) and introduce the timescale

$$\tau = \frac{2m\sigma_0^2}{\hbar} = \frac{\sigma_0}{v_s}, \tag{2.38}$$

associated with the relative spreading of the wave packet, as seen above. Depending on the relationship between t and τ, we can identify three dynamical regimes [22]:

1. The very short-time or *Ehrenfest-Huygens regime*, $t \ll \tau$, where the wave packet remains almost spreadless: $\sigma_t \approx \sigma_0$.
2. The short-time or *Fresnel regime*, $t \ll \tau$, where the spreading increases nearly quadratically with time: $\sigma_t \approx \sigma_0 + \left(\hbar^2/8m^2\sigma_0^3\right)t^2$.
3. The long-time or *Fraunhofer regime*, $t \gg \tau$, where the Gaussian wave packet spreads linearly with time: $\sigma_t \approx \left(\hbar/2m\sigma_0\right)t$.

This scheme results very useful in order to determine a way to elucidate which process, translational motion or spreading, is going to dominate the future evolution of the wave packet. Let us recast Eq. (2.31) in terms of v_s and consider that t is the time required for the centroidal position to cover a distance $d = v_0 t \approx \sigma_0$. Substituting the latter expression into Eq. (2.31), we obtain

$$\sigma_t = \sigma_0\sqrt{1+\left(\frac{v_s}{v_0}\right)^2}. \qquad (2.39)$$

Accordingly, using only information about the initial preparation of the wave packet, we can infer information about its subsequent dynamical behavior. More specifically, if $v_s \ll v_0$, the wave-packet spreading along the transverse direction will be relatively slow, this being equivalent to the condition in time $t \ll \tau$. On the other hand, if $v_s \gg v_0$, the wave packet will spread very rapidly compared to its propagation along x, which corresponds to the time condition $t \gg \tau$.

2.2.4   BOHMIAN FORMULATION

The Bohmian formulation of quantum mechanics [18, 23], proposed in 1952 by Bohm as a way to circumvent the conclusions of Von Neumann's theorem on the impossibility of hidden variables in quantum mechanics [24, 25], provides us with a fair description of quantum systems in terms of both waves and (quantum) trajectories. The wave function Ψ provides with the dynamical information about the whole available configuration space to quantum particles, which will evolve accordingly as if they are "guided"—in this way, the quantum motion displayed by particles reflects the evolution of the wave function. The fundamental Bohmian equations of motion are usually derived from the standard version of quantum mechanics through the transformation $(\Psi, \Psi^*) \to (\rho, S)$, where Ψ and Ψ^* are generally complex-valued functions of the position $(\mathbf{r})$ and time (t), and ρ and S are real-valued functions of the same variables. More explicitly, the transformation relation between both types of functions (or fields) for a particle of mass m is given by

$$\Psi(\mathbf{r},t) = \rho^{1/2}(\mathbf{r},t)e^{iS(\mathbf{r},t)/\hbar}, \qquad (2.40)$$

(and its complex conjugate). After substituting Eq. (2.40) into the time-dependent Schrödinger equation, Eq. (2.2), two real coupled partial differential equations are obtained,

$$\frac{\partial \rho}{\partial t} + \frac{1}{m}\nabla(\rho\nabla S) = 0, \tag{2.41a}$$

$$\frac{\partial S}{\partial t} + \frac{(\nabla S)^2}{2m} + V_{\text{eff}} = 0, \tag{2.41b}$$

which come from the imaginary and real parts, respectively, of the resulting equation. The former is the continuity equation, which accounts for the probability conservation, while the latter is the *quantum Hamilton-Jacobi equation*, with

$$V_{\text{eff}}(\mathbf{r},t) = V(r) - \frac{\hbar^2}{2m}\frac{\nabla^2\rho^{1/2}(\mathbf{r},t)}{\rho^{1/2}(\mathbf{r},t)}, \tag{2.42}$$

being an *effective* total potential. The last term in the right-hand side of Eq. (2.42) is the so-called *quantum potential*,

$$Q \equiv -\frac{\hbar^2}{2m}\frac{\nabla^2\rho^{1/2}(\mathbf{r},t)}{\rho^{1/2}(\mathbf{r},t)} = \frac{\hbar^2}{4m}\left[\frac{1}{2}\left(\frac{\nabla\rho}{\rho}\right)^2 - \frac{\nabla^2\rho}{\rho}\right], \tag{2.43}$$

which, as well as ρ, depends on both $\mathbf{r}$ and t. This term is regarded as a potential because, like V, it also rules the quantum particle dynamics. However, its nature is fully quantum mechanical due to its dependence on the quantum state via ρ. Note that, since, statistically, ρ describes the evolution of a swarm of identical noninteracting particles, the dependence of Q on ρ means that the dynamics of a single particle from the swarm is going to be influenced by the behavior of the other. That is, quantum particle dynamics are *nonlocal*. Rather than being a particular feature of Bohmian mechanics, this property is inherent to quantum mechanics in general, which manifests through the kinetic operator, $\widehat{\mathcal{K}} = -(\hbar^2/2m)\nabla^2$, in its standard version. It is well known that, from a computational viewpoint, in order to evaluate the action of $\widehat{\mathcal{K}}$ accurately, one has to consider a very good representation of it. In this other sense, since Q arises from the action of this operator on Ψ after considering Eq. (2.40), i.e.,

$$\widehat{\mathcal{K}}\Psi = -\frac{\hbar^2}{2m}\nabla^2\Psi = \frac{(\nabla S)^2}{2m} - \frac{\hbar^2}{2m}\frac{\nabla^2\rho^{1/2}}{\rho^{1/2}}, \tag{2.43}$$

it could also be associated with a sort of nonlocal kinetic energy [17]. Nonlocality only disappears when $Q = 0$. Then, the particle dynamics become fully classical.

Since Eq. (2.41b) is a (quantum) Hamilton-Jacobi equation, paths along which quantum particles travel may be defined according to the guidance condition,

$$\mathbf{v} = \frac{\nabla S}{m} = \frac{\hbar}{m}\text{Im}\left(\frac{\nabla\Psi}{\Psi}\right), \tag{2.45}$$

where $\dot{\mathbf{r}} = \mathbf{v}$, in analogy to classical mechanics. Equations (2.41) and (2.45) form a closed set of coupled equations, which describes the evolution of a swarm of identical particles under the "guidance" of the quantum state at each time. It is important in this regard to call the attention on the fact that the Bohmian momentum associated with the velocity field defined by Eq. (2.45), $\mathbf{p} = m\mathbf{v}$, should not be misunderstood either with the variable used to describe the wave function in the momentum representation or the momentum that appears in the Wigner representation. Unlike these momenta, the Bohmian one is totally defined in the configuration space and, therefore, it is position dependent, while the other two types of momentum are independent of the position coordinate.

Closely related to Bohm's ideas, although predating them, in 1926 Madelung proposed an alternative reformulation to Schrödinger's wave mechanics in terms of streamlines and that nowadays we denote as quantum hydrodynamics [26]. This interpretation or formulation is directly connected to some relevant phenomena in quantum mechanics, e.g., superconductivity [27] or Bose-Einstein condensation [28]. As in Bohmian mechanics, one also starts with a polar formulation of the probability amplitude Eq. (2.40), but focussing on

$$\rho = R^2 = \Psi^* \Psi, \tag{2.46a}$$

$$\mathbf{J} = \rho \mathbf{v} = R^2 \frac{\nabla S}{m}, \tag{2.46b}$$

where $\rho(\mathbf{r}, t)$ is the probability density, $\mathbf{J}(\mathbf{r}, t)$ is the quantum probability density current, and v is the velocity field Eq. (2.45), which describes the flow of the latter. Taking this into account, Eq. (2.41) can be again expressed as

$$\frac{\partial \rho}{\partial t} + \nabla \cdot \mathbf{J} = 0, \tag{2.47a}$$

$$\frac{d\mathbf{v}}{dt} = \frac{\partial \mathbf{v}}{\partial t} + (\mathbf{v} \cdot \nabla)\mathbf{v} = -\frac{1}{m} \nabla(V + Q), \tag{2.47b}$$

which constitute the formal basis of quantum hydrodynamics and have a direct correspondence with those of classical fluid mechanics if m is identified with the mass of a piece of fluid separated from the rest by a closed surface, $m\rho$ is the fluid density, and $\mathbf{v}$ is the velocity field of the flow. However, unlike classical fluids, quantum fluids correspond to probability flows, with no material structure. That is, they only characterize statistical events at each point in space and time, in spite of the fact that the time evolution of these events can be better understood when compared with the motion of ordinary fluids. Moreover, whereas the classical concept of fluid can be applied to describe the statistical behavior of a macroscopic ensemble of particles, in quantum mechanics it is applied to single particles. Within this formulation, Q can be recast as

$$Q = -\frac{\hbar^2}{4m} \left[\frac{1}{2}(\nabla \ln \rho)^2 + \nabla^2 \ln \rho \right]. \tag{2.48}$$

In order to understand the relationship between the expectation value of a quantum operator and the statistical Bohmian description, consider $\hat{A}$ is a Hermitian operator, which can be a function of the position and momentum operators, $\hat{\mathbf{r}}$ and $\hat{\mathbf{p}} = -i\hbar\nabla$, i.e., $\hat{A} = \hat{A}(\hat{\mathbf{r}}, -i\hbar\nabla)$. The expectation value of this operator is defined as

$$\langle \hat{A} \rangle = \langle \Psi | \hat{A} | \Psi \rangle = \frac{\int \Psi^*(\hat{A}\Psi)d\mathbf{r}}{\int \Psi^*\Psi d\mathbf{r}}, \tag{2.49}$$

where $\Psi(\mathbf{r},t) = \langle \mathbf{r} | \Psi(t) \rangle$ is the wave function in the system configuration representation and

$$\left[\hat{A}\Psi\right](\mathbf{r},t) \equiv \left\langle \mathbf{r} \left| \hat{A} \left(\int |\mathbf{r}'\rangle\langle\mathbf{r}'| d\mathbf{r}' \right) \right| \Psi(t) \right\rangle = \int \hat{A}(\hat{\mathbf{r}}, \hat{\mathbf{r}}')\Psi(\mathbf{r}',t)d\mathbf{r}', \tag{2.50}$$

with $\hat{A}(\hat{\mathbf{r}}, \hat{\mathbf{r}}') \equiv \langle \mathbf{r} | \hat{A} | \mathbf{r}' \rangle$. For Hermitian operators, $\hat{A}$, only its real part has to be taken into account in the calculation of its expectation value. Hence, Eq. (2.49) can be expressed as

$$\langle \hat{A} \rangle = \frac{\text{Re}\left\{\int \Psi^*(\hat{A}\Psi)d\mathbf{r}\right\}}{\int \Psi^*\Psi d\mathbf{r}} = \frac{\int \text{Re}\left\{\Psi^*(\hat{A}\Psi)\right\} d\mathbf{r}}{\int \Psi^*\Psi d\mathbf{r}}. \tag{2.51}$$

Moreover, one can also consider the quantity

$$A \equiv \frac{\text{Re}\left\{\Psi^*(\hat{A}\Psi)\right\}}{\Psi^*\Psi} \tag{2.52}$$

to represent the *local* value of the operator $\hat{A}$, given in terms of the associated field function $A(\mathbf{r},t)$. In other words, the quantity Eq. (2.52) can be interpreted as the property A for a given particle.

For example, consider the position, momentum, and energy operators in the configuration representation,

$$\hat{\mathbf{r}}(\mathbf{r},\mathbf{r}') = \mathbf{r}\delta(\mathbf{r} - \mathbf{r}'), \tag{2.53a}$$

$$\hat{\mathbf{p}} = -i\hbar\delta(\mathbf{r} - \mathbf{r}')\nabla, \tag{2.53b}$$

$$\hat{H} = \delta(\mathbf{r} - \mathbf{r}')\left[-\frac{\hbar^2}{2m}\nabla^2 + V(\mathbf{r})\right], \tag{2.53c}$$

respectively. The associated field functions are

$$\mathbf{r}(\mathbf{r},t) = \frac{\text{Re}\left\{\Psi^*\hat{\mathbf{r}}\Psi\right\}}{\Psi^*\Psi} = \mathbf{r}(t), \tag{2.54a}$$

$$\mathbf{p}(\mathbf{r},t) = \frac{Re\{\Psi^*(-i\hbar\nabla)\Psi\}}{\Psi^*\Psi} = \nabla S, \tag{2.54b}$$

$$E(\mathbf{r},t) = \frac{Re\{\Psi^*\left(-\dfrac{\hbar^2}{2m}\nabla^2 + V\right)\Psi\}}{\Psi^*\Psi} = \frac{(\nabla S)^2}{2m} + V_{\text{eff}}, \tag{2.54c}$$

which will provide us with the position, momentum, and energy of a Bohmian particle when they are evaluated along its trajectory. Indeed, in this case, note that Eq. (2.54a) corresponds precisely to the equation of motion Eq. (2.45), thus being a solution of Eq. (2.54b).

If instead of a particle there is a statistical ensemble of them (or, equivalently, some set of initial conditions has to be sampled) distributed according to $\rho(\mathbf{r},t)$, the average value of A can be computed as in classical mechanics,

$$\langle A(t)\rangle = \int \rho(\mathbf{r},t)\, A(\mathbf{r},t)\, d\mathbf{r}. \tag{2.55}$$

Thus, sampling Eqs. (2.53) over ρ,

$$\bar{\mathbf{r}} = \int \rho \mathbf{r}\, d\mathbf{r} = \int \Psi^*\hat{\mathbf{r}}\Psi\, d\mathbf{r} = \langle\hat{\mathbf{r}}\rangle, \tag{2.56a}$$

$$\bar{\mathbf{p}} = \int \rho\nabla S\, d\mathbf{r} = \int \Psi^*(-i\hbar\nabla)\Psi\, d\mathbf{r} = \langle\hat{\mathbf{p}}\rangle, \tag{2.56b}$$

$$\bar{E} = \int \rho\left[\frac{(\nabla S)^2}{2m} + V_{\text{eff}}\right] d\mathbf{r} = \int \Psi^*\left[-\frac{\hbar^2}{2m}\nabla^2 + V\right]\Psi\, d\mathbf{r} = \langle\hat{H}\rangle, \tag{2.56c}$$

which coincide with the corresponding expectation values obtained from standard quantum mechanics, this showing the equivalence at a predictive level of both approaches. Obviously, from a trajectory viewpoint, i.e., when the associated local field functions are evaluated along trajectories, Eqs. (2.56) reads as

$$\bar{\mathbf{r}}_B = \frac{1}{N}\sum_{i=1}^{N} w_i \mathbf{r}_i(t), \tag{2.57a}$$

$$\bar{\mathbf{p}}_B = \frac{1}{N}\sum_{i=1}^{N} w_i \nabla S(\mathbf{r}_i(t)), \tag{2.57b}$$

$$\bar{E}_B = \frac{1}{N}\sum_{i=1}^{N} w_i \left\{\frac{[\nabla S(\mathbf{r}_i(t))]^2}{2m} + V_{\text{eff}}(\mathbf{r}_i(t))\right\}, \tag{2.57c}$$

where N is the total number of trajectories considered, w_i is the associated weight—if the trajectories are initially sampled according to ρ_0, then $w_i = 1$ for all trajectories, otherwise $w_i \approx \rho(\mathbf{r}(t_0))$—and the subscript B means that these average values are computed from a sampling of Bohmian trajectories. As in classical statistical treatments, provided the sampling of initial conditions is properly carried out according to some initial distribution function (this role is played here by $\rho(0)$), in the limit $N \rightarrow \infty$, the quantities given by Eqs. (2.57) will approach the corresponding value of their quantum homologous, given by Eqs. (2.56). Taking this into account, one readily notes that the uncertainty principle can be directly related to a statistical result instead of to an inherent impossibility to measure positions or momenta—the source for this impossibility would be rather associated with the way how things happen (interact) at quantum scales. In this sense, the inequality

$$\Delta r_i \Delta p_i \geq \frac{\hbar}{2}, \tag{2.58}$$

expresses the relationship between two statistical quantities (in this case, position and momentum) in quantum mechanics, where

$$(\Delta r_i)^2 = \overline{r_i^2} - \overline{r_i}^2 \approx \overline{r_{B,i}^2} - \overline{r_{B,i}}^2 , \tag{2.59a}$$

$$(\Delta p_i)^2 = \overline{p_i^2} - \overline{p_i}^2 \approx \overline{p_{B,i}^2} - \overline{p_{B,i}}^2 , \tag{2.59b}$$

with $i = x, y, z$.

2.2.4.1 Application: Single-Slit Diffraction from a Bohmian Perspective

Following the analysis of the single-slit diffraction introduced at the end of the previous section, here we are now going to reconsider it at a Bohmian level. Thus, first, it is important to note that transverse propagation (if the incident wave carried a transverse momentum) and spreading are independent of one another. To understand how the "compete," first we consider the phase of the time-dependent wave function, Eq. (2.33), which reads as

$$S(x,t) = p_0 \left(x - x_t \right) + \frac{\hbar^2 t}{8m\sigma_0^2 \sigma_t^2} \left(x - x_t \right)^2 + E_0 t - \frac{\hbar}{2} \tan^{-1} \left(\frac{\hbar t}{2m\sigma_0^2} \right). \tag{2.60}$$

When this expression is substituted into the guidance equation, determined by the velocity field Eq. (2.45), we obtain

$$\dot{x} = \frac{1}{m} \frac{\partial S}{\partial x} = p_0 + \frac{\hbar^2 t}{4m\sigma_0^2 \sigma_t^2} \left(x - x_t \right). \tag{2.61}$$

Then, integrating over time (as it can readily be seen, this integration is analytical), it is found that

$$x(t) = x_t + \frac{\sigma_t}{\sigma_0} \left[x(0) - x_0 \right]. \tag{2.62}$$

Equation (2.62) defines the family of Bohmian trajectories whose dynamics is ruled by the free Gaussian wave packet. As seen in this equation, the decoupling between translational motion and spreading also manifests in the corresponding Bohmian trajectories. When considering the three time regimes mentioned in p. 92, we find that [22]

1. Ehrenfest-Huygens regime: The Bohmian trajectories remain nearly parallel to one another and also to the centroidal one, x_t, i.e., $x(t) \approx x(0) + v_0 t$.
2. Fresnel regime: The Bohmian trajectories undergo a kind of uniformly accelerated motion, expressible as $x(t) \approx x(0) + v_0 t + a_{\text{eff}} t^2 / 2$, with $a_{\text{eff}} = x(0) / \tau^2$.
3. Fraunhofer regime: The Bohmian trajectories display a kind of uniform rectilinear motion, with an effective velocity given by $v_{\text{eff}} = v_0 + (x_0 / \sigma_0) v_s$, i.e., $x(t) \approx v_{\text{eff}} t$.

The latter case is interesting because it defines the asymptotic state that will be eventually observed and that can give rise to very different outcomes. For example, let us assume that the maximum values of $|x_0|$ are of the order of σ_0 (or a few times σ_0, as much). Therefore, $|x_0| / \sigma_0 \sim 1$ and $v_{\text{eff}} \approx v_0 + v_s$. The long-time dynamics will then be governed by the larger of these two contributions. If v_0 is dominant, the asymptotic motion is basically classical-like (no significant spreading in comparison with the distances covered by particles). However, if v_s is dominant, the asymptotic motion is seemingly classical, though determined by the nonclassical rate $v_s x_0 / \sigma_0$, which means that every particle from the swarm will propagate with a different asymptotic velocity. This velocity increases as x_0 increases, which is in correspondence with the fact that higher-frequency components will propagate faster—in Bohmian mechanics, this high-frequency components affect the trajectories in the foremost part of the wave packet, whereas lower-frequency ones affect its rearmost part, as will be seen below (see Figure 2.1). The fact of having very different values for v_0 and v_s leads to very interesting effects.

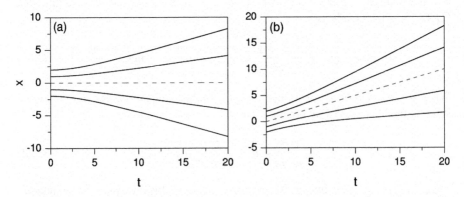

FIGURE 2.1 Set of Bohmian trajectories associated with the propagation of a free Gaussian wave packet with $v_0 = 0$ (a) and $v_0 \neq 0$ (b). In each panel, the dashed line denotes the instantaneous position of the centroid of the wave packet. Such classical trajectory also corresponds to the Bohmian trajectory, which is launched with initial condition $x(0) = x_0$.

After a closer inspection to Eq. (2.62), one readily notices that any trajectory separates from its neighbors at a constant rate regardless of the initial value of the group velocity v_0. For example, given two initial conditions, $x_1(0)$ and $x_2(0)$, the (time-dependent) distance between the two corresponding trajectories will be

$$d(t) = |x_2(t) - x_1(t)| = \frac{\sigma_t}{\sigma_0} d(0), \tag{2.63}$$

irrespective of the value of v_0 and where $d(0) = |x_2(0) - x_1(0)|$. Taking into account this distance, Eq. (2.62) can be recast as

$$x(t) = x_t + sign[\delta x(0)] d(t) \tag{2.64}$$

if we identify $x_2(t) = x(t)$ and $x_1(t) = x_t$, and $\delta x(0) = x_2(0) - x_1(0)$. In Figure 2.1, we can see this effect for: (a) $v_0 = 0$ and (b) $v_0 \neq 0$. As can be noticed in this figure, initially, in the Fresnel region, the trajectories are basically parallel, while they start diverting and displaying a rectilinear motion asymptotically, at the Fraunhofer region. In panel (a), since $v_0 = 0$, the lower-frequency components are near the centroidal trajectory, while higher-frequency components can be found at the outmost part of the diffusive beam (these components having positive and negative sign). On the other hand, in panel (b), the higher-frequency components are at the forefront part of the wave packet, while the lower-frequency ones appear in its rearmost part.

2.3 NUMERICAL METHODOLOGIES IN REAL TIME

Apart from output data, numerical simulations in real time allow us to investigate and analyze how the neutron flow is affected or develops from one space region at a given time to another subsequent space region. In the particular case we are dealing with here, this means a reliable method to understand how interference and diffraction phenomena emerge. This thus goes beyond standard theoretical optical-type approaches based on the use of auxiliary rays and integrals that mask all the process.

In Section 2.2, we have shown that simple models can be used in the place of more refined ones in order to understand the physics (phenomenology) implicit in neutron diffraction. This allows us to skip the particular shape of interaction potentials, and rather focus on the initial conditions, given in terms of the particular choice for the neutron wave function. Of course, different levels of refinement can be considered here, but this discussion goes beyond the scope of this chapter (for particular descriptions, the interested reader may consult other chapters in this book).

Here we are going to focus on two particularly interesting numerical methods when dealing with paraxial working conditions. One is based on Heller's frozen-Gaussian approach (see Section 2.3.1), where the diffracted plane wave is approximated by a linear combination of Gaussian wave packets. As time evolves, all these wave packets start developing a phase, which eventually gives rise to an interference process that nicely reproduces the features provided by the standard diffraction theory. Because it is a very intuitive approach, directly connected to the well-known Ehrenfest's theorem and quantum-classical correspondence, as well as it is very simple to implement (the evolution of the wave packets is fully analytical), it is the method that has been used to produce the results presented in Section 2.4. Regarding the Bohmian formulation of the problem, this method also introduces an additional nice feature: it provides us with an analytical nonlocal expression for the guidance equation.

The second method, more sophisticated, solves exactly Schrödinger's equation by discretizing both space and time. The main algorithm is quite standard in atomic and molecular physics, condensed matter physics, or chemical physics, and has its roots in the resolution of optical problems. This is a beam propagation method nicely combines the split-operator algorithm (based on an application of the so-called Baker-Campbell-Hausdorff formula) with the Fourier transform, which takes advantage of the dual coordinate-momentum representation of the wave function. As in the case of the frozen-Gaussian method, here we also get an analytical nonlocal expression for the guidance equation, although this is in terms of the discrete plane-wave basis set arising from the discretization and limitation of the configuration (coordinate) space.

2.3.1 FROZEN-GAUSSIAN BASIS SETS

2.3.1.1 General Framework

Perhaps, one of the very first concepts we all learn in quantum mechanics is that the direct analog of a classical point-like system (i.e., a system or particle described by a well-specified point in the phase space) is a localized wave packet, typically a Gaussian one [29]. Following Ehrenfest's theorem, the spatial spreading with time of these wave packets is relatively slow and negligible compared with the space variations of the potential functions describing the interactions acting on the system. This implies, first, that they nearly preserve their initial shape and hence their initial width over reasonable timescales. Second, their centroid essentially follows a trajectory describable by Hamilton's equations, according to Ehrenfest's theorem [30].

In 1975, just when the first wave-propagation methods were already being considered, Heller considered such nice features to implement a simple propagation method based on the semiclassical harmonic approximation [31, 32]: whenever the space variations associated with the interaction potential are smooth enough compared with the space extension of a Gaussian wave packet (i.e., for relatively narrow, localized wave packets), it can be shown that the latter's dynamics is governed by the local harmonic approximation of such a potential and hence a set of simple ordinary differential equations can be used in the place of Schrödinger's equation. In other words, under such working conditions, Gaussian wave packets remain Gaussian all through their evolution in time. And, by means of the superposition principle, this rule can also be extended to other types of localized wave functions, as it is seen below.

Thus, following Heller's proposal, consider an N-dimensional Gaussian ansatz,

$$\Psi(\mathbf{r},t) = \exp\left\{\frac{i}{\hbar}\left[(\mathbf{r}-\mathbf{r}_t)A_t(\mathbf{r}-\mathbf{r}_t)^T + \mathbf{p}_t \cdot (\mathbf{r}-\mathbf{r}_t)^T + \gamma_t\right]\right\}, \qquad (2.65)$$

which is a functional of time-dependent functions, vectors ('T' stands for transpose) and matrices:

1. The instantaneous (vector) position $\mathbf{r}_t$ and momentum $\mathbf{p}_t$ of the wave-packet centroid. If the ansatz Eq. (2.65) is a normalized, then $\langle \hat{\mathbf{r}} \rangle = \mathbf{r}_t$ and $\langle \hat{\mathbf{p}} \rangle = \mathbf{p}_t$.
2. A complex-valued $N \times N$ covariance matrix, A_t, accounting for the wave-packet shape (Gaussian-type) distortions: the diagonal elements describe spreading along each direction or degree of freedom, while the off-diagonal elements deal with the correlations between different directions.
3. A complex-valued parameter, γ_t, which ensures the norm conservation at any time (imaginary part) and introduces a global time-dependent phase factor (real part). This latter factor actually corresponds to the Gouy phase in Gaussian-beam optics.

For simplicity in the notation and to avoid misunderstandings, in this section a subscript t shall be used to denote time dependence [for instance, $\mathbf{r}_t \equiv \mathbf{r}(t)$]. From Eq. (2.65), it can be seen that the norm is explicitly described by the quantity

$$\operatorname{Im}(\gamma_t) = -\frac{\hbar}{4} \ln \left\{ \left(\frac{2}{\pi\hbar} \right)^N \Delta[\operatorname{Im}(A_t)] \right\} = -\frac{\hbar}{4} \operatorname{Tr} \left\{ \ln \left[\left(\frac{2}{\pi\hbar} \right)^N \operatorname{Im}(A_t) \right] \right\}, \quad (2.66)$$

where $\Delta(A_t)$ is the determinant of A_t and $\exp\{\operatorname{Tr}[\ln(A_t)]\} = \Delta(A_t)$. Hence, in order to ensure the norm preservation with time, the initial value γ_0 has to be chosen as

$$\operatorname{Im}(\gamma_0) = -\frac{\hbar}{4} \operatorname{Tr} \left\{ \ln \left[\left(\frac{2}{\pi\hbar} \right)^N \operatorname{Im}(A_0) \right] \right\}. \quad (2.67)$$

For all practical purposes, γ_0 can be chosen to be a pure imaginary number, i.e., $\gamma_0 = i\operatorname{Im}(\gamma_0)$.

The set of equations of motion accounting for the time evolution of the different parameters ruling the behavior of the ansatz Eq. (2.65) can easily be determined by substituting the latter into the time-dependent Schrödinger's equation after considering the harmonic approximation of the potential function. Thus, consider the potential function $V(\mathbf{r})$ is Taylor-expanded around the classical trajectory $\mathbf{r}_t$ up to second order, which renders the effective time-dependent potential function

$$V_t^{\text{eff}}(\mathbf{r}) = V_t + \mathbf{V}_t' \cdot (\mathbf{r} - \mathbf{r}_t)^T + \frac{1}{2}(\mathbf{r} - \mathbf{r}_t) V_t''(\mathbf{r} - \mathbf{r}_t)^T, \quad (2.68)$$

where

$$V_t = V(\mathbf{r} = \mathbf{r}_t), \quad (2.69a)$$

$$\mathbf{V}_t' = \nabla V(\mathbf{r})|_{\mathbf{r} = \mathbf{r}_t}, \quad (2.69b)$$

$$V_t'' = \left. \frac{\partial^2 V(\mathbf{r})}{\partial r_i \, \partial r_j} \right|_{\mathbf{r} = \mathbf{r}_t}, \quad (2.69c)$$

with $i, j = 1, 2, \ldots, N$. When this effective potential function is substituted into the time-dependent Schrödinger equation, the latter turns into an effective equation that describes the local evolution of the wave packet around the instantaneous position $\mathbf{r}_t$,

$$i\hbar \frac{\partial \Psi}{\partial t} = \hat{H}_t^{\text{eff}} \Psi, \qquad (2.70)$$

by means of the effective, time-dependent Hamiltonian

$$\hat{H}_t^{\text{eff}} \equiv -\frac{\hbar^2}{2m} \nabla^2 + V_t^{\text{eff}}(\mathbf{r}), \qquad (2.71)$$

which is correct up to second-order corrections in the position.

Substitution of the ansatz Eq. (2.65) into the effective Schrödinger's equation Eq. (2.70), and then gathering factors in terms of powers of same order of $(\mathbf{r} - \mathbf{r}_t)$, leads to the set of ordinary differential equations

$$\dot{\mathbf{r}}_t = \left. \frac{\partial H}{\partial \mathbf{p}} \right|_{\mathbf{p}=\mathbf{p}_t} = \frac{\mathbf{p}_t}{m}, \qquad (2.72a)$$

$$\dot{\mathbf{p}}_t = \left. \frac{\partial H}{\partial \mathbf{r}} \right|_{\mathbf{r}=\mathbf{r}_t} = -\mathbf{V}'(\mathbf{r}_t), \qquad (2.72b)$$

$$\dot{\mathbf{A}}_t = -\frac{2}{m} \mathbf{A}_t^2 - \frac{1}{2} V''(\mathbf{r}_t), \qquad (2.72c)$$

$$\dot{\gamma}_t = \frac{i\hbar}{m} \text{Tr}(\mathbf{A}_t) + \mathcal{L}_t = \frac{i\hbar}{m} \text{Tr}(\mathbf{A}_t) + \mathbf{p}_t \cdot \dot{\mathbf{r}}_t^T - E, \qquad (2.72d)$$

where the first two equations are just the classical Hamilton's equations for the centroid of the wave packet (as mentioned above, it evolves as an ordinary classical trajectory). Actually, the close analogy with a classical motion is also manifest in the last term of Eq. (2.72d), in terms of the classical Lagrangian $\mathcal{L}_t$ in the first equality, and the classical total energy in the second equality,

$$E = \frac{\mathbf{p}_t^2}{2m} + V(\mathbf{r}_t). \qquad (2.73)$$

which is constant along the trajectory $\mathbf{r}_t$. When Eq. (2.72d) is integrated over time, L_t gives the classical action (apart from a normalization factor), which is consistent with the semiclassical nature of the method.

Even though in an approximated fashion, the set of ordinary differential equations, Eqs. (2.72), simplifies the search of numerical solutions of the Schrödinger's equation. These solutions will be relatively accurate provided third- and higher-order derivatives of the potential function can be neglected, which means that deviations from the true solution are smaller than

$$\sigma^2 \left(\frac{\partial^3 V / \partial r_i\, \partial r_j\, \partial r_k}{\partial V / \partial r_l} \right)_{\mathbf{r}=\mathbf{r}_t}, \tag{2.74}$$

where σ denotes the effective width of the wave packet and $i, j, k = 1, 2, \ldots, N$. Conversely, for polynomial-type potential functions of second or lesser order, the time evolution obtained from this set of equations is exact and so the evolution displayed by the ansatz, Eq. (2.65), which is fully analytical. The same can also be applied in the case of dissipative systems, such as the Caldirola-Kanai Hamiltonian model [33].

As for the guidance equation, notice that it is analytical (even though its solutions is, in general, numerical),

$$\dot{\mathbf{r}}_t = \frac{\mathbf{p}_t}{m} + \frac{2\,\mathrm{Tr}(\mathbf{A}_t)}{m} \cdot (\mathbf{r} - \mathbf{r}_t)^T. \tag{2.75}$$

This equation shows that the evolution of Bohmian trajectories is essentially ruled by two competing types of motion, where one obeys the classical laws of motion, while the other is linked to the spreading of the wave packet [21].

2.3.1.2   From Localized Gaussians to Diffracted Plane Waves

The fact that the superposition principle still holds for the approximated Schrödinger's equation allows to recast any localized wave packet as a bare sum of Gaussian wave packets. This is particularly interesting in the case of collimated beams, where the extension and shape of the diffracted density (wave function) are strongly connected to the action of the transfer function of the opening. Thus, let us consider the diffraction problem within the general framework of scattering theory (after all, the physics is exactly the same). Following standard notation from matter-wave scattering theory [34], consider an incident or incoming monochromatic plane wave,

$$\Psi_{\mathrm{in}}(\mathbf{r}, t = 0) = \frac{1}{\sqrt{k_{z,0}}}\, e^{i\mathbf{k}_0 \cdot \mathbf{r}} = \frac{1}{\sqrt{k_{z,0}}}\, e^{-ik_{z,0}z + i\mathbf{k}_0 \cdot \mathbf{R}}, \tag{2.76}$$

where $\mathbf{k}_0 = (\mathbf{K}_0, k_{z,0})$ is the incident wave vector, $\mathbf{R}$ is a position vector parallel to the plane where the opening is allocated, such that $\mathbf{r} = (\mathbf{R}, z)$, and the normalization is taken with respect to the incident flux. The wave, which propagates along the $\mathbf{k}_0$ direction, perpendicular to the aperture plane, represents an incident neutron beam with well-defined energy, i.e., with mean

$$\bar{E} = \langle \hat{H} \rangle = \frac{\hbar^2 \mathbf{k}_0^2}{2m} \tag{2.77}$$

where $\mathbf{p}_0 = \hbar\mathbf{k}_0$, and zero energy dispersion, $\Delta E = \sqrt{\langle \hat{H}^2 \rangle - \hat{H}^2}$.

Computationally, the propagation of the incoming plane wave Eq. (2.76) can be tackled as follows. First, at $t = 0$, it is recast in terms of a coherent superposition of source plane waves acted by a Gaussian spatial propagator, as

$$\Psi_{in}(\mathbf{r}) = \int \mathcal{K}(r - r_0) e^{i\mathbf{k}_0 \cdot \mathbf{r}_0^T} d\mathbf{r}_0, \tag{2.78}$$

just as a Huygens' integral, where the propagator is given by the three-dimensional form of the ansatz Eq. (2.65) at $t = 0$. Accordingly, the wave function, Eq. (2.78), reads explicitly as

$$\Psi_{in}(\mathbf{r}) \sim \int e^{i(\mathbf{r}-\mathbf{r}_0)A_0(\mathbf{r}-\mathbf{r}_0)^T/\hbar + i\mathbf{k}_0 \cdot (\mathbf{r}-\mathbf{r}_0)^T + i\gamma_0/\hbar} e^{i\mathbf{k}_0 \cdot \mathbf{r}_0^T} d\mathbf{r}_0, \tag{2.79}$$

(without loss of generality, the normalization prefactor has been removed). For practical (computational) purposes, this wave function can be recast as a coherent superposition of a finite number (N) of frozen Gaussians,

$$\Psi_{in}(\mathbf{r}) \sim \sum_{n=1}^{N} e^{i(\mathbf{r}-\mathbf{r}_0^n)A_0^n(\mathbf{r}-\mathbf{r}_0^n)^T/\hbar + i\mathbf{k}_0^n \cdot (\mathbf{r}-\mathbf{r}_0^n)^T + i\gamma_0^n/\hbar} e^{i\mathbf{k}_0^n \cdot \mathbf{r}_0^{n^T}}, \tag{2.80}$$

with consecutive centroids separated a distance $\Delta\mathbf{r}_0$. The monochromaticity of Eq. (2.80) will increase as $\Delta\mathbf{r}_0$ gets smaller, the number of wave packets increases, and the extension covered by them increases. Due to the linearity of the Schrödinger's equation, the time evolution of Eq. (2.80) is now straightforwardly obtained by replacing each wave packet with its time-evolved form, given by Eq. (2.65), i.e.,

$$\Psi(\mathbf{r}) \sim \sum_{l=1}^{\ell} e^{i(\mathbf{r}-\mathbf{r}_t^l)A_t^l(\mathbf{r}-\mathbf{r}_t^l)^T/\hbar + i\mathbf{k}_t^l \cdot (\mathbf{r}-\mathbf{r}_t^l)^T + i\gamma_t^l/\hbar} e^{i\mathbf{k}_0 \cdot \mathbf{r}_0^{l^T}}, \tag{2.81}$$

(the subscript 'l' in the last phase factor has been removed because it is the same for all Gaussians). That is, generally speaking, instead of solving the time-dependent Schrödinger's equation, the time evolution of the (quasi-plane) wave function, Eq. (2.80), is obtained after solving a set of $4N$ ordinary differential equations.

2.3.1.3 Computation of Bohmian Trajectories

Regarding the Bohmian trajectories, substituting the wave function, Eq. (2.81), into the Bohmian guidance equation, we find

$$\dot{\mathbf{r}} = \frac{1}{2m|\Psi|^2} \left\{ \Psi^* \sum_{l=1}^{\ell} \left[\mathbf{p}_t^l + 2A_t^l(\mathbf{r}-\mathbf{r}_t^l)^T \right] \Psi^l + \Psi \sum_{l=1}^{\ell} \left[\mathbf{p}_t^l + 2(A_t^l)^\dagger(\mathbf{r}-\mathbf{r}_t^l)^T \right] \Psi^{l,*} \right\}, \tag{2.82}$$

By defining the vector

$$\mathbf{d} = \mathbf{p}_t + 2A_t(\mathbf{r} - \mathbf{r}_t)^T, \tag{2.83}$$

Equation (2.82) can be recast as

$$\dot{\mathbf{r}} = \frac{1}{m|\Psi|^2} \sum_{l=1}^{\ell} \left[\text{Re}(\mathbf{d}^l)\text{Re}(\Psi^*\Psi^l) - \text{Im}(\mathbf{d}^l)\text{Im}(\Psi^*\Psi^l) \right], \tag{2.84}$$

Although this is an already compact form, one can further proceed in order to obtain a particularly more suitable functional form regarding the derivation of analytical results. To this end, first each Gaussian wave packet is recast in the usual polar form, with $R = \rho^{1/2}$, so that

$$\text{Re}(\Psi^l\Psi^{l',*}) = R^l R^{l'} \cos\omega_{ll'}, \tag{2.85a}$$

$$\text{Im}(\Psi^l\Psi^{l',*}) = R^l R^{l'} \sin\omega_{ll'}, \tag{2.85b}$$

$$|\Psi|^2 = \sum_{l,l'=1}^{l} \Psi^l\Psi^{l',*} = \sum_{l,l'=1}^{l} R^l R^{l'} \cos\omega_{ll'}, \tag{2.85c}$$

where $\omega_{ll'} = (S^l - S^{l'})/\hbar$. Accordingly, substituting these expressions into Eq. (2.84), this equation of motion acquires the more compact form

$$\dot{\mathbf{r}} = \frac{1}{m} \frac{\sum_{l,l'=1}^{\ell} R^l R^{l'} \left[\text{Re}(\mathbf{d}^l)\cos\omega_{ll'} - \text{Im}(\mathbf{d}^l)\sin\omega_{ll'} \right]}{\sum_{l,l'=1}^{\ell} R^l R^{l'} \cos\omega_{ll'}}. \tag{2.86}$$

2.3.2 SPECTRAL BASIS SETS

There are problems of interest, where the exact solution can be recast in terms of coherent superpositions of energy eigenstates, which constitute a suitable working basis set in the same way that Gaussian wave packets has also served for the purpose in the previous sections. This is the case, for instance, when dealing with neutron diffraction through wave guides, the modes in the guide plus some continuum contributions constitute an excellent basis set [35]. In the case of diffraction through openings, it is also possible to devise optimal basis sets after arising from the Fourier transform of the opening [17]. In either case, as in Heller's method, this avoids dealing explicitly with the exact solution of the time-dependent Schrödinger equation.

For simplicity and without loss of generality, consider a one-dimensional application, although the generalization to more dimensions is straightforward. To start with, consider the decomposition of the wave function in its Fourier components or, equivalently, its representation in the momentum space,

$$\Psi(x,t) = \frac{1}{\sqrt{2\pi\hbar}} \int c(p,t) e^{ipx/\hbar}\, dp, \tag{2.87}$$

where

$$c(p,t) = \frac{1}{\sqrt{2\pi\hbar}} \int \Psi(x,t) e^{-ipx/\hbar}\, dp \tag{2.88}$$

denotes the wave function in the momentum space (the notation c is used here instead of $\tilde{\Psi}$ to stress the role of this quantity as a complex-valued weight associated with the plane wave $e^{ipx/\hbar}$ rather than its role as a amplitude probability in the momentum space). This representation is just a recast of the wave function in terms of a basis set formed by plane waves—in bound problems, these waves recombine to form a discrete basis set of stationary waves, i.e., energy eigenstates of the Hamiltonian according to the conventional textbook picture that we have of bound systems (see below). In principle, we are not going to set any condition on $c(p,t)$, so we assume the general case of a complex-valued function, just as $\Psi(x,t)$. Therefore, we can express it in the usual polar form, $c(p,t) = |c(p,t)| e^{i\varphi(p,t)}$. Accordingly, we find that

$$\Psi^* \nabla \Psi = \frac{1}{\sqrt{2\pi\hbar}} \frac{1}{\hbar} \iint p |c(p,t)| |c(p',t)| e^{i\Delta k_{pp'} x + i\delta_{pp'}}\, dp\, dp', \tag{2.89a}$$

$$|\Psi|^2 = \frac{1}{\sqrt{2\pi\hbar}} \iint |c(p,t)| |c(p',t)| e^{i\Delta k_{pp'} x + i\delta_{pp'}}\, dp\, dp', \tag{2.89b}$$

where $\Delta k_{pp'} = k_p - k_{p'} = (p - p')/\hbar$ and $\delta_{pp'}(t) = \varphi(p,t) - \varphi(p',t)$. After substituting these two quantities into the guidance condition and rearranging terms, we obtain

$$\dot{r} = \frac{1}{m} \frac{\iint p |c(p,t)| |c(p',t)| \cos\left(\Delta k_{pp'} x + \delta_{pp'}\right) dp dp'}{\iint |c(p,t)| |c(p',t)| \cos\left(\Delta k_{pp'} x + \delta_{pp'}\right) dp dp'}, \tag{2.90}$$

which is pretty similar to Eq. (2.86). As it can be seen, Eq. (2.90) is an analytical expression that allows us to compute the Bohmian trajectories avoiding the inconvenience of the partial derivatives of the wave function if the Fourier decomposition of the wave function can be determined at each time step. In principle, this may seem a hard task, but there are many situations where this calculation can be done straightaway. Moreover, it results very convenient from the viewpoint of the accuracy of the simulation, since usually Ψ is only known at certain discrete positions on a grid (see below), since Eq. (2.90) does not use explicitly the value of Ψ, but the position x along the trajectory, the time at which it is evaluated, and the superposition of a set of basis functions, $c(p,t)$.

2.3.3   PSEUDOSPECTRAL BASIS SETS

2.3.3.1   Space and Time Discretization

Another situation where we can find similar analytical expressions for the Bohmian
guidance equation arises when we consider the fast Fourier transform (FFT) [30]
to solve the spatial part of Schrödinger's equation (i.e., to compute the action of the
Laplacian over the wave function) in numerical grids [37]. As will be seen below,
although the Fourier transform renders a continuous function, because of the discretization of the latter in a grid, the Fourier transform will also be discrete, leading
to a form similar to Eq. (2.86).

To start with, notice that the evolution of the wave function from t_1 to t_2, with
$t_2 > t_1$, can be formally expressed as

$$\Psi(\mathbf{r},t_2) = \hat{U}(t_2,t_1)\Psi(\mathbf{r},t_1), \tag{2.91}$$

where

$$\hat{U}(t_2,t_1) = e^{-i(t_2-t_1)\hat{H}/\hbar} \tag{2.92}$$

is the *time-evolution operator*. The difference among the various grid methods arises
from how the action of this operator over the wave function is computed. More specifically, there are two steps here: (1) the time evolution and (2) the implementation
of the Laplacian operator in $\hat{H}$. The accuracy of each method thus relies directly on
the algorithm devised to compute the action of $\hat{U}$.

Consider the time is discretized in time steps δ relatively small. Then, Taylor
expanding the evolution operator, Eq. (2.92), up to first order, one finds

$$\Psi^{n+1}(\mathbf{r}) \approx \left(1 - \frac{i\delta}{\hbar}\hat{H}\right)\Psi^n(\mathbf{r}). \tag{2.93}$$

That is, the wave function at time $t_{n+1} = (n+1)\delta$ is simply and *explicitly* computed
from its value at $t_n = n\delta$. However, despite the simplicity of this first-order approximation, it is numerically unstable. This instability can be removed by considering an
implicit scheme, where

$$\Psi^n(\mathbf{r}) \approx \left(1 + \frac{i\delta}{\hbar}\hat{H}\right)\Psi^{n+1}(\mathbf{r}). \tag{2.94}$$

However, this other scheme has an important flaw: it is not unitary, which is important to preserve conservation of the probability. Although $\hat{U}$ and $\hat{U}^\dagger$ are unitary
(because of the Hermiticity of $\hat{H}$), the same does not happen with their power-series
approximations. To avoid this drawback, though, one can use the so-called Cayley
operator,

$$\hat{U}(\delta) \approx \frac{1 - i\delta\hat{H}/2\hbar}{1 + i\delta\hat{H}/2\hbar}. \tag{2.95}$$

This operator constitutes the core of the *Crank-Nicholson method* [38, 39], which results very convenient in the simulation of one-dimensional systems, because of its stability and accuracy for relatively long times.

At first-order approximation, it is also possible to obtain another reliable scheme based on the time-symmetric relationship

$$\Psi(\mathbf{r}, t+\delta) - \Psi(\mathbf{r}, t-\delta) \approx \left(e^{-i\delta\hat{H}/\hbar} - e^{i\delta\hat{H}/\hbar} \right) \Psi(\mathbf{r}, t). \tag{2.96}$$

Substituting $\hat{U}$ and $\hat{U}^{\dagger}$ by their first-order approximations into Eq. (2.105), and discretizing the resulting expression, yields

$$\Psi^{n+1}(\mathbf{r}) \approx \Psi^{n-1}(\mathbf{r}) - \frac{2i\delta}{\hbar} \hat{H} \Psi^{n}(\mathbf{r}). \tag{2.97}$$

This method is the so-called *second-order difference scheme* [37], with a degree of accuracy similar to that of the schemes based on Eq. (2.94) or Eq. (2.95), although more advantageous computationally, since it is simpler to implement. The only extra requirement is that, in order to solve Eq. (2.97), it is necessary to know previously Ψ^{0} and Ψ^{1}—the latter can be obtained by solving Eq. (2.93) with a simple second-order Runge-Kutta scheme, for example.

There are other more sophisticated methods, which go to a higher level of accuracy. One of the simplest and most popular ones is the so-called split-operator method [40], which consists in representing Eq. (2.92) as

$$\hat{U} \approx e^{-i\delta\hat{T}/2\hbar} e^{i\delta\hat{V}/\hbar} e^{-i\delta\hat{T}/2\hbar}, \tag{2.98}$$

i.e., splitting the Hamiltonian in such a way that one carries the evaluation of the potential and kinetic operators at different steps. This essentially implies an emphasis on the evaluation of the action of the Hamiltonian, since the time integration can be performed in any standard way given the accuracy of the representation of $\hat{U}$.

2.3.3.2 The Fourier Method

In order to improve the performance in the evaluation of the kinetic operator, a second type of methods is also used, which is based on a spectral (Fourier) decomposition of the wave function in the momentum representation [40, 41]. In this sense, the most efficient method is the so-called FFT [36], which presents an exponential convergence to the actual value of the derivative—the algorithm to compute the Fourier transform in this way is faster than a conventional one and requires a lower-computational effort. This scheme is thus based on the great effectiveness of Fourier analysis to compute the action of the kinetic operator. More specifically, taking advantage of the diagonal form of the kinetic operator in the momentum representation, we have that

$$\hat{T}\tilde{\Psi}(\mathbf{p}, t) = -\frac{\hat{\mathbf{p}}^2}{2m} \tilde{\Psi}(\mathbf{p}, t) = -\frac{\mathbf{p}^2}{2m} \tilde{\Psi}(\mathbf{p}, t), \tag{2.99}$$

where $\tilde{\Psi}(\mathbf{p},t)$ is the wave function in the momentum representation. The action of the Hamiltonian on the wave function is then given by

$$\hat{H}\Psi(\mathbf{r},t) = \left(\frac{1}{2\pi\hbar}\right)^{3/2} \int \frac{\mathbf{p}^2}{2m} \tilde{\Psi}(\mathbf{p},t) e^{i\mathbf{p}\cdot\mathbf{r}/\hbar} \, d\mathbf{p} + V(\mathbf{r})\Psi(\mathbf{r},t) \qquad (2.100)$$

(with Ψ properly discretized). The Fourier transform to obtain $\tilde{\Psi}$ and the inverse Fourier transform in Eq. (2.100) are carried out by making use of the FFT algorithm [36], which allows to compute those transforms in faster manner than a conventional algorithm and with a smaller computational effort.

In the FFT algorithm, though, the spatial discretization leads to discretization of the possible momenta, which are connected to the maximum and minimum dimensions of the spatial grid. For example, consider one dimension. The grid covers an extension L and has N sampling points (where $N = m2^n$, with $m = 1$, 3, or 5, and n being a nonzero positive number), such that $x_i = i\varepsilon$, with $i = 0$, 1, 2, ..., $N-1$, and $\varepsilon = L/N$. With this, the distance between two adjacent points in the momentum space is given by $p_{min} = 2\pi\hbar/L$, while the largest momentum available (in absolute value) is $p_{max} = 2\pi\hbar/2\varepsilon$, which corresponds to the Nyquist critical frequency [36]. The momentum grid then goes between $-p_{max}$ and p_{max}, or, equivalently, given the periodicity of the Fourier transform, from 0 to $(N-1)p_{min}$, taking into account that $p_j = jp_{min}$, with $j = 0$, 1, 2, ..., $N-1$. The (discrete) spectral Fourier decomposition of the wave function at the point x_i is then given by

$$\Psi_i = \frac{1}{N} \sum_{j=0}^{N-1} \tilde{\Psi}_j e^{ik_j x_i}, \qquad (2.101)$$

with $k_j = jp_j/\hbar$ and

$$\tilde{\Psi}_j = \frac{1}{N} \sum_{i=0}^{N-1} \Psi_i e^{ik_j x_i} \qquad (2.102)$$

Now, taking into account the periodicity of the Fourier decomposition, Eq. (2.101), we should notice that:

1. The zero momentum component corresponds to $j = 0$.
2. Positive momentum components, $0 < p < p_{max}$, correspond to $1 \le j \le N/2 - 1$.
3. Negative momentum components, $-p_{max} < p < 0$, correspond to $N/2 + 1 \le j \le N - 1$.
4. The value $j = N/2$ corresponds to both $\pm p_{max}$.

With this in mind, the discrete version of Eq. (2.100) is then given by

$$\hat{H}\Psi_i^n = \sum_{j=0}^{N-1} \frac{k_j^2 \hbar^2}{2m} \tilde{\Psi}_j^n e^{ik_j x_i} + V(x_i)\Psi_i^n, \qquad (2.103)$$

where the subscript n denotes the time step $t = n\delta$. Here, the FFT algorithm is not in the form how Eq. (2.103) is written, but in the form how it is computed. In order to get a good accuracy, it is important to ensure that the physical momenta involved in the problem under study are contained within the range limited by the maximum and minimum momenta—or, equivalently, within the corresponding energetic range [37].

Finally, it is worth mentioning the fact that many times the propagation may lead to reflections at the boundaries of the grid considered. In this case, it is common to resort to *absorbing boundaries* [42, 43] or *mask functions* [44, 45], which smoothly "swallow" (or act like "swallowing") the part of the wave function that reaches the grid boundaries, thus avoiding the contamination of the simulation with nonphysical reflections.

Taken into account the way how the wave function is propagated, there are also two possibilities to compute the evolution of the Bohmian trajectories. One of them is locally, i.e., taking into account the values of the wave function and its derivatives at each grid point and, by means of some fair interpolator, to propagate numerically the Bohmian guidance equation. The other one is globally and follows from the spectral (Fourier) decomposition of the wave function. This method is, in principle, more accurate, since one does not need of any interpolator, but just to know the components of the wave function at each time step. Then, by using the discretized form of Eq. (2.87),

$$\Psi_i^n \sim \sum_j \tilde{\Psi}_j^n e^{ik_j x_i}, \tag{2.104}$$

and proceeding in a similar manner, we find the discrete (Fourier) version of Eq. (2.90), which reads as

$$\dot{\mathbf{r}} = \frac{1}{m} \frac{\sum_{j,l} p_j |\tilde{\Psi}_j^n| |\tilde{\Psi}_l^n| \cos(\Delta k_{lj} x + \delta_{lj})}{\sum_{j,l} |\tilde{\Psi}_j^n| |\tilde{\Psi}_l^n| \cos(\Delta k_{lj} x + \delta_{lj})}, \tag{2.105}$$

where $\tilde{\Psi}_j^n = |\tilde{\Psi}_j^n| e^{i\varphi_j^n}$, $\Delta k_{lj} = k_l - k_j = (p_l - p_j)/\hbar$, and $\delta_{lj} = \varphi_l^n - \varphi_j^n$. Notice in Eq. (2.105) that the grid point x_i is substituted by any general value x, since Eq. (2.104) is also general and can be used to compute the value of the wave function at any spatial point x due to the global character of the Fourier decomposition. The time evolution, from x^n to x^{n+1}, can be now performed by using a simple Runge-Kutta algorithm.

2.4 NUMERICAL SIMULATIONS

2.4.1 SINGLE-SLIT DIFFRACTION

Although Feynman himself regarded the two-slit experiment as having "in it the heart of quantum mechanics" [46], the essence of this theory already appears in a simpler phenomenon: slit diffraction. As it happens with any type of wave, diffraction arises whenever a wave meets an object. If distances typically of several multiple numbers of the wavelength, we will then observe this wave phenomena, which consists of a series of dark and light fringes going into the region of *geometric shadow*. As shown by the above experiments, this phenomenon occurs for massive particles as well as for light.

Let us therefore analyze the physics associated with matter-wave diffraction, which essentially consists in studying the effects of boundaries on matter waves. Just to work on realistic conditions, we are going to consider the experiments carried out by Zeilinger et al. [2] with cold neutrons—i.e., with a de Broglie wavelength of the order of a few Angstroms (in this case, $\lambda = 19.26$ Å). Assuming different timescales along the directions perpendicular and parallel to the slit plane allows us some convenient simplifications in the treatment of this phenomenon. We shall start this description from the diffracted neutron beam at $t = 0$, which we consider to be a wave packet with the following shape. Along the perpendicular (z) direction, the wave packet can be well accounted for by a Gaussian wave packet, which is preferable against a simple plane wave just to keep the wave function spatially localized along this direction. In the x-direction, the wave packet can be assumed to be nearly constant along the extension covered by the slit and zero everywhere else. Accordingly, the whole (initial) wave function can be considered to be a linear combination of Gaussian wave packets (see Section 2.3.1), relatively narrow along x and very wide along z. From the discussion in Sections 2.2.1 and 2.2.2, if we choose $\sigma_{0,x} \ll w$ and $\sigma_{0,z} \gg w$, where w is the slit width, the Gaussians will only display a relevant spreading along the x-direction (it will be meaningless along z for the whole duration of the time propagation). This is important, because the overlapping or interference of all these wave packets will give rise to the appearance of the corresponding diffraction features in a similar fashion to Huygens' view of diffraction; i.e., this phenomenon arises as the interference of a number of coherent sources. Moreover, since there is no coupling between the motion along each direction—within this simple model, $V(x,z) = 0$ everywhere—the spreading along x is independent of the translational motion along z. Hence, the time evolution for all these wave packets is analytical, being described in Section 2.2.3.

A series of snapshots of the probability density, $\rho(x,z)$, describing the diffraction process are displayed in Figure 2.2 in order to illustrate its time evolution. In this case, the slit width is 92.1 μm, which corresponds to one of the cases considered in the experiment [2], and ρ is shown at different values of its centroidal position, z_t, along the z-direction—note that this is equivalent to considering different times according to the simple relation

$$z = \left(\frac{2\pi\hbar}{m\lambda} \right) t. \tag{2.106}$$

As it can be noticed, while ρ spreads along the x-direction, developing a series of maxima (for the sake of clarity, the maximum value in all plots has been set up at a 10% of the probability density maximum), its width along the z-direction remains essentially throughout the whole time evolution.

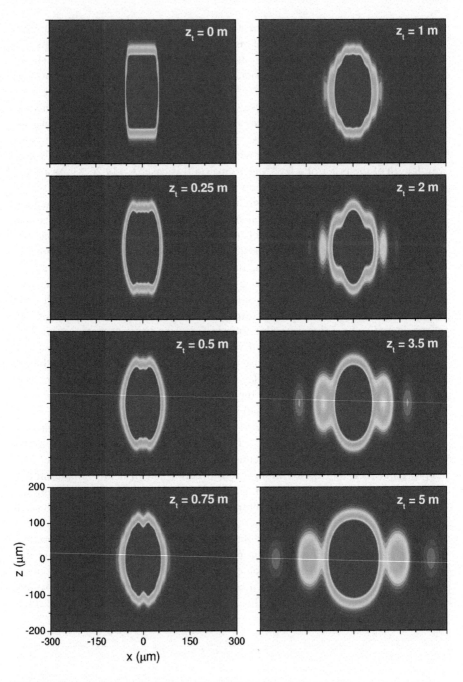

FIGURE 2.2 Time evolution of the probability density associated with a wave packet simulating the diffraction of a cold neutron beam ($\lambda = 19.26$ Å) by a single slit ($w = 92.1 \, \mu$m). Each snapshot is labeled according to the wave-packet centroid position along the z-axis, z_t. In order to better appreciate the diffraction effects, only contours below 10% of the probability density maximum are displayed.

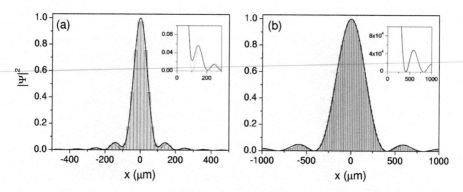

FIGURE 2.3 Intensity patterns for single-slit diffraction of cold neutrons ($\lambda = 19.26$ Å) by slits with widths: (a) 92.1 μm and (b) 23 μm. The solid line indicates a standard quantum-mechanical calculation, while histograms have been obtained by box-counting trajectory arrivals at different positions along the x-directions (with bins of a width of 20 μm) and a distance $z_t = 5$. To compare with, the single-slit Fraunhofer diffraction pattern given by Eq. (2.107) is also displayed with dotted line.

In Figure 2.3, the single-slit diffraction patterns for the two experimental slit widths [2] are shown: (a) $w = 92.1$ μm and (b) $w = 23$ μm. These patterns, computed at a distance $z_t = 5$ m from the slits, show the agreement between standard quantum mechanics (solid line) and Bohmian mechanics (histogram bars). As the initial wave function is described in a good approximation by a constant amplitude along the slit width and zero everywhere else, the shape of the patterns fits pretty well the well-known Fraunhofer single-slit diffraction formula,

$$\rho_\infty(x) = \left| \psi_\infty(x) \right|^2 \sim \left[\frac{\sin(\pi w x / \lambda z_t)}{\pi w x / \lambda z_t} \right]^2, \qquad (2.107)$$

where the subscript '∞' denotes the fact that the pattern is described far beyond the slit, within the Fraunhofer regime (i.e., with z_t being relatively large compared to characteristic transverse distances, e.g., the size spanned by the pattern itself). The slight deviation between the computed pattern and this formula (see inset in Figure 2.3a) comes from the fact that, for $z_t = 5$ m, the diffracted beam has not yet reached the Fraunhofer regime. A simple guess allows us to understand this fact. Fraunhofer diffraction is reached at distances from the slit such that $z_t \gg z_F \equiv \pi w^2 / 4\lambda$. This value amounts to $z_F \approx 3.5$ m for the case represented in Figure 2.3a and $z_F \approx 0.2$ m for the case of Figure 2.3b.

A set of representative Bohmian trajectories illustrating the dynamics associated with these patterns is displayed in Figure 2.4 for the diffraction through the slit with width $w = 23$ μm. Bearing in mind a quantum hydrodynamic context, we observe that these trajectories evolve in an essentially laminar way—i.e., without displaying very convoluted or intricate motions—as the stream of a nonturbulent fluid. Though simple, this example allows us to notice how Bohmian trajectories provide a clear and intuitive physical picture of the quantum probability density

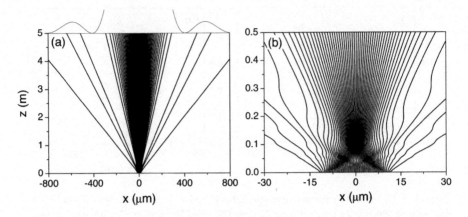

FIGURE 2.4   (a) Bohmian trajectories corresponding to diffraction through a slit of width 23 μm. At the top of the figure, intensity diffraction pattern at $z_t = 5$ m. (b) Enlargement of part (a) to show the topology of the quantum trajectories in the Fresnel region.

current or quantum flux, a quantity that is not commonly used in standard quantum mechanics, since it is not an *observable*. The regimes mentioned in Sections 2.2.3 and 2.2.4, namely the Fresnel and the Fraunhofer regime, emerge clearly here. At relatively low values of z_t (up to $z_t \sim z_F$), in the Fresnel regime, a kind of "wiggling" motion is apparent. On the other hand, beyond z_F (i.e., $z_t \gg z_F$), the trajectories become more rectilinear, this being a signature of the Fraunhofer regime. Within this regime, the asymptotic motion becomes stationary, i.e., the probability density does not change its shape and Bohmian motion is uniform.

Although it is enough to only consider the wave function and Bohmian trajectories in order to fully characterize the quantum system dynamics, just for sake of completeness it is also interesting to have a look at the associated quantum potential, which is displayed in Figure 2.5. In this figure, each frame keeps a one-to-one correspondence with those from Figure 2.2 for the probability density. To better understand the quantum potential dynamics, let us proceed as follows. Due to the uncoupling between the degrees of freedom x and z, the wave function is factorizable at any time, i.e., $\Psi(x,z,t) = \psi_x(x,t)\psi_z(z,t)$, and therefore Q is also expressible as a sum of two components all the way through,

$$Q(x,z,t) = Q_x(x,t) + Q_z(z,t). \tag{2.108}$$

Now, since ψ_z is nearly a Gaussian wave packet at any time,

$$Q_z \approx \frac{\hbar^2}{4m\sigma_{z,t}^2}\left(1 - \frac{z^2}{\sigma_{z,t}^2}\right), \tag{2.109}$$

i.e., Q_z displays the form of an inverted parabola that widens as t increases. Because $\sigma_{z,0}$ has been chosen relatively large in this model, the width of the Gaussian remains almost unaffected for a long time, and therefore Eq. (2.109) will remain essentially constant.

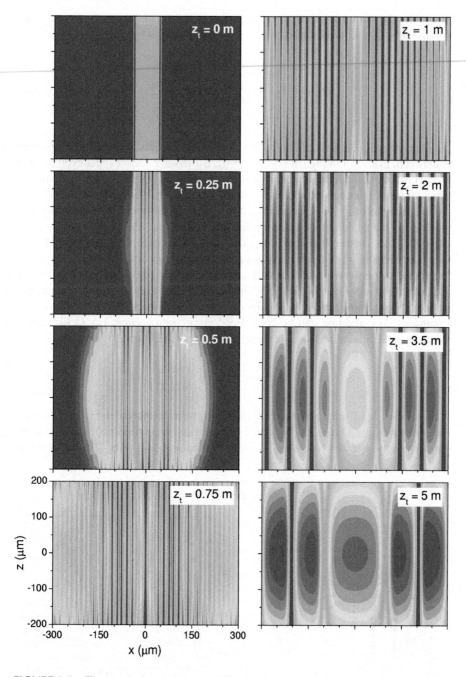

FIGURE 2.5 Time evolution of the probability density associated with a wave packet simulating the diffraction of a cold neutron beam ($\lambda = 19.26$ Å) by a single slit ($w = 92.1$ µm). Each frame corresponds to each snapshot of the probability density displayed in Figure 2.2 (labels have the same meaning). In order to better appreciate the appearance of the diffraction channels, only contours up to $Q = -2.5 \times 10^{-32}$ meV are shown.

Regarding the variations along the x-direction, since ψ_x approaches a square function initially, Q_x can be described in a good approximation by

$$Q_x \sim \begin{cases} 0, & |x| \leq w/2 \\ \delta(x \mp w/2), & x = \pm w/2, \\ -\infty, & |x| \geq w/2 \end{cases} \tag{2.110}$$

i.e., Q_x essentially consists of two δ-functions at the borders of the slit and a plateau in between, as shown in Figure 2.5 for $z_t = 0$. As the wave packet starts evolving and Fresnel diffraction features appear, Q starts developing a series of parallel stripes, which eventually merge into the well-known Fraunhofer diffraction channels in the far field (see plots for $z_t = 3.5$ m and $z_t = 5$ m). This gradual evolution can be followed by studying the profile of the full quantum potential, Eq. (2.108), at subsequent positions of z_t, as shown in Figure 2.6. In this figure, we can see the evolution of Q_x for the wider slit along z_t (which is equivalent to observe its time evolution due to the linear relation between z_t and t) as well as its contour-plot representation. As it can be noticed, the passage from the Fresnel regime to the Fraunhofer one is quite remarkable, the latter being characterized by a series of diffraction channels ruling the asymptotic dynamics. This structure can be determined taking into account Eq. (2.107), from which Q_x reads in the far field (see panel for $z_t = 5$ m in Figure 2.5) as

$$Q_x^\infty \approx 2 - \frac{4}{(\alpha x)^2} + \frac{4}{(\alpha x)^2} \frac{\cos(\alpha x)}{\mathrm{sinc}(\alpha x)}, \tag{2.111}$$

with $\alpha = \pi w / \lambda z_t$. Comparing the Bohmian trajectories displayed in Figure 2.4a with the contour plot of Figure 2.6, we readily notice how these trajectories evolve along the seemingly plateau structures, avoiding the canyon-like ones. The latter come

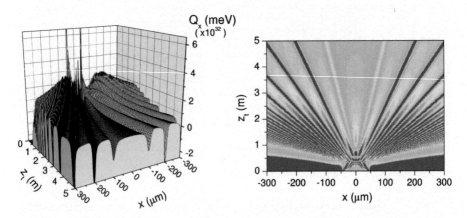

FIGURE 2.6 Quantum potential ruling the dynamics of the Bohmian trajectories displayed in Figure 2.4a. Evolution of the slab $Q_x(x, z_t, t)$ (left) and its contour-plot representation (right) as a function of z_t.

from the nodes that appear in the wave function between consecutive diffraction peaks (see Figure 2.3)—quantum forces are relatively strong and repulsive along in the vicinity of these regions. The appearance of these plateaus explains both the uniform motion described by trajectories in the far field and the fact that trajectories traveling along one of them never go into another (due to the strong quantum forces developed along the nodal lines).

2.4.2 YOUNG'S TWO SLITS

In the previous section, we have analyzed the case of single-slit diffraction assuming that the initial wave function resulted from the effect produced by the diffracting object (in that case, a single aperture) on some incoming wave function. This fact stresses the direct connection between diffraction and boundary conditions. In Young's two-slit experiment, this combination results particularly interesting: the superposition of the two waves coming from each slit gives rise to an interference pattern, which is modulated by the diffraction pattern associated with these slits.

In order to illustrate the relationship between diffraction and boundary effects in Young's two-slit experiment, we are going to consider hard-wall slits, as in Section 2.4.1, using as working parameters those provided by Zeilinger et al. [2] for two-slit interference with slow neutrons. In this case, the de Broglie wavelength was $\lambda = 18.45$, which means an associated (subsonic) velocity $v = 214.4$ m/s and a total time of flight $\tau_f = 2.26 \times 10^{-2}$ s (detectors are placed at a distance of 5 m from the screen containing the two slits). Regarding the two slits, the width for both slits is $w = 22.2$ μm (we assume identical slits for simplicity, but without loss of generality), and the distance between their centers is $d = 126.3$ μm (the separation between the inner borders of the slits is 104.1 μm). As for the initial wave functions, we have considered two types. One assumes that the slits are described by hard walls, with total transmission along their widths and zero everywhere else. Therefore, the transmitted (initial) wave functions are quasi-plane waves along the x-direction and a relatively wide Gaussian wave packet along the z-direction, as before, although taking into account that now we have a coherent superposition of two of such wave packets. In the other case, the slits are assumed to have a Gaussian transmission, thus producing Gaussian wave packets. These wave packets have the same width along the z-direction than the previous ones, and along the x-direction their width is $\sigma_{0,x} = w / 4$, such that only a tiny tail goes beyond the boundaries of the (hard wall) slits (the probability associated with these tails is meaningless). The corresponding initial probability densities are displayed in Figure 2.7a.

When the wave functions considered are let to freely evolve until their centroids in the z-direction reach the final or *flight* distance $z_f = 5$ m, the interference patterns that we find along the x-direction are those displayed in Figure 2.7b. As it can be noticed, the interference pattern (along the x-direction) associated with the hard-wall slits is in agreement with the two-slit interference formula for this type of slit in the Fraunhofer regime,

$$\rho_\infty(x) = |\psi_\infty(x)|^2 \sim \left[\frac{\sin(\pi w x / \lambda z_f)}{\pi w x / \lambda z_f}\right]^2 \cos^2(\pi d x / \lambda z_f). \qquad (2.112)$$

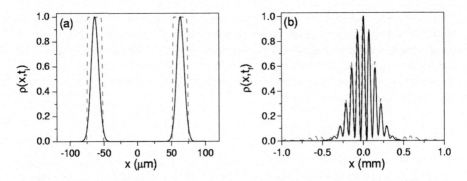

FIGURE 2.7 (a) Initial probability densities associated with hard-wall slits (dashed line) and Gaussian slits (black solid line). (b) Young-type probability density patterns produced by the interference of the wave packets illustrated in panel (a) at a flight distance $z_f = 5$ m.

That is, we have interference fringes of width $\delta x = \lambda z_f / d = 73$ μm and centered at $x_n = n\lambda z_f / d$, with $n = 0, \pm 1, \pm 2, \ldots$ These fringes are modulated by the diffraction factor, which displays maxima at $x = 0$ and each time that $x_\ell^{max} = (\ell + 1/2)\lambda z_f / w$, and minima whenever $x_\ell^{min} = \ell\lambda z_f / w$. In this sense, notice that if $n / \ell = d / w$, the ℓ-th diffraction minimum will cancel out the n-th interference fringe. For example, in our case, we have $d / w = 5.69$, which means that the $\ell = 3$ diffraction minimum will almost cancel out the $n = 17$ interference fringe.

In the case of the Gaussian slits, the diffraction envelope is not a sinc function, but a Gaussian (diffraction by Gaussian slits produces wave functions that remain always Gaussian), as shown in Figure 2.7b. This means that the diffraction term in Eq. (2.112) has to be replaced by a Gaussian. As it has been seen in previous sections, the long-time limit of the probability density will be a Gaussian centered at $x_0 = 0$ and $p_0 = 0$, and with a width $\sigma_{x,0} = \sigma_0$. Accordingly, the pattern produced by two Gaussian slits will be described[1] by

$$\rho_\infty(x) = |\psi_\infty(x)|^2 \sim e^{-x^2/2\bar{\sigma}^2} \cos^2(\pi dx / \lambda z_f), \qquad (2.113)$$

where $\bar{\sigma} \equiv \lambda z_f / 4\pi\sigma_0$. Again here the interference orders appear at the same places, since the distance between the centers of the slits has not changed. However, contrarily to what happens with the previous case, the diffraction envelope cancels any interference feature beyond the reach of the Gaussian, i.e., for a few times the width $\bar{\sigma}$ (e.g., for $x = 3\bar{\sigma} \approx 0.4$ mm the probability density is almost vanished).

In order to visualize the process that connects the initial distributions displayed in Figure 2.7a with the corresponding final interference fringes observed in Figure 2.7b, in Figure 2.8 we show the associated Bohmian trajectories. In the case of hard-wall slits (see Figure 2.8a), we notice how the trajectories with outermost initial conditions (with respect to the margins of the slits) undergo a faster transversal motion (see the enlargement of the Fresnel regime in Figure 2.8b, thus leading to a faster overlapping between the two outgoing or diffracted waves than in the Gaussian case. In this latter case, represented in Figure 2.8c, single-slit diffraction proceeds more slowly

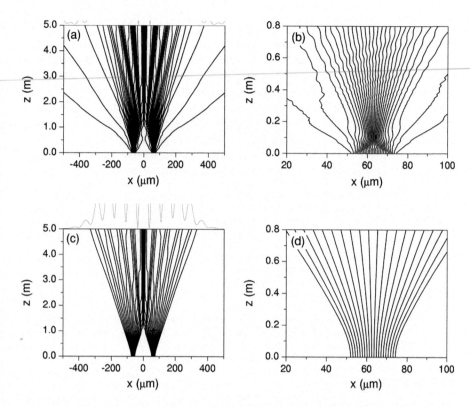

FIGURE 2.8 Bohmian trajectories evenly distributed along the two slits for the two cases considered in Figure 2.7: hard-wall slits (upper row) and Gaussian slits (lower row). On top of the left-hand side panels, final probability density up to a 10% of the maximum value. Enlargements showing the Fresnel regime in both cases are shown in the corresponding panels on the right-hand side.

and gradually, as shown in Figure 2.8d. Actually, this is the reason why the reach of the outermost Bohmian trajectories is much shorter than in the previous case and, eventually, why the associated interference pattern decreases monotonically to zero as the distance from $x = 0$ increases: there are no trajectories than can travel further away, as in the case of the hard-wall slits.

2.5 FINAL REMARKS

As mentioned in introductory Section 2.1, this chapter has been prepared with the intention that it can serve as a tutorial to acquire some skills and tools about the numerical modeling and simulation necessary or of interest in neutron optics. Apart from the fact that neutrons are neutral particles and have no mass, differences with conventional optical phenomena are minor, so that all the tools available in optics should also be of some help in neutron optics. Such differences, however, must be kept in mind, because they imply some subtleties implicit in the analytical modeling.

The procedure here considered thus offers an alternative route to tackle neutron-optics problems to more conventional ones, which combines both standard quantum tools (wave functions) with other tools of interest, namely Bohmian trajectories. These trajectories explicitly show how the neutron flow evolves in configuration space, thus providing a more intuitive picture of the development of diffraction and interference features in configuration space. Of course, this does not mean that real neutrons travel along Bohmian trajectories, but only that their average flow follows such paths. Furthermore, it has also been shown that, far from been another speculative interpretation of quantum mechanics, there are elements in the theory that can be truly tested, such as the transverse momentum, already measured for photons [47].

ACKNOWLEDGMENTS

Financial support from the Spanish MINECO (Grant No. FIS2016-76110-P) is acknowledged.

ENDNOTE

1 Similar results can be obtained if one considers a coherent superposition of two wave packets, with the centers at $x_0 = \pm d/2$ and $p_0 = 0$, and then computes the corresponding long-time limit expression.

REFERENCES

1. A. G. Klein and S. A. Werner, "Neutron optics," *Rev. Prog. Phys.*, vol. 46, pp. 259–335, 1983.
2. A. Zeilinger, R. Gähler, C. G. Shull, W. Treimer, and W. Mampe, "Single- and double-slit diffraction of neutrons," *Rev. Mod. Phys.*, vol. 60, pp. 1067–1073, 1988.
3. H. Rauch and S. A. Werner, *Neutron Interferometry. Lessons in Experimental Quantum Mechanics.* New York: Oxford University Press, 2000.
4. H. Rauch and S. A. Werner, *Neutron Interferometry: Lessons in Experimental Quantum Mechanics, Wave-Particle Duality, and Entanglement.* New York: Oxford University Press, 2015.
5. K. Durstberger-Rennhofer and Y. Hasegawa, "Energy entanglement in neutron interferometry," *Physica B*, vol. 406, pp. 2373–2376, 2011.
6. T. Denkmayr, J. Dressel, H. Geppert-Kleinrath, Y. Hasegawa, and S. Sponar, "Weak values from strong interactions in neutron interferometry," *Physica B*, vol. 551, pp. 339–346, 2018.
7. C. Dewdney, Ph. Gueret, A. Kyprianidis, and J. P. Vigier, "Testing wave–particle dualism with time-dependent neutron interferometry," *Phys. Lett.*, vol. 102A, pp. 291–294, 1984.
8. C. Dewdney, A. Garuccio, A. Kyprianidis, and J. P. Vigier, "Energy conservation and complementarity in neutron single crystal interferometry," *Phys. Lett.*, vol. 104A, pp. 325–328, 1984.
9. C. Dewdney, A. Garuccio, Ph. Gueret, A. Kyprianidis, and J. P. Vigier, "Time-dependent neutron interferometry: evidence against wave packet collapse?," *Found. Phys.*, vol. 15, pp. 1031–1042, 1985.
10. C. Dewdney, "Particle trajectories and interference in a time-dependent model of neutron single crystal interferometry," *Phys. Lett.*, vol. 109A, pp. 377–384, 1985.

11. C. Dewdney, P. R. Holland, and A. Kyprianidis, "A quantum potential approach to spin superposition in neutron interferometry," *Phys. Lett. A*, vol. 121, pp. 105–110, 1987.
12. C. Dewdney, "The quantum potential approach to neutron interferometry experiments," *Physica B*, vol. 151, pp. 160–170, 1988.
13. H. R. Brown, C. Dewdney, and G. Horton, "Bohm particles and their detection in the light of neutron interferometry," *Found. Phys.*, vol. 25, pp. 329–347, 1995.
14. B. T. M. Willis and C. J. Carlile, *Experimental Neutron Scattering*. Oxford: Oxford University Press, 2009.
15. R. P. Feynman and A. R. Hibbs, *Quantum Mechanics and Path Integrals*. New York: McGraw-Hill, 1965.
16. E. Hecht, *Optics*. 4th ed., New York: Addison-Wesley Longman, 2002.
17. A. S. Sanz, M. Davidović, and M. Božić, "Full quantum mechanical analysis of atomic three-grating Mach-Zehnder interferometry," *Ann. Phys.*, vol. 353, pp. 205–221, 2015.
18. A. S. Sanz and S. Miret-Artés, *A Trajectory Description of Quantum Processes. I. Fundamentals, vol. 850 of Lecture Notes in Physics*. Berlin: Springer, 2012.
19. A. S. Sanz and S. Miret-Artés, "A causal look into the quantum Talbot effect," *J. Chem. Phys.*, vol. 126, pp. 234106-1–234106-11, 2007.
20. A. S. Sanz and S. Miret-Artés, *A Trajectory Description of Quantum Processes. II. Applications, vol. 831 of Lecture Notes in Physics*. Berlin: Springer, 2014.
21. A. S. Sanz and S. Miret-Artés, "A trajectory-based understanding of quantum interference," *J. Phys. A: Math. Theor.*, vol. 41, pp. 435303-1–435303-23, 2008.
22. A. S. Sanz and S. Miret-Artés, "Quantum phase analysis with quantum trajectories: a step towards the creation of a Bohmian thinking," *Am. J. Phys.*, vol. 80, pp. 525–533, 2012.
23. P. R. Holland, *The Quantum Theory of Motion*. Cambridge: Cambridge University Press, 1993.
24. D. Bohm, "A suggested interpretation of the quantum theory in terms of "hidden" variables. i," *Phys. Rev.*, vol. 85, pp. 166–179, 1952.
25. D. Bohm, "A suggested interpretation of the quantum theory in terms of "hidden" variables. ii," *Phys. Rev.*, vol. 85, pp. 180–193, 1952.
26. E. Madelung, "Quantentheorie in hydrodynamischer form," *Z. Phys.*, vol. 40, pp. 322–326, 1926.
27. F. London, "Planck's constant and low temperature transfer," *Rev. Mod. Phys.*, vol. 17, pp. 310–320, 1945.
28. A. L. Fetter and A. A. Svidzinsky, "Vortices in a trapped dilute Bose-Einstein condensate," *J. Phys.: Condens. Matter*, vol. 13, pp. R135–R194, 2001.
29. P. A. M. Dirac, *Principles of Quantum Mechanics*. Oxford: Oxford University Press, 1930.
30. L. I. Schiff, *Quantum Mechanics*. 3rd ed., Singapore: McGraw-Hill, 1968.
31. E. J. Heller, "Time-dependent approach to semiclassical dynamics," *J. Chem. Phys.*, vol. 62, pp. 1544–1555, 1975.
32. E. J. Heller, "Frozen Gaussians: a very simple semiclassical approximation," *J. Chem. Phys.*, vol. 75, pp. 2923–2931, 1981.
33. A. S. Sanz, R. Martínez-Casado, H. C. Peñate-Rodríguez, G. Rojas-Lorenzo, and S. Miret-Artés, "Dissipative Bohmian mechanics within the Caldirola-Kanai framework: a trajectory analysis of wave-packet dynamics in viscid media," *Ann. Phys.*, vol. 347, pp. 1–20, 2014.
34. A. S. Sanz and S. Miret-Artés, "Selective adsorption resonances: quantum and stochastic approaches," *Phys. Rep.*, vol. 451, pp. 37–154, 2007.
35. J. Tounli, A. Alvarado, and A. S. Sanz, "Boundary bound diffraction: a combined spectral and Bohmian analysis," *Phys. Scr.*, vol. 94, pp. 035202-1–035202-15, 2007.
36. W. H. Press, S. A. Teukolsky, W. T. Vetterling, and B. P. Flannery, *Numerical Recipes in Fortran 77: The Art of Scientific Computing*, vol. 1st, 2nd ed., Cambridge: Cambridge University Press, 1992.

37. C. Leforestier, R. H. Bisseling, C. Cerjan, M. D. Feit, R. Friesner, A. Guldberg, A. Hammerich, G. Jolicard, W. Karrlein, H.-D. Meyer, N. Lipkin, O. Roncero, and R. Kosloff, "A comparison of different propagation schemes for the time dependent Schrödinger equation," *J. Comp. Phys.*, vol. 94, pp. 59–80, 1991.

38. A. Goldberg, H. M. Schey, and J. L. Schwartz, "Computer-generated motion pictures of one-dimensional quantum-mechanical transmission and reflection phenomena," *Am. J. Phys.*, vol. 35, pp. 177–186, 1967.

39. I. Galbraith, Y. S. Ching, and E. Abraham, "Two-dimensional time-dependent quantum-mechanical scattering event," *Am. J. Phys.*, vol. 52, pp. 60–68, 1984.

40. M. D. Feit, J. J. A. Fleck, and A. Steiger, "Solution of the Schrödinger equation by a spectral method," *J. Comput. Phys.*, vol. 47, pp. 412–433, 1982.

41. D. Kosloff and R. Kosloff, "A Fourier method solution for the time dependent Schrödinger equation as a tool in molecular dynamics," *J. Comp. Phys.*, vol. 52, pp. 35–53, 1983.

42. R. Kosloff and D. Kosloff, "Absorbing boundaries for wave propagation problems," *J. Comp. Phys.*, vol. 63, pp. 363–376, 1986.

43. D. E. Manolopoulos, "Derivation and reflection properties of a transmission-free absorbing potential," *J. Comp. Phys.*, vol. 117, pp. 9552–9559, 2002.

44. R. Heather and H. Metiu, "An efficient procedure for calculating the evolution of the wave function by fast Fourier transform methods for systems with spatially extended wave function and localized potential," *J. Comp. Phys.*, vol. 86, pp. 5009–5017, 1986.

45. P. Pernot and W. A. Lester Jr., "Multidimensional wave-packet analysis: splitting method for time-resolved property determination," *Intern. J. Quantum Chem.*, vol. 40, pp. 577–588, 1991.

46. R. P. Feynman, R. B. Leighton, and M. Sands, *The Feynman Lectures on Physics*, vol. 3. Reading, MA: Addison-Wesley, 1965.

47. S. Kocsis, B. Braverman, S. Ravets, M. J. Stevens, R. P. Mirin, L. K. Shalm, and A. M. Steinberg, "Observing the average trajectories of single photons in a two-slit interferometer," *Science*, vol. 332, pp. 1170–1173, 2011.

3 Neutron Confinement and Waveguiding

*Ignacio Molina de la Peña, Maria L. Calvo,
and Ramón F. Álvarez-Estrada*

CONTENTS

3.1 INTRODUCTION

This chapter is motivated by several facts: (i) the worldwide program implementing the use and applications of slow neutrons, (ii) the resulting construction and development of small nuclear reactors and accelerator-based and spallation sources for producing them, and (iii) persistent research in the last decades and in the present one toward improving the control and focalization of slow neutrons, in order to improve (i). Our analysis will be based upon the interactions of slow neutrons with the atomic nuclei of certain elements, such as titanium, carbon, boron, and silicon, which play a central role (other atomic nuclei being also considered, as we proceed).

A general reference on slow neutrons (n) and their properties (and, updated up to about 1995, various applications) is [1]. See Refs. [2–4] for further very useful information. For an account, updated up to about 2010, regarding the relevance of slow neutrons and their usefulness, see the comments in Ref. [5] and, for their focusing, see Ref. [6]. In this connection, it may be adequate to quote from Ref. [5]:

> "Because of their unique properties, neutrons are used to investigate a growing number of research areas, in both traditional and new fields and from fundamental science to technology; no end to this growth can be foreseen. Currently, 4,000 to 5,000 European researchers are using neutron scattering for their scientific work. This demand drives the improvement of neutron instrumentation which, to a large extent, is related to neutron optics ... the next decade will see the increased application of focusing and polarization devices ...."

In order to present our study and proposals in a comprehensive way, we shall outline, in successive subsections in this section, various general properties about slow neutrons (their interactions, sources, and optics, say, their optical-like behavior). Their transportation by means of hollow guides with large transverse cross sections along relatively long distances will be considered in Section 3.2. A simple introduction to confined propagation along waveguides of small transverse cross sections and first proposals for slow neutrons will be treated in Section 3.3. Previous work on waveguides of small transverse circular cross sections (fibers) is presented in Section 3.4. Section 3.5 deals with film waveguides: the first experiments and some proposals and subsequent comments. Section 3.6 contains conclusions and discussions. Several subjects regarding fundamentals, to be dealt with here in Subsections 3.1.1 through 3.1.3, are treated in greater length and depth (and with wider references) in Chapter 1. However, it seemed adequate to overview some of them swiftly here, so as to be able to enter quickly in our main subject in this chapter.

3.1.1 SOME GENERAL FEATURES

In a broad sense, (nonrelativistic) neutrons with kinetic energies E less than or, at most, about a few tens of kilo-electron volts will be referred here to as slow neutrons. In a similar sense, slow neutrons with E ranging from about 0.003 eV up to about 0.5 eV are named thermal, although the more specific designation "thermal neutrons" is usually attributed to those having E about 0.025 eV. Slow neutrons with E in the ranges 5×10^{-5} eV to 0.025 eV and 0.5 eV to a few tens of kilo-electron volts are also named cold and epithermal, respectively.

Two key properties of beams of slow neutrons, central for our purposes here, refer to their behavior as they propagate through a material medium: (i) they suffer relatively small absorption and attenuation (to be characterized quantitatively by the linear coefficient μ, usually expressed in cm^{-1}), (ii) as a consequence, they are relatively non-destructive. Property (i) arises from the electrical neutrality of the neutron, so that it does not interact with electrons bound in atoms, but only with atomic nuclei, occupying tiny regions inside those atoms. In magnetic materials, there are additional interactions of the neutron magnetic moment with various magnetic moments in the atom. In the present study, we do not consider such neutron magnetic properties. Property (ii) has a qualified limitation in certain cases, arising from specific nuclear reactions (which imply the disappearance of the incoming neutron). On the contrary, X-rays beams have, precisely, properties opposite to (i) and (ii) above, as they propagate through matter: X-rays are strongly absorbed (for instance, by atomic electrons) and also are quite destructive, in general. On the other hand, slow neutron fluxes are usually much smaller than photon fluxes. See Subsection 1.7.4.

Let an incoming slow neutron beam, with flux F_0 (neutrons/(cm$^2 \cdot$ s)), penetrate into a material medium with linear coefficient μ (Table 3.2) and propagate into it along a distance z. Then, the beam is attenuated due to its interaction with the atomic nuclei in the medium and the flux F (neutrons/(cm$^2 \cdot$ s)) at the distance z is, according to Eq. (3.1) and Figure 3.1:

$$F(z) = F_0 e^{-z\mu}. \qquad (3.1)$$

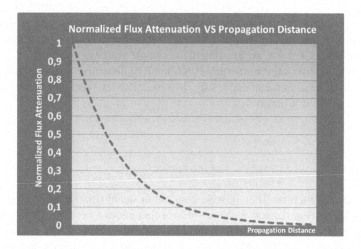

FIGURE 3.1   Flux attenuation with distance, where an exponential decay is exhibited.

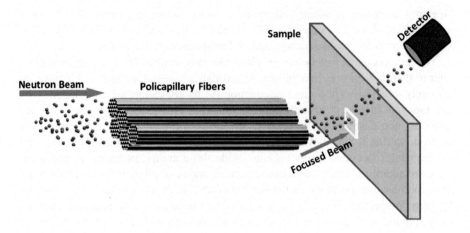

FIGURE 3.2   Schematic representation of neutron beams used for "analysis of matter."

There are growing research and technologically relevant areas in which slow (in particular, thermal and epithermal) neutrons are employed. We shall limit here to quote two areas (a much wider perspective being given in other chapters in this book):

1. A very broad and diversified one, referred to "analysis of matter" or the investigation of materials, which includes a wide variety of areas of interest such as analysis of solids (structure of crystals, defects, residual stress, …), liquids, ultracold atomic systems, and biomolecules to name a few. Figure 3.2 displays, in essence, this application: a neutron beam propagates through some concentrating and focalizing (solid-material) device (a set of parallel tubes in Figure 3.2), which enable to focus the former onto a sample of the material to be studied. After traveling through and interacting with the sample, neutrons impinge in a detector: they carry information on the structure and properties of the sample, arising from that interaction. Thus, the analysis of the detected neutrons makes that information of the sample available.
2. Boron neutron capture therapy (BNCT) [7, 8] and current clinical applications, to be discussed very briefly in Subsection 3.5.3 and, at greater length and detail, in Chapter 6.

3.1.2   Neutron Sources

There are three broad classes of neutron sources:

1. **Nuclear reactors:** They provide constant fluxes of slow neutrons. The most intense nuclear reactors in the world have continuous (constant) neutron fluxes of order 10^{15} (neutrons/(cm$^2 \cdot$ s)). An example is the one located at the Institute Laue-Langevin (Grenoble, France), the continuous neutron flux provided by it being 1.5×10^{15} (neutrons/(cm$^2 \cdot$ s)) with a thermal power of 58.3 W. There are also ongoing programs for research and applications based upon small or

medium nuclear reactors, with smaller continuous fluxes smaller by several orders of magnitude, for instance, about 10^9 (neutrons/(cm$^2 \cdot$ s)). An example is the Tehran Research Reactor in Iran, built and set into operation for application to BNCT [9]. See Subsection 1.2.4.1 and Chapter 6.

2. **Accelerator-based neutron sources:** Typically, they are based upon accelerating protons (or deuterons), at energies about MeV of tens of mega-electron volts, in cyclotrons. At a later stage, those protons collide with suitable nuclear targets (lithium, beryllium) and the resulting nuclear reactions yield pulses of fast neutrons, with energies in the MeV range. Subsequently, the fast neutron pulses are moderated (thereby, losing energy) and become slow neutron pulses. An example, with flux about 10^9 (neutrons/(cm$^2 \cdot$ s)) (not continuous, but approximately constant for some limited time intervals) will be considered in Subsection 3.5.3.2. See Chapter 6.

3. **Spallation sources:** Accelerator-driven pulsed proton beams (using synchrotrons, cyclotrons), with energies of the order of hundreds of mega-electron volts (higher than in the above accelerator-based sources), knock suitable nuclei like U_{92}^{238}. The resulting high energy nuclear collisions in the target produce many fast neutrons which, after being moderated, give rise to beams of slow neutrons. See Figure 3.3 for details. Spallation sources provide nonstationary (non-constant) neutron fluxes. A spallation source can generate several tens of epithermal neutrons per proton and pulsed beams containing 10^{16} (neutrons/s) and even more. See Subsection 1.2.4.3. Table 3.1 lists some important spallation sources currently under full operation and under construction [10]. In particular, we remark that BNCT is planned to be carried out in the European Spallation (ESS) source in Lund (Sweden) (currently under construction).

3.1.3 NEUTRON OPTICS

Slow neutrons are microscopic sub-atomic particles and, so, as they propagate in space, are described by wave functions and probability amplitudes and are subject to

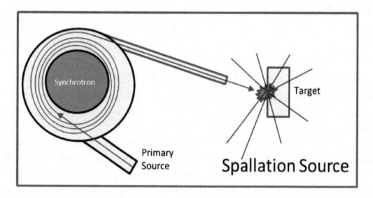

FIGURE 3.3 Schematic diagram of a synchrotron-based spallation source. About 40 neutrons per proton could be generated.

TABLE 3.1

Spallation Sources

KENS (KEK, Japan)

SINQ (PSI, Switzerland)

ISIS (Rutherford Lab, UK)

CNR-SNS (Oak Ridge, USA)

ESS (Lund, Sweden): operation expected since 2019

SNS (Guandong, China)

FRM (Munich, Germany)

ILL (Grenoble, France)

interference and diffraction phenomena, which go under the name of neutron optics [1, 3, 4]. If the neutron with energy E moves in vacuum (or in air, as a first approximation), its de Broglie wavelength λ_{db} is given by:

$$E = \frac{2\pi^2\hbar^2}{m_n\lambda_{db}^2},$$ (3.2)

where $\hbar$ is Planck's constant and m_n is the neutron mass. Let a slow neutron propagate through a material medium having (positive) density ρ (nuclei/cm^3) (of order 10^{22} (nuclei/cm^3) for the materials of interest here (and about three orders of magnitude smaller, for air). The wave function $\Psi(x, t)$ for a slow neutron at the position x (used here, generically, to denote one, two or three spatial dimensions) at time t, as the former propagates through the medium, satisfies the Schrödinger equation:

$$\left[\frac{-\hbar^2}{2m_n}\nabla^2 + V(x)\right]\Psi(x, t) = i\hbar\frac{\partial\Psi(x, t)}{\partial t}.$$ (3.3)

∇^2 is the Laplacian and $V(x)$ is the potential due to the nuclei in the medium on the slow neutron. In neutron optics, and as a first approximation, $V(x)$ can be approximated by the effective ("optical"):

$$V(x) = \frac{2\pi\hbar^2}{m_n}b\rho.$$ (3.4)

See Section 1.3.1 and, in particular, Subsection 1.3.1.1. Although, in principle, $V(x)$ could depend on x, in practice it is allowed to regard the former as constant in the spatial region of interest, in which the slow neutron is propagating (as the right-hand side of Eq. (3.4) displays). In turn, b is the (coherent) neutron-nucleus scattering amplitude at low energy. b determines the interaction seen by the slow neutron as it propagates through the medium and it will be crucial for our studies here. In principle, b has both a real (Re(b)) and an imaginary part (Im(b)), determined by the absorption of a slow neutron by the corresponding nucleus. To the extent that nuclear absorption be a small effect (to be taken into account duly at some stage, if required), and as a first approximation, we shall regard b as real, in what follows. The

TABLE 3.2

Materials and Related Neutron Properties

Material	b (10^{-12} cm)	μ (cm^{-1})
Ti	−0.337	0.45
Si	0.42	0.11
Al	0.35	0.098
B	−	60
Air	−	5.7×10^{-4}
Mn	−0.37	0.6
V	−0.05	0.2

order of magnitude of b is about 10^{-13} cm. b is positive for the nuclei of almost all elements in the Periodic Table, except for a few: b is negative only for the naturally occurring H, Li, Ti, V, Mn ("naturally occurring" meaning that the averaging over all isotopes of the corresponding element has carried out). See Table 3.2. b is also negative for certain specific isotopes, not to be considered here. In Table 3.2, μ is the linear coefficient (see Eq. (3.1)).

For boron, the value of μ and $\mathrm{Im}(b)$ are so large that $\mathrm{Re}(b)$ has been regarded as negligible, as a first approximation. The order of magnitude of $V(x)$, for typical material media of interest here, is of the order of 10^{-7} to 10^{-8} eV. For air, the value of $V(x)$ is about three orders of magnitude smaller.

For certain purposes, one can bypass the use of the Schrödinger equation by using the geometrical-optics standpoint. In the latter, one introduces the neutron refractive index:

$$n^2 = 1 - \frac{2\pi\hbar^2}{m_n}\frac{b\rho}{E}. \tag{3.5}$$

If a slow neutron moves in vacuum with wave-vector k, then $V(x) \equiv 0$, $n = 1$, and $E = E_k = \hbar^2 k^2 / (2m_n)$. Let such a slow neutron, having moved initially in vacuum, approach a material medium with $V(x) \neq 0$ and $\mathrm{Re}(b) > 0$, $\mathrm{Im}(b) \simeq 0$ (that is, nuclear absorption being neglected). Consequently, it follows that for most materials: $n < 1$. Then, from the standpoint of neutron optics (and upon recalling the analogy with ordinary optics of light, where the refractive index is usually larger than 1) such a medium could be regarded as less dense than vacuum. The phenomena which occur due to such an approach to that material medium have been studied in Section 1.5: see, in particular, Subsection 1.5.2.

3.2 GUIDES (WITH LARGE CROSS SECTION)

Routinely, slow neutrons are transported from the sources to the locations where they will be applied by using hollow guides (tubes) [1, 3, 11–14]. See also Chapter 6. The latter have transverse dimensions of the order of several cm and allow transporting the neutron beam along lengths of order tens of meters.

At later stages, suitable (and, technologically, more sophisticated) guides enable to focalize slow neutron beams onto smaller transverse scale, say, down to mm for various purposes (and, as we shall see later when treating polycapillary glass fibers (PGFs), even, down to tens of microns).

For thermal neutrons, the beam size and the aimed spot size (scale of focalization) were [7] about 100 mm and 1 mm, respectively [1, 3]. It was then expected that work on improved techniques for neutron focusing would constitute a promising development.

They are formed by an inner cylinder-like (the core, along the axis of the guide), surrounded by a cylinder-like one (clad or cladding) composed by a different material with a smaller refractive index (or by several ones, disposed concentrically). The core may be empty or made up by a suitable material.

Confined propagation of slow neutrons in the core occurs if the refractive index of the clad is smaller than that of the core:

$$n_{core} > n_{clad} \tag{3.6}$$

Let us consider briefly a slow neutron approaching one end of a hollow guide, within the geometrical-optics approximation. Let θ be the angle formed by the incoming neutron momentum k with the axis (z) of the waveguide. Let φ be the angle between k and the normal to the limiting surface separating the core (vacuum, with $b = 0$) from the clad (with $b > 0$). One has $\varphi + \theta = \pi / 2$. See Subsection 1.5.2. Let $\varphi = \varphi_{cr}$ fulfill $sin\ \varphi_{cr} = n$. Let us introduce the acceptance or critical glancing angle $\theta_{cr} = \pi / 2 - \varphi_{cr}$. Then, according to standard geometrical-optics-like arguments, the neutron suffers total reflection back into vacuum core if $\theta \le \theta_{cr}$ so that $\varphi_{cr} \le \varphi$, and it does not penetrate into the clad. Then, total reflection occurs for $\theta \le \theta_{cr}$. For typical values of $\theta(> 0)$ and ρ, n is a bit smaller than $+1$, φ_{cr} is a bit smaller than $\pi / 2$, and the following approximation holds:

$$\theta_{cr} \simeq \frac{\hbar \left[4\pi\rho_{cl} b_{cl} \right]^{1/2}}{\left[2m_n E \right]^{1/2}} = \lambda_{dB} \left[b\rho / \pi \right]^{\frac{1}{2}}, \tag{3.7}$$

with $E = E_K = \hbar^2 k^2 / (2m_n)$. So, there exists a certain characteristic angle θ_{cr} (see Figure 3.4, with air as core), such that the neutron does propagate confined

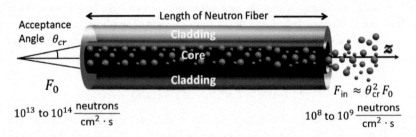

FIGURE 3.4 Critical angle θ_{cr} representation for a simple hollow waveguide. Incoming flux F_0 and exit flux F_{in} are represented.

along the core if: $\theta < \theta_{cr}$. Then, total internal reflection of the neutron in the inner surface (core-cladding) occurs. For typical materials of interest here, θ_{cr} is of the order of 10^{-2} to 10^{-3} radians. Thus, the geometrical-optics-like phenomenon of total reflection has provided the basis for the hollow neutron guides (with suitably large transverse dimension about several centimeters), enabling the channeling of a slow neutron beam along relatively long distances. The neutrons propagate through the inner empty space of the guide and, if $\theta \le \theta_{cr}$ holds, they suffer multiple total reflections on the walls, which are made up by some suitable material with $b > 0$. See Figure 3.4.

Let F_0 be the flux of an incoming beam of slow neutrons approaching that end of a guide. The flux F_{in} of neutron beams which have entered into and propagate confined along the core of the guide is of the order of $\theta_{cr}^2 F_0$, within the geometrical-optics approximation. Typically, the influence of the acceptance angle reduces the propagating flux F_{in} by a factor of order 10^{-4}, compared to F_0. Notice that this kind of reduction in the propagating flux, due to θ_{cr} value, is independent on (and has a different physical origin from) that arising from the beam attenuation represented by the linear coefficient μ (see Figure 3.4).

In practice, one deals with neutron beams which are not monochromatic but have a spectrum of de Broglie wavelengths [14]. Typically, the walls of hollow neutron guides are made up of nickel-coated boron glass. Their cross sections may be rectangular, having sides about 10 cm, and their lengths may be about 80 m. They are not straight, but have some curvature radius R_{cu}. Those designed to transport thermal neutron beams [1] have R_{cu} about one to several hundred meters. Since typical transverse dimensions of these guides are much larger (by several orders of magnitude) than the de Broglie wavelengths for thermal neutrons, the latter has been described through geometrical-optics approaches.

One important feature (and application) of such hollow neutron guides, which makes them technologically interesting, is that they can be connected directly to nuclear reactors. Then, the guides are employed to extract slow neutron beams from the latter, and to channel and transport such beams along relatively long distances for various purposes (investigation of condensed matter and biological matter samples) [1, 3]. See Figure 3.2. Microguides for thermal neutrons have been realized experimentally (using a suitable evaporation technique) [15]. In that work, one microguide consisted on a sandwich of alternate Ni and Al layers: it contained 100 double layers, with total thickness 0.037 mm and radius of curvature 15 cm. Thus, the distance between walls in the microguide was about 320 mm. They were also employed to study the deflection of neutrons [15]. There are further and interesting developments on neutron guides (using certain devices named super-mirrors). For further developments, see Refs. [6, 16, 17] and references therein.

3.3 NEUTRON WAVEGUIDES

They are formed by an inner cylinder-like region made up by a suitable material (the core, along the axis of the waveguide), surrounded by a cylinder-like one (clad or cladding) composed by a different material with a smaller refractive index (or by several ones, disposed concentrically).

3.3.1 SOME GENERAL FEATURES

The confined propagation of slow neutrons occurs mainly along the core of the waveguide and is described by means of certain specific solutions $\Psi(x, t)$ of the Schrödinger equation, named propagation modes. Specifically, a propagation mode $\Psi(x, t) = \varphi(x)e^{-iEt/\hbar}$ is a solution of Eq. (3.3) (with time-dependence $e^{-iEt/\hbar}$, E being the neutron energy)

$$\left[\frac{-\hbar^2}{2m_n} \nabla^2 + V(x) \right] \varphi(x) = E\varphi(x). \tag{3.8}$$

In Eq. (3.8), $\varphi(x)$ is an eigenfunction of $\left[\frac{-\hbar^2}{2m} \nabla^2 + V(x) \right]$ with eigenvalue E. $\varphi(x)$ oscillates and, so, describes a wave along certain crucial inner region (core) in the propagation direction (z), while it decays rapidly (say, exponentially) along those transverse directions in which confinement occurs, far from the core. Those transverse directions reduce to one (say, the x direction) for planar waveguides (two spatial dimensions), while they include both transverse directions (x and y) for waveguides with fiber-like structure (three spatial dimensions), as we shall discuss in the following subsections. Figure 3.4 displays a fiber-like waveguide. For brevity, we shall discuss here several features in terms of fiber-like waveguides (having z as the propagation direction and x, y as transverse directions), the extension to planar waveguides being left for Subsection 3.5.

Confined propagation of slow neutrons (with energy E) occurs mainly (but not exclusively) along the core in the z-direction. The condition for that confined propagation to occur is:

$$\Delta(b\rho) = (b\rho)_{\text{clad}} - (b\rho)_{\text{core}} > 0, \tag{3.9}$$

which, in the geometrical-optics approximation, yields Eq. (3.6). If this condition holds, as we assume, there is always one propagation mode (the so-called fundamental one). An important issue is whether other propagation modes exist, besides the fundamental one. Let A be the total area of the core, in the (x, y) plane. In general, the total number of propagation modes increases if $A \cdot \Delta(b\rho)$ increases. For thermal neutrons, $\left(\frac{V_{\text{clad}} - V_{\text{core}}}{E} \right)$ is of the order of about 10^{-6}. If there are further propagation modes, besides the fundamental one, $A^{1/2}$ (which is of order of the transverse dimension of the core, in the (x, y) plane) has to be larger than the de Broglie wavelength λ_{db}. Two interesting examples are:

1. A core with $b < 0$, the clad being air (having negligible $b\rho$) or a material with $b > 0$.
2. A hollow waveguide, with a (hollow) core formed just by air, the clad being one concentric cylinder with $b > 0$ (and, possibly, more concentric cylinders with $b > 0$ or $b = 0$).

3.3.2 FIRST THEORETICAL PROPOSALS: OVERVIEW

The possibility of guided slow neutron waves propagating along thin films was proposed and discussed in Ref. [18]. Figure 3.12 displays a planar waveguide (a thin

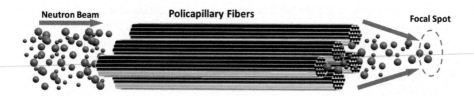

FIGURE 3.5 Representation of a PGF (five parallel glass fibers) focusing a neutron beam. For simplicity, only 19 HCCs are displayed in each glass fiber. This figure is also a schematic representation of the experiments in Refs. [28, 29].

film). In it, there are lateral confinements along one of the two transverse directions (say, x), orthogonal to the direction of propagation (z). This proposal was based upon the quantum mechanical properties of the neutron.

Proposals formulating the possible confined propagation of thermal neutrons along waveguides (neutron fibers) with small transverse dimensions (small circular cross section) were studied in Refs. [19, 20], by exploiting analogies with light propagation along optical fibers [21]. In these proposals, based on quantum mechanics [22], there would be lateral confinement along both transverse directions (x and y), orthogonal to the direction of propagation (z). See Figure 3.4 and Subsection 3.4.1.

PGFs have been an important proposal and development, as we shall discuss. See Ref. [23] for a comprehensive review, in which confined propagation of radiation is described in a purely geometrical-optics standpoint. See Figure 3.5 for details. A typical PGF may have an overall diameter in the millimeter range and a length about a few tens of centimeters. One PGF may contain about a few thousand individual, approximately parallel, HCC (hollow capillary channels), separated among themselves by silica glass. PGFs may have some curvature, although that possibility is not displayed, for simplicity, in Figure 3.5. Each single HCC with an internal diameter about a few microns constitutes one hollow waveguide (its core being just air) for the confined propagation and focusing of neutrons (and, actually, of X-rays, which have attracted the largest part of activity on PGFs). Notice that there is lateral confinement along both transverse directions, orthogonal to the propagation direction. In short, PGFs and bundles thereof yield devices for the experimental confined propagation and focusing of radiation, known as capillary optics. See Subsection 3.4.2.

See Subsection 3.4.3 for a short overview of conjectures regarding possible confined propagation of slow neutron in shorter scales (nanotubes).

3.4 NEUTRON FIBERS

Based upon an analogy with light propagation along optical fibers (thin solid, say, glass-like, or plastic dielectric waveguides) [21], the possible confined propagation of slow neutrons along waveguides of small transverse cross section (neutron fibers) and some speculations about possible applications (to neutron radiotherapy) were discussed some time ago [19, 20]. There are some interesting differences between the propagation of ordinary and that of slow neutrons. Thus: (i) the interaction giving rise to the phenomena (electromagnetic for light, versus strong nuclear force for neutrons), (ii) the ranges of wavelengths (about some thousands of Angstroms, Å, for light, versus an interval about 1Å for thermal neutrons, or even shorter for epithermal ones).

Before entering those proposals, recall (Subsection 3.1.3 and Table 3.2 and Section 1.3) that, for a few isotopes and naturally occurring elements, the coherent low-energy neutron-nucleus scattering amplitude b fulfills $\mathrm{Re}(b) < 0$, that is, $\mathrm{Re}(n^2) > 1$. Those materials can be regarded as more dense than vacuum, from the neutron optics point of view. One proposal [19] was the possible existence of confined propagation modes of thermal neutrons in fully solid (say, non-hollow) waveguides having small transverse cross section (fibers), made up by suitable materials such that the real part of the corresponding b be negative $(\mathrm{Re}(b) < 0$, so that $\mathrm{Re}(n^2) > 1)$ and nuclear absorption be adequately small. Such a proposal was based on the wave phenomena associated to thermal neutrons and on various properties of suitable (propagation mode) solutions of the associated Schrödinger Eq. (3.8).

Among the various nuclei with $\mathrm{Re}(b) < 0$, the proposal [19] concentrated on Ti (73.8 per cent of natural Ti is Ti_{22}^{48}, which is a spinless nucleus having $\mathrm{Re}(b) = -5.84 \times 10^{-13}$ cm), with some additional discussion of Ni_{28}^{62} (which is also spinless, but with an abundance of 3.7 per cent only in natural Ni). In their simplest version, those neutron fibers could be unclad and surrounded by air, due to the fact that $\mathrm{Re}(n^2) > 1$ (and that $|\rho b|$ for air is about three orders of magnitude smaller than for the core). By employing the approximate Eq. (3.1), the thermal flux $F(z)$ transmitted in the case of a Ti neutron fiber with length about $z = 5$ cm was estimated [20] to be about 10 per cent of F_0 while that for a Ni_{28}^{62} one can be estimated to be higher.

Another possibility [19, 20] was a "cladded neutron fiber," namely, a cylindrical solid waveguide (core) of an element like those mentioned above, surrounded by an outer coaxial cylinder (the clad) of finite thickness, which in turn, would be surrounded by air. See Figure 3.4. The element making up the clad would be required to have positive nuclear scattering amplitude $(\mathrm{Re}(b) > 0)$ and small nuclear absorption. The possible confined propagation of slow neutrons along magnetized waveguides (say, those containing manganese) was also discussed briefly [19]. Then, spin (and birefringence) effects for neutrons would play a role. In particular, in a $\mathrm{Mn}_2\mathrm{N}$ fiber, magnetized along the fiber $(z-)$ axis, thermal neutrons would propagate confined only if their spins were anti-parallel to the net magnetization.

Some improving alternatives for cladded neutron fibers were analyzed [20]: there, the aim was to allow for confined neutron propagation along reasonable lengths longer (with smaller scattering and absorption losses) than those for a Ti fiber. In those improved cladded fibers (see also Figure 3.4), the core (subscript "co") could be either Al or Si or air, while the clad (of course, also of finite thickness) could be either Ni or Fe. Now, $\rho_{co}\mathrm{Re}(b_{co}) \geq 0$ and $\rho_{cl}\mathrm{Re}(b_{cl}) > 0$, with $\rho_{co}\mathrm{Re}(b_{co}) < \rho_{cl}\mathrm{Re}(b_{cl})$. Again, the actual cladded fiber is surrounded by air, for which $\rho\mathrm{Re}(b)$ is also positive and much smaller.

3.4.1   Single Fiber: Quantum-Mechanical Analysis

3.4.1.1   Propagation Modes

In the following treatment, we consider the propagation modes for slow neutrons in neutron fibers. The use of Eqs. (3.4) and (3.8) will amount to omit the neutron spin. This simplification suffices provided that magnetized fibers be excluded.

Let us consider, in three-dimensional space, a very lengthy and straight neutron fiber with zero curvature and finite transverse cross section T. The latter lies in the (x, y)-plane, so that the z-axis is parallel to any axis of the fiber. T may have an arbitrary shape. See Figure 3.6. Three- and two-dimensional vectors will be denoted by over-bars and boldface symbols, respectively, so that $\bar{x} = (x, y, z) = (\boldsymbol{x}, z)$. The confined propagation of slow neutrons along the fiber can be modeled through Eq. (3.8), where $V(\bar{x}) = V(\boldsymbol{x})$ fulfills: $V(\boldsymbol{x}) = V_{co} = 2\pi\hbar^2 b_{co}\rho_{co} / m_n$ for $\boldsymbol{x}$ inside the inner part of T (the core, represented by the subscript co), while $V(\boldsymbol{x}) = V_{cl} = \frac{2\pi\hbar^2 b_{cl}\rho_{cl}}{m_n}$, for $\boldsymbol{x}$ outside T (the infinite outer medium or clad, with subscript cl). Here, b_{co} and b_{cl} stand for the average coherent amplitudes for the low-energy scattering of a neutron by an atomic nucleus belonging to the core and the clad, respectively. The possible absorptions of slow neutrons by both the core and the clad are taken as negligible, so that both b_{co} and b_{cl} are assumed to be real. The inclusion of nuclear absorption has also been taken into account [19, 20]. The numbers of nuclei per cm^3 in the core and in the clad are ρ_{co} and ρ_{cl}, respectively. It is necessary that $V_{co} < V_{cl}$, in order that confined neutron propagation along the fiber occurs.

The thermal neutron propagation modes, with total energy E and propagation constant β_α (in an interval about one $\mathring{A}^{-1}$ or about 10 nm^{-1}), are the physically relevant solutions of Eq. (3.8). They bear the following structure:

$$\varphi = \phi(\boldsymbol{x})_\alpha \cdot \exp(i\beta_\alpha z) \cdot \exp(-iEt / \hbar). \tag{3.10}$$

With $\int d^2\boldsymbol{x} |\phi(\boldsymbol{x})_\alpha|^2 = 1$ (normalization). α denotes a set of additional quantum numbers, also required in order to specify uniquely, together with either E or β_α, the

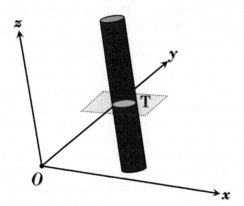

FIGURE 3.6   Unbent three-dimensional fiber having (small) transverse cross section T. The axis of the fiber is parallel to the z-axis.

solution of Eq. (3.8). As for optical waveguides, a neutron fiber is named monomode if there is only one possible choice of values for α. Otherwise, the neutron fiber is called multimode.

As both b_{cl} and b_{co} are real, then so is β_α, and we take $\beta_\alpha > 0$. As the kinetic energy is positive, one has: $E > V_{cl}$. Eqs. (3.8) and (3.10) yield:

$$\left[-\frac{\hbar^2}{2m_n}\Delta_T + V(x) - V_{cl}\right]\phi(x)_\alpha = -\frac{\hbar^2}{2m_n}\chi_\alpha^2\phi(x)_\alpha. \tag{3.11}$$

$$E = V_{cl} + \frac{\hbar^2}{2m_n}\beta_\alpha^2 - \frac{\hbar^2}{2m_n}\chi_\alpha^2, \tag{3.12}$$

where $\Delta_T = \partial^2/\partial x^2 + \partial^2/\partial y^2$. Moreover, since both b_{cl} and b_{co} are real, χ_α is also real and fulfills $\chi_\alpha > 0$. In particular, these properties hold for a hollow fiber with a clad in which absorption of neutrons be negligible. Eq. (3.11) has, and fully determines, a finite set of real negative eigenvalues $-\chi_\alpha^2$. Then, for given E and for the finite set of allowed values of χ_α^2, Eq. (3.12) determines a finite set of values of β_α only. Notice that $V(x) - V_{cl}$ vanishes or equals a negative constant, depending on whether x lies outside or inside T, respectively. Both ϕ_α and κ_α can be obtained by solving the time-independent Schrödinger Eq. (3.11) with the following standard boundary conditions: both ϕ and its normal derivative have to be continuous across all walls, and $\phi(x)$ has to vanish exponentially as the modulus of x tends to $+\infty$. Moreover, ϕ_α has to be finite for any x. As the Schrödinger Eq. (3.11) refers to two spatial dimensions and involves an attractive potential [19, 20], the fundamental mode (or ground state) always exists, but there may be or not further propagation modes (see later). In particular, $\kappa_\alpha/\beta_\alpha \leq 10^{-2}$ holds for thermal neutrons in typical cases. The following bound holds [24].

$$\chi_\alpha \leq \left[4\pi\left(b_{cl}\rho_{cl} + b_{co}\rho_{co}\right)\right]^{1/2} \equiv \chi_{max}. \tag{3.13}$$

The largest value of $\chi_\alpha \,(> 0)$ allowed by Eq. (3.11) and compatible with the bound in Eq. (3.13) corresponds, precisely, to the fundamental mode. The values of $\chi_\alpha \,(> 0)$ associated to higher modes decrease in magnitude.

For an unbent fiber, the number of allowed propagation modes, $N_{pm}\left(\chi_0^2\right)$ in the range $0 \leq \chi_0^2 \leq \chi_\alpha^2 \leq \chi_{max}^2$ is the number of independent solutions $\phi(x)_\alpha$ of Eq. (3.11) as α varies, having χ_α^2 in that range (say, excluding higher modes with $0 \leq \chi_\alpha^2 \leq \chi_0^2$). Provided that $N_{pm}\left(\chi_0^2\right)$ be appreciably larger than unity (the neutron fiber being multimode, then), the former can be estimated easily in quasi-classical approximation. In turn, the previous condition will be fulfilled provided that transverse dimension of the fiber (say, the square root of the area $A(T)$ of T) be adequately (or much) larger than β_α^{-1}. General studies about quasi-classical approximations in quantum mechanics [24] provide the foundations for the approximate formula for $N_{pm}\left(\chi_0^2\right)$ to be employed. It reads [20, 24]:

$$N_{pm}\left(\chi_0^2\right) = \frac{A(T)}{2\pi}\left[4\pi\left(b_{cl}\rho_{cl} + b_{co}\rho_{co}\right) - \chi_0^2\right]. \tag{3.14}$$

The total number of propagation modes having χ_α^2 in the range $0 \le \chi_\alpha^2 \le \chi_{max}^2$ (say, with $\chi_0^2 = 0$) is, also in quasi-classical approximation:

$$N_{pm} = N_{pm}\left(\chi_0^2 = 0\right) = 2A(T)\left(b_{cl}\rho_{cl} + b_{co}\rho_{co}\right). \tag{3.15}$$

As an illustrative example [20], we shall consider a very long and straight absorptionless fiber with circular cross section T (a circle of radius R in the (x, y) plane). The origin $\vec{x} = (0,0,0)$ is specifically chosen to be the center of one of those cross sections T. Let the standard polar coordinates r and φ ($x = r\cos\varphi$, $y = r\sin\varphi$) be introduced. Then, the clad and the core correspond to $r > R$ and $r < R$, respectively. In cylindrical coordinates (r, φ, z), the solution of Eq. (3.11), which is finite at $r = 0$ and continuous at $r = R$, is:

$$\phi(x)_\alpha = \exp(iM\varphi) \cdot J_{|M|}(K_1 r), \; r < R. \tag{3.16}$$

$$\phi(x)_\alpha = \frac{J_{|M|}(K_1 R)}{H_{|M|}^{(1)}(i\chi_\alpha R)} \cdot \exp(iM\varphi) \cdot H_{|M|}^{(1)}(i\chi_\alpha r), \; r > R. \tag{3.17}$$

$$K_1^2 + \chi_\alpha^2 = 4\pi\left(b_{cl}\rho_{cl} + b_{co}\rho_{co}\right), \tag{3.18}$$

with $M = 0,1,2\ldots$ Here, α denotes M and the $K's$, collectively. $H_{|M|}^{(1)}$ and $J_{|M|}$ are the standard Hankel and Bessel functions [25]. One still has to impose one additional boundary condition, namely, that $\partial\phi(x)_\alpha / \partial r$ be continuous, at $r = R$ for any φ. It reads:

$$\left[J_{|M|}(K_1 R)\right]^{-1} \left[\frac{dJ_{|M|}(K_1 r)}{dr}\right]_{r=R} = \left[H_{|M|}^{(1)}(i\chi_\alpha R)\right]^{-1} \left[\frac{dH_{|M|}^{(1)}(i\chi_\alpha r)}{dr}\right]_{r=R}. \tag{3.19}$$

The allowed values of χ_α (for given M) turn out to be the solutions of Eq. (3.19). Then, the allowed values of K_1 follow from Eq. (3.18). The ground state of Eq. (3.11) corresponds to its lowest eigenvalue, that is, to the largest χ_α (which is close to, but slightly smaller than, χ_{max}): the corresponding K_1 is very small. That ground state is independent on φ (as $M = 0$), does not vanish for any finite r and decreases as $r^{-1/2}\exp[-\chi_\alpha r]$ if $r \to +\infty$. The remaining propagation modes are excited ones: they correspond to decreasing values of χ_α (with either $M = 0$ or increasing integer $|M|$), decrease as $r^{-1/2}\exp[-\chi_\alpha r]$ if $r \to +\infty$ and develop an increasing number of zeroes for finite values of r. Further results and details for those propagation modes have been given in precedent works [19, 20, 26]. In particular, some approximate formulas [20] for K_1 or χ_α (for both $M = 0$ and $M \ne 0$), constitute the counterparts of related formulas for optical waveguides. For $M = 0$, other useful approximate formulas together with some detailed numerical solutions for the 16 lowest eigenvalues of Eq. (3.11) (except for the ground state) have been obtained [27].

3.4.1.2 Curved Fiber

We now turn to curved (bent) fiber: see Figure 3.7, as an example.

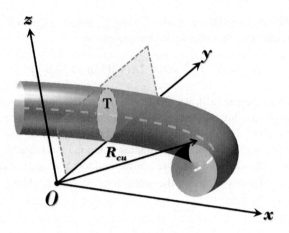

FIGURE 3.7 Curved (bent) three-dimensional fiber.

Let us now consider a very lengthy three-dimensional fiber and, for definiteness, let us suppose that: (i) the fiber is perfectly straight from some $z = z_-$ up to some z_0, (ii) the fiber has a large curvature radius R_{cu}, for $z_0 < z < z_1$. Both z_- and z_1 are finite: the straight part of the fiber starts at z_-, while the curved one ends at z_1. The length $z_0 - z_- = l$ is assumed to be quite large (as a first approximation, one could set $z_- = -\infty$). R_{cu} is much larger than any neutron wavelength and then the transverse dimensions of the fiber (for instance, $R_{cu} \geq 0.1\ m$). $z_1 - z_0 = l$ can be regarded as the length of the curved part of the fiber.

We suppose a confined beam of slow neutrons, propagating initially along the straight part of the fiber. Let $F_{in,tot}$ be the (statistical average of the) incoming quantum-mechanical probability flux across the transverse section of the straight part (see Chapter 1). As those neutrons enter into the curved part of the fiber, there is a non-vanishing probability for them to have no longer confined propagation but to escape outside the fiber. The tendency of the initially confined neutrons to propagate unconfined, toward increasing separations from the fiber, is named curvature or bending loss. Let $F(z_1)$ be the quantum-mechanical probability flux for neutrons to propagate still confined across the transverse cross section of the fiber, by the end z_1 of the curved part of the fiber. One has the approximate three-dimensional formula [24]:

$$F(z_1) \simeq F_{in,tot} \Upsilon. \tag{3.20}$$

Υ can be interpreted as its transmission coefficient which assesses the importance of bending losses. The following approximate formula displaying how Υ depends on R_{cu} and on the length l of the curved part of the fiber has been justified [24]:

$$\Upsilon \simeq \exp\left[-a_{cu}l \ \exp\left[-b_{cu}R_{cu} \ \right] / R_{cu}^{\frac{1}{2}} \right]. \tag{3.21}$$

Eq. (3.21), which holds in three spatial dimensions, describes the decrease of the probability flux of the confined neutron propagating along the curved fiber, due to curvature losses. The constants a_{cu} and b_{cu} depend, in principle, on whether the waveguide is multimode or monomode. It can be justified [24] that $b_{cu} = \left(\frac{2}{3}\right)\beta_0 \left[\frac{\chi_0}{\beta_0}\right]^3 (> 0)$. χ_0 is the smallest in the set of all χ_α which are excited effectively in the unbent part of the fiber: hence, the corresponding $\beta_\alpha \equiv \beta_0$ is maximum. Thus, we expect that all modes such that $\chi_0 \leq \chi_\alpha \leq \chi_{max}$ do get excited effectively. With this χ_0^2, the number of modes which propagate effectively can be estimated as $N_{pm}\left(\chi_0^2\right)$ (Eq. (3.14)). To find a_{cu} is more difficult: see Subsection 3.4.2.2. In order to realize the physical consistency of the approximate Eq. (3.21), we notice the following. Let R_{cu} become very large. Then, Υ tends quickly toward unity. As R_{cu} tends to zero, Υ goes quickly to zero. Previously, a theoretical study of bending losses for a curved neutron fiber in the simpler case of two spatial dimensions had been carried out [20]. For the analysis of bending losses for electromagnetic radiation, see Ref. [21]. A geometrical optics treatment of bending losses has been applied for X-rays and neutrons [23].

3.4.2 CONFINED PROPAGATION IN MULTICAPILLARY GLASS FIBERS

In principle, PGFs could be obtained by stretching glass tubes adequately, at suitably high temperatures. A typical polycapillary fiber (Figures 3.5 and 3.9) may have a diameter in the millimeter range (a bit smaller, eventually) and a length about a few tens of centimeters. Each PGF contains many (say, about 10^3) individual, essentially parallel, HCC. Single HCC with an internal diameter about a few microns have indeed been achieved experimentally, and we shall concentrate on them. Then, the former can be regarded, potentially, as a hollow waveguide for the confined propagation of radiation with wavelengths adequately smaller than that internal diameter. Consequently, X-rays can propagate confined along each single HCC. Bundles of such PGFs are currently employed for the focusing of X-rays, hereby providing basic devices for performing experiments in capillary optics of X-rays.

For a survey of the experimental development of capillary optics in the 1980s, first and mostly for X-rays (together with short comments about possible applications for slow neutrons), see M. A. Kumakhov and F. F. Komarov [23]. For other review, dealing with neutron capillary optics, see Ref. [4]. It appears that M. A. Kumakhov developed a patent for experimental capillary optics in 1984 [23]. Since then, experimental capillary optics has been considerably advanced and applied for X-rays, in several institutions. We quote [26]: the I. V. Kurchatov Institute in Moskow (Russia), the Center for X-Ray Optics in State University of New York at Albany, Albany, New York (USA), the Institute for Roentgen Optics (based on the Kurchatov Institute, mentioned above), the company Unisantis S. A. (www.unisantis.com/) in Geneva and the Institute for Advanced Instruments (IfG) GmbH (www.ifg-adlershof.de) in Berlin (Germany), mostly for X-rays. X-rays capillary optics constitutes nowadays a very active research field with important and expanding applications.

3.4.2.1 Experiments with Slow Neutrons

Confined propagation of slow (specifically, cold) neutrons has been experimentally established in PGFs (made of lead-silica glass) earlier in 1992 [28, 29]. Figure 3.5

could also serve as a (very simplified) description of the experiments presented in Refs. [28, 29].

Bundles of PGFs, as described in Subsection 3.3.2, were employed specifically for the confined propagation and focusing of slow neutrons [28, 29]. Any of the polycapillary fibers employed in the experiments [28, 29] has, typically, a diameter $d_{pf} = 0.4$ mm and a length between, say, 150 mm and 200 mm and it contains more than $N_{pf} = 1000$ parallel individual HCC. In turn, each single HCC in a PGF had an internal diameter $d_{hcc} = 6$ μm and could be regarded, quite reasonably, as a hollow waveguide of very small cross section, along which the confined neutrons propagate. See Figures 3.9 and 3.10.

Each polycapillary fiber employed in the first experiments [28, 29] was made of lead-silica glass. Accordingly, each of the above HCC was surrounded by a lead-silica clad (with $\text{Re}(n^2) < 1$): the latter had a width about 5 μm, so that, beyond such a distance, the cladding region ended and another neighboring parallel hollow channel was met. That is, two neighboring (essentially parallel) HCC were separated, on average, by a lead-silica clad of width about 5 μm.

The fibers may be straight or bent: in the latter case, they have curvature radius $R_{cu} \geq 0.1$ m [28, 29]. See Figure 3.8.

Those devices enabled, in principle, to increase the flux of confined neutrons which emerge from the exit end of a parallel assembly of fibers and concentrate onto a small region (focal spot), located at the focal distance. Specifically [29], a parallel assembly of 721 polycapillary fibers with an overall diameter of 15 mm at the exit end has provided an amplification of the neutron flux by a factor 7, at a focal distance equal to 104 mm [22]. The diameter of the focal spot was 1 mm. See Figure 3.10.

The 1992 experiments [28, 29] were analyzed by means of geometrical optics. Further related, and quite detailed, studies about neutron transmission have been carried out later using also geometrical optics [30–34]. See Ref. [35] for further developments, Ref. [36] for an overview, and Refs. [26, 37] for other presentations.

Straight Fiber

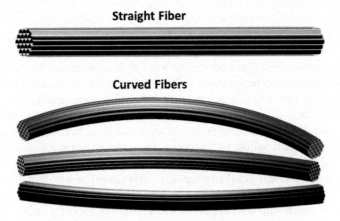

Curved Fibers

FIGURE 3.8   Various examples of polycapillary glass fibers. Bundle has:

    i. Overall Transverse diameter about 15 mm
    ii. About 700 single PGF

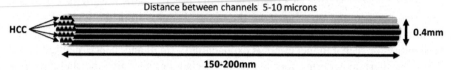

FIGURE 3.9 Single polycapillary glass fiber: quantitative details.

3.4.2.2 Estimates

It appears [24] that a single individual HCC with small cross section is conceptually analogous to, and can be regarded as, the single neutron fiber conjectured theoretically [19, 20]. This will be our understanding here, like in Ref. [24]. In order to avoid confusion, neither a single PGF nor a bundle will be regarded to be a single neutron fiber here.

Let the average energy of a thermal neutron, in a typical beam coming from a nuclear reactor, be about 10^{-2} eV, so that the average neutron velocity v is about 10^3 ms^{-1}. Let F_0 denote some average thermal neutron flux (per unit surface and time). Typically, each neutron is represented by a statistical mixture of wave functions, so that F_0 should be interpreted as, and replaced by, $\langle F_0 \rangle$. The average smallest separation between two neighboring thermal neutrons propagating in the beam is about $d_{n-n} = (v / F_0)^{1/3}$. For the highest fluxes, F_0 is about 10^{15} neutrons cm^{-2}s^{-1}. Then, d_{n-n} is of the order of 10^4 Å to 10^5 Å. The thermal neutron beam is, typically, non-monochromatic, so that a spectrum of wavelengths is met: for instance, it ranged between 2 Å and 9 Å in the first experiments [28, 29]. Since the wavelengths in such intervals are smaller than d_{n-n}, one can reasonably neglect the overlaps between the wave packets associated to different neutrons in the beam and regard each confined neutron as propagating independently.

The average distance, d_{nhc}, between two neighboring (parallel) HCC can be estimated from $d_{pf}^2 \simeq d_{nhc}^2 N_{pf}$, with the fiber diameter $d_{pf} = 0.4$ mm. One finds $d_{nhc} \simeq 5$ to 10 μm.

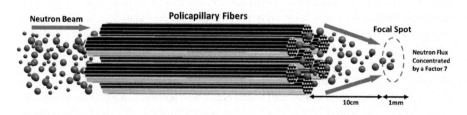

FIGURE 3.10 Focalization of a neutron beam by means of a bundle of polycapillary glass fibers. See Refs. [28, 29].

A priori, one could entertain the possibility that one slow neutron, propagating initially confined along certain HCC (denoted by *hcc*1), could escape, by transmission through the surrounding clad due to quantum-mechanical tunnel effect, to one of the neighboring parallel HCC and, eventually, escape finally toward the externally surrounding medium (air). Let d be the radial distance from the internal surface limiting the *hcc*1 to some point in its surrounding clad, so that $0 \leq d \leq 5$ μm. The probability for such a tunnel effect is exponentially small and, hence, negligible, for almost all thermal neutrons, for the lead-silica clad, provided that the clad thickness be about or larger than $d \geq 0.5$ μm $= d_{tu}$ [20]. Consequently, since d_{nhc} is about one order of magnitude larger than d_{tu}, one concludes that such a tunnel effect is negligible. One could simply say that neutrons propagate confined along any individual HCC as if the latter were surrounded by a clad of infinite width (say, as if $d_{nhc} = \infty$): this practical point of view can be adopted in thermal neutron propagation in PGFs.

The description of thermal neutrons, propagating confined along the above hollow capillary guides, with small transverse cross section, appears to require the quantum-mechanical approach and the associated Schrödinger Eq. (3.11) [24, 27].

Let us concentrate on a typical HCC, like those in the first experiments guiding thermal neutrons [28, 29]. E can be taken to be about 0.025 eV, as $V(x)$ and V_{cl} are much smaller. For such a channel ($b_{co}\rho_{co} = 0$), and for values of the number (ρ_{cl}) of nuclei per cm^3 and of the neutron-nucleus scattering amplitude b_{cl}^2 of a typical clad, one expects that the right-hand side of Eq. (3.13) ranges from about 10^{-3} Å to about 10^{-2} Å. Then, for typical values of β_α ranging from, say, 0.7 Å^{-1} to 3 Å^{-1}, $\chi_\alpha / \beta_\alpha$ for the highest modes (say, the fundamental one and the first few ones above it) varies from 10^{-2} to 10^{-4}. The HCC behaves naturally as a multimode waveguide and the total number of propagation modes N_{pm}, allowed in principle, varies between 4.5×10^2 and 4.5×10^4. Of course, for a given incoming slow neutron beam, it is not warranted, a priori, that all allowed propagation modes will be effectively excited. A reasonable expectation seems to be that, as neutrons enter into the HCC, they find more favorable to propagate into lower mode, say, the fundamental one and those having smaller values of χ_α. An interesting problem is to provide an estimate of the number of propagation modes which do get excited effectively: see Subsection 3.4.1.2.

Even if the inner diameter of a HCC ($d_{hcc} = 6$ μm) in a PGF is three to four orders of magnitudes larger than the average de Broglie wavelength of a thermal neutron, and although the total number of allowed propagation modes for the latter is rather large, one should not rely entirely on a geometrical optics description. In fact, a quantum-mechanical description for slow neutrons, propagating confined of a HCC in a PGF, should be entertained, as least for a suitable fraction of those neutrons. Thus, for certain effectively propagation modes (namely, the fundamental one and a subset of higher modes above it) $\chi_\alpha d_{hcc} / 2$ may be of order unity and even larger, and, hence, quantum effects may be important. In fact, for values typical of the actual HCC, Eq. (3.13) implies that $\chi_\alpha d_{hcc}$ has an upper bound which varies between 2.8 and 28. Stated into other terms, slow neutrons propagating confined along capillary channels with small transverse cross section can give rise to interference phenomena. The detailed and very interesting analysis [27] on neutron propagation along capillary guides (under certain specific conditions met in experiments) has provided further arguments supporting the physical necessity of treatments of such a phenomenon by employing quantum-mechanical wave functions.

We now turn to bending losses in multicapillary glass fibers. Polycapillary fibers were not straight. When initially confined beams of slow neutrons propagate in curved multicapillary glass fibers, the flux decreases due, in particular, to bending losses. Those bending losses have been experimentally measured (that is, the associated transmission coefficient Υ) [29]. Those experimental data have also been analyzed theoretically [29], by means of a geometrical optics formula for Υ [23].

A new theoretical (quantum-mechanical) analysis of bending losses in three dimensions, specifically addressed to the experimental results [29], has been carried out [24]. In outline, the latter analysis proceeds by applying Eq. (3.21) to those curvature losses, accepting that the HCC behave as multimode waveguides. An appealing fact was that the dependence of Υ on R_{cu}, as given by Eq. (3.21), appeared to allow for an approximate description of the experimental data. For multimode waveguides, a_{cu} has been estimated through a fit to the experimental data [24]. The behavior in Eq. (3.21), with $a_{cu}l \simeq 0.95$ m$^{1/2}$ and $b_{cu} \simeq 0.51$ m^{-1} is approximately consistent [24] with the experimental data [29] for the multimode case, as R_{cu} varies in, say, 0.1 m $< R_{cu} <$ 13 m. The order of magnitude of b_{cu} seems physically reasonable and essentially consistent with the above arguments (and with estimates of $(2/3)\beta_\alpha \left[\chi_\alpha / \beta_\alpha\right]^3$), for typical average values of β_α about 1 Å^{-1} and of $\chi_\alpha / \beta_\alpha$ between 10^{-3} and 10^{-4}). For $b_{cu} \simeq 0.51$ m^{-1} and a maximum propagation constant for the effectively propagating modes (namely, β_0) about 3 Å^{-1}, one gets $\chi_0 / \beta_0 \simeq 2.9 \times 10^{-4}$. If the right-hand side of Eq. (3.13) is about 10^{-3} Å, so that $N_{pm} \simeq 5.4 \times 10^{-2}$, the total number of effectively excited modes is estimated to be $N_{pm}\left(\chi_0^2\right) \simeq 108$. If, on the other hand, the right-hand side of Eq. (3.13) is about 10^{-2} Å, so that $N_{pm} \simeq 4.5 \times 10^4$, the total number of effectively excited modes is $N_{pm}\left(\chi_0^2\right) \simeq 4.2 \times 10^4$. See also Refs. [24, 26].

The case with N (more or less) parallel HCC (with $N \gg 1$), which occurs in reality with polycapillary fibers (so that $N = N_{pf}$) and with bundles thereof, can be treated similarly, and leads, approximately, to the same Υ [24].

3.4.3 NANOTUBES: CONJECTURES ON CONFINED PROPAGATION

3.4.3.1 General Aspects

We recall that the minimum transverse size for a X-rays beam (whatever the focusing device) is about 10 nm [38].

Carbon nanotubes (CNTs) play an increasingly important role in nanotechnology, since their discovery in 1991 by means of high-resolution microscopy observations [39]. CNTs turn out to be very interesting, by virtue of several specific properties: their very small transverse sizes, at the nanometer or nm scale (1 nm = 10^{-3} μm = 10^{-9} m), their unique structure, their possibility of being metallic or superconducting (depending on their geometric structure), their remarkable capabilities for allowing for ballistic transport, their extremely high thermal conductivity and optical polarizability, their possibilities for reaching high structural perfection. For accounts about CNTs see Refs. [40–42].

We consider a CNT with length L (1 μm $\leq L \leq$ 100 μm). This CNT may be (i) a single-wall carbon nanotube (SWCNT), that is, one ($N = 1$) single layer composed by carbon atoms folded into a hollow cylinder (the hollow or "empty" interior containing air), or (ii) a multiwall carbon nanotube (MWCNT), that is, a set of $N(\geq 2)$ carbon layers wrapped coaxially. That is, the MWCNT resembles a SWCNT coated by $N - 1$ additional concentric walls, also made by carbon atoms, so that there are $N - 1$ (> 0) additional "empty" interiors (also filled by air) between successive walls. Adequate figures displaying adequately the structures of SWCNT and MWCNT can be seen, for instance, in Ref. [41]. Regarding the preparation of CNTs, we shall limit ourselves to quote chemical vapor deposition (CVD): it appears to be adequate for producing large amounts of SWCNTs and it is also employed to obtain MWCNTs.

For a SWCNT, typical dimensions are: 10 to 17 nm of overall transverse diameter and diameter d of the internal "empty" cylinder about 5 to 7 nm. Then, the thickness w of the wall of a SWCNT is about 2.5 to 5 nm. For a MWCNT, we shall take 50 nm as some average, or typical, overall transverse diameter and the distance Δ between two successive coaxial walls to be about 0.35 nm. As a first approximation, we shall assume that the diameter of the innermost central "empty" cylinder and the thickness of any wall in a typical MWCNT are of the same order as in a SWCNT. Then, one may estimate that, in such a MWCNT, N varies between 4 and 8. Here, we shall concentrate on essentially straight CNTs.

Let A_{tot} be the area of the total transverse cross section of the nanotube (in nm^2). Also, let A ($< A_{tot}$ and in nm^2 as well) be the total area of all transverse cross sections of all N "empty" interiors of the CNT ("effective area"). As examples, we consider: (i) a SWCNT with $d = 6$ nm and $A(1) \simeq 28$ nm^2, and (ii) a MWCNT, with $N = 6$ and overall transverse diameter $\simeq 50$ nm, in which $A(6) \simeq 207$ nm^2 (the total area of its transverse cross section being $A_{tot} \simeq 2044$ nm^2).

Here, we shall overview a conjecture on the possible confined propagation of thermal neutrons along CNTs behaving as waveguides with transverse dimensions at nm scales. In so doing, we shall follow a proposal made originally by G. F. Calvo, and developed later [43] to which we refer for a detailed presentation and discussion. Such an analysis will play, for thermal neutrons, a role similar to that already carried out about the focusing of X-rays at the nm scales [38].

Let us consider an incoming thermal neutron flux F_0 (neutrons/(cm$^2 \cdot$ s)), which approaches one end of the CNT from outside. The flux of confined, which have entered into and propagate confined along the CNT, can be roughly estimated to be $F_{in} \simeq \theta_{cr}^2 \times F_0$, with the same θ_{cr} as in Eq. (3.7) (compare with Figure 3.4). Then, θ_{cr}^2 reduces the neutron flux by a factor about 10^{-5}. See Figure 3.11. For

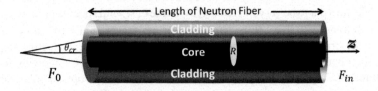

FIGURE 3.11   Incoming flux F_0 in and exit flux F_{in} out of a nanotube. (schematic representation not displaying its specific structure).

$F_0 \simeq 10^{14}$ neutrons/cm$^2 \cdot$s, one estimates $F_{\text{in}} \simeq 10^9$ neutrons/cm$^2 \cdot$s. The number of confined neutrons transmitted across the area A during τ hours is $\simeq 3.6 \times 10^{-2} A \times \tau$, which we shall now estimate. For a SWCNT ($A = A(1)$), we get $\simeq 1$ neutron per hour. For a MWCNT, with $N = 6$ and $A(6) \simeq 207$ nm^2, one has $\simeq 6$ neutrons per hour. One would expect that the above estimates of the number of confined neutrons transmitted across A during τ hours hold, in order of magnitude, in a quantum description of the thermal neutron propagation. The necessity of such a description will be discussed below.

3.4.3.2   Conjectures

The total area A of all transverse cross sections of all "empty" interiors of the CNT is very small. Then, a quantum-mechanical description of the confined propagation of thermal neutrons along CNTs is necessary. The following arguments will support this expectation.

1. Let N_{pm} be the total number of propagation modes for neutrons moving confined along a waveguide, in quasi-classical approximation. The general formula for N_{pm} given in Eq. (3.15) gives for a CNT (either SWCNT or MWCNT): $N_{\text{pm}} \simeq 2b\rho A$ (the factor 2 counting both spin projections for the neutron). Here, $b\rho(\equiv b_{cl}\rho_{cl})$ refers to any carbon wall (which constitutes the clad). Notice that the hollow interiors (the cores) contain air, for which we take, as a first approximation, $b_{co}\rho_{co} \simeq 0$. Using $b\rho \simeq 7.5 \times 10^{-4}$ nm^{-2} for the carbon walls and the estimates for A for both SWCNTs and MWCNTs given in the previous subsection, one finds $N_{\text{pm}} < 1$. Then, the reliability of the quasi-classical approximation for the computation of the total number of propagation modes [22] cannot be trusted.
2. For thermal neutrons, the de Broglie wavelength is about 10^{-1} nm to 1 nm, to be compared to the transverse sizes of the CNT.

The above arguments indicate that quantum effects are relevant below the μm scale and that the possible description of thermal neutron confined propagation in CNTs requires quantum mechanics. Due to the smallness of A, one may anticipate that only one propagation mode (namely, the fundamental one) would, at best, be relevant.

A study of the propagation modes in CNTs with $N(\geq 1)$ "empty" regions has been carried out [43], to which we refer for details. We shall assume, for a while: (i) the most external wall (that after the $n = N$-th empty space) of the CNT has infinite thickness. It will also be assumed, for simplicity, that: (ii) all walls and "empty" regions are concentric and have circular cross sections.

We consider a SWCNT ($N = 1$) first. According to several arguments and estimates [43], the fundamental mode of a confined slow neutron in the SWCNT appears to spread far outside the waveguide, in the transverse plane. Loss effects would upset quite dramatically the probability distributions determined by that propagation mode in the transverse plane and, hence, its confined propagation along the SWCNT. Then, it appears very unlikely that the SWCNT could support, in an effective way, the fundamental mode: of course, excited propagation modes are even more unlikely, not to say fully excluded.

We now consider a MWCNT with $N > 1$ concentric coaxial "empty" regions, having spatial sizes as in Subsection 3.4.3.1 and behaving as a monomode waveguide. For this MWCNT, which has a larger "effective area" $A(N)$, the fundamental mode becomes spatially more concentrated in the $N(> 1)$ successive "empty" interiors than it was for $N = 1$ (as $A(1) < A(N)$). For the monomode MWCNT, estimates indicate that, for suitable N and under assumption (i), the probability distribution for the fundamental mode in the transverse plane would be essentially concentrated into the waveguide, without extending significantly beyond the most external wall.

The later conclusion will now be reconsidered, upon turning to the following additional physical effects [43], not taken into account above and yielding losses in the confined propagation of thermal neutrons along CNTs, that is, making their flux to decrease. They are: nuclear absorption and diffuse scattering, bending (or curvature) losses and tunneling (the most influential effect). Both nuclear absorption and diffuse scattering are negligible both in the carbon walls and in the "empty" spaces. Bending losses are expected to be much smaller (and, tolerable) for MWCNTs than for SWCNTs. Regarding tunneling, and no longer following assumption (i), we shall assume the most external $(n = N)$ carbon wall has also finite thickness w. Then, a thermal neutron, propagating initially confined along the CNT, could escape, upon performing tunnelings across the various carbon walls and, eventually, across the outermost (N-th) one, toward the external medium (air) surrounding the CNT. In principle, this loss effect is more relevant for CNTs in the nanoscale than for PGFs. We shall summarize the results of estimates in Ref. [43]. For $N = 6$, $d \simeq 6$ nm, $w \simeq 3.5$ nm, 30 μm $< L < 50$ μm, it has been estimated that tunneling losses would seem to be tolerable and would not spoil completely the confined propagation (although they would affect it). Thus, the fact that $N > 1$ makes confined neutron propagation along MWCNTs more likely and less sensible to loss effects than in SWCNTs. These conclusions [43] seem to hold also for other shapes of the transverse cross sections, that is, the above assumption (ii) does not seem necessary.

In conclusion, for confined propagation of thermal neutrons (mostly, in the fundamental mode) at the nanoscale, MWCNTs would be a more favorable possibility than SWCNTs. The nanoscale seems the lower limit in which neutron confined propagation could occur. Further devices may be needed in the process of focalizing neutrons between, say, the μm scale and the nm one: for such a purpose, PGFs may be adequate.

3.5　NEUTRON FILM WAVEGUIDES

3.5.1　PREVIOUS EXPERIMENTS

Guided slow neutron waves in thin planar waveguides were established experimentally in 1994 [44]. See Figure 3.12 for a description of the device. Neutrons propagated along z and there was lateral confinement only along one transverse direction (x). The thin film was a TiO$_2$/Ti/Si layer. The guiding layer (core) was a thin titanium (Ti) planar waveguide with thickness 120 nm along the x direction and length

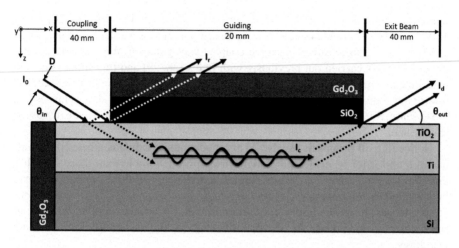

FIGURE 3.12   Experiment arrangement in $TiO_2/Ti/Si$ thin planar waveguides (as in Ref. [44].)

20 mm along z. The upper film in the clad was formed by three successive layers (TiO_2, SiO_2, and Gd_2O_2) above the core. The SiO_2 layer had a thickness of 600 nm. The lower film in the clad was a Si layer below the core. More specifically, the experimental setting formed by the three successive layers above the core enabled the coupling of the incoming neutron wave (with flux I_0) to the propagation mode along the Ti core. In practice, the SiO_2 layer could be regarded as if it were infinitely thick. Sizes along the y direction were about a few tens of millimeters. The ratio of the thermal neutron flux detected after the waveguiding effect in the Ti layer over the measuring incoming flux (namely, I_d/I_0, as displayed in Figure 3.12) was about 3.5×10^{-4}, within the order of magnitude expected. The confined neutron flux along z was strongly attenuated beyond 20 mm.

Waveguiding effects for polarized slow neutrons in magnetic thin films have also been observed experimentally [45–48]. In particular, the experiment [45] made use of a magnetic CoZr /Al/CoZr trilayer, deposited successively on $Gd_3 Ga_5O_{12}$: polarized slow neutrons propagated confined in the Al layer (core), with thickness 280 nm and length about 6 mm.

3.5.2   PROPOSALS FOR CONFINED NEUTRON PROPAGATION

We shall outline some proposals [49], based upon two coupled waveguides (TCW), for generating slow neutron beams which, possibly, would propagate along longer distances (say, about 1 m) in the z-axis and be confined in transverse directions down to about a fraction of a micron. The first proposal (two coupled waveguides (1) or TCW1) will be based upon the successful experiment using a thin Ti film, reviewed in Subsection 3.5.1, and include an additional suitable waveguide in order to reduce neutron absorption. The second proposal (two coupled waveguides (2) or TCW2), although not based upon experiments carried out previously, will constitute a natural improving extension of TCW1 and of previous proposals in Refs. [19–20]. Both TCW1 and TCW2 could possibly have some applications, as we shall comment later.

The third proposal, involving three coupled waveguides (CW3) and based upon the experiments carried out previously with PGFs [28–29] and thin films [44], will constitute a natural extension of them and of the previous proposal in Refs. [26, 37] toward the controlled exploration of shorter scales with focalizing slow neutrons in a broad sense. So, for the sake of completeness, CW3 may well deserve to be discussed shortly here.

Notice that all proposals involve Ti as an adequate candidate, since: (i) the values of b and μ for it seem, in a global sense, more favorable than those for other possibilities (see Table 3.2), (ii) its natural abundance, (iii) pure samples of it can be prepared in a laboratory, (iv) its interesting mechanical, chemical, and physical properties. It will be important to remind that, for air, the value of $b\rho$ is about three orders of magnitude smaller than for the typical elements used in cores and claddings.

3.5.2.1 Two Coupled Waveguides Model 1 (TCW1)

Two planar waveguides are successively coupled along the propagation direction (z). See Figures 3.13 and 3.14.

The first waveguide has a Ti film (core) and a clad formed by two films (SiO_2 and Si). The whole device is placed in air: that is, there is air above SiO_2 and below Si. Confined propagation along z occurs first in the Ti core, like in Ref. [44]. Transverse sizes along the x and y directions could be similar to those in the experiment in Feng et al. [44], in Figure 3.12. Anyway, we shall suppose that each of the SiO_2 and Si films in the first waveguide have widths along x larger than 0.5 μm: see the comment below regarding the widths of the SiO_2 and Si films for the second waveguide. For simplicity in a first discussion, the clad above the Ti core reduces just to a SiO_2 film (compared to Figure 3.12). Notice that here, and contrary to the experiment in Ref. [44], one is not interested in detecting neutrons leaving the Ti core. Anyway, there is one important difference regarding the first waveguide in Figure 3.13: its length

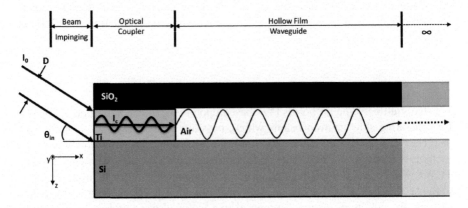

FIGURE 3.13 Schematic representation of the device in the TCW1 proposal.

along z is chosen to be definitely shorter than that in Figure 3.12 (say, less than 1 cm), so as to reduce attenuation in the Ti core.

The second waveguide has air as core (with transverse dimensions similar to those of the first waveguide). The clad is similar to that of the first waveguide: two films (SiO$_2$ and Si). The confinement in the core of the second waveguide follows from Eq. (3.9), the very small value of $b\rho$ for air (core) and the larger positive value of $b\rho$ for the SiO$_2$ and Si claddings (by about three orders of magnitude). However, the latter confinement is not strict, due to the finite widths of the SiO$_2$ and Si claddings. In order that losses across the clad (by quantum mechanical tunneling across it toward air) be negligible, it is required that each of the SiO$_2$ and Si films have widths along x larger than 0.5 μm, as we shall now justify. Let any of the two films in the clad of the second waveguide have density ρ, coherent scattering amplitude $b > 0$, and transverse width along x equal to d. The probability for tunneling of a slow neutron from the air core of the second waveguide across a distance d in one of its claddings is roughly proportional to $e^{-4(4\pi b\rho)^{1/2}d}$, which is adequately small for $d > 0.5$ μm.

The overall length of the second waveguide along z could be several tens of centimeters or even about 1 m. Confined slow neutrons from the Ti core of the first waveguide would enter, possibly with very small attenuation, into the air core of the second waveguide. Since air has a small linear coefficient μ (see Table 3.2), a direct application of Eq. (3.1) shows that there would be small attenuation in the core of the second waveguide and, hence, at its exit there would be an appreciable neutron flux. See Figure 3.13.

At this stage, the structure of the first waveguide could be modified by illuminating laterally the Ti guide with the incoming neutron beam so as to become similar to that in Figure 3.12. This would facilitate the coupling of an incoming neutron to the propagation mode along the Ti core. See Figure 3.14.

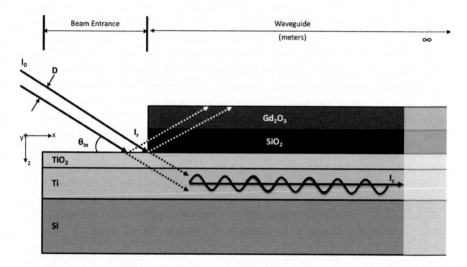

FIGURE 3.14 Schematic representation of the modified TCW1 proposal, with lateral neutron beam illumination improvement.

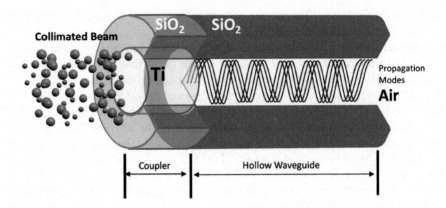

FIGURE 3.15   Schematic representation of the TCW2 proposed device.

We stress that the neutron beam, as it propagates along z, is confined only along one transverse direction, namely, the x direction. There is no confinement in the other transverse direction (y). The following proposal TCW2 will be aimed to achieve confinement along y as well, still keeping confinement along x, wherever possible.

3.5.2.2   Two Coupled Waveguides Model 2 (TCW2)

It also contains two coupled waveguides. See Figure 3.15 for this device, which is placed in air. The first waveguide would no longer be a thin film, like in the previous proposal TCW1, but a single solid cylinder (with circular cross section), with Ti as inner core and SiO_2 as surrounding clad. The diameter of the Ti core would be of the order of 1 μm and the width of the SiO_2 clad would be of the same order. The length of the first waveguide should be smaller than 1 cm. The second waveguide would be a hollow cylinder (with circular cross section), having air as inner core and SiO_2 as surrounding clad as well. The width of the clad of the second waveguide could be similar to that of the first waveguide. In the coupling, the axis (z) in the Ti core of the first waveguide should coincide with the axis of the second waveguide. The length of the second waveguide could be some tens of centimeters and, possibly, about 1 m. Neutrons, propagating confined along the Ti core of the first waveguide, would enter into the air core of the second waveguide and propagate along the latter without appreciable attenuation along distances similar to those in the proposal TCW1, as another use of Eq. (3.1) shows. The (strict) confinement in the core of the first wave-guide follows from Eq. (3.5) and the values of b in Table 3.2. The confinement in the core of the second waveguide follows from Eq. (3.5), the very small value of $b\rho$ for air (core) and the larger positive value of $b\rho$ for the SiO_2 clad (by about three orders of magnitude). However, the latter confinement is not strict, due to the finite width of the SiO_2 clad. Losses in the propagating neutron flux across the surrounding SiO_2 clad of the second waveguide (by quantum mechanical tunneling across it toward air) would be negligible, since its width is larger than 0.5 μm: the justification is similar to the one for the clad of the second waveguide in the proposal TCW1.

In the present proposal TCW2, there would be confinement along both transverse directions x and y.

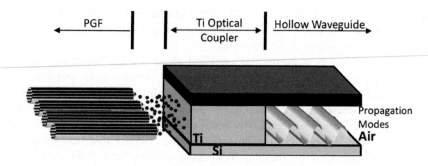

FIGURE 3.16   Schematic representation of the CW3 proposal, coupling PGFs to TCW1.

3.5.2.3   Three Coupled Waveguides Model 3 (CW3)

We shall complement the previous TCW1 and TCW2 by discussing the following natural possibility (CW3), formed by three successively coupled waveguides, for the controlled exploration of short scales using slow neutrons. See Figure 3.16 for a schematic representation.

The first waveguide is a suitable bundle of parallel PGFs, with their axes parallel to the z axis, directly coupled to the second waveguide, as we shall describe. The second and third waveguides are very similar to (and similarly oriented as) the ones in the previous proposal TCW1 (Figure 3.13): then, they have similar sizes along the directions y and z, while their thickness along x is now taken to be larger than that of a HCC, say, about a few microns. In each PGF in the first waveguide, any single HCC should have an internal diameter (a few microns) matching as much as possible the thickness of the second waveguide along x. Since a typical PGF has an overall diameter in the mm range and the dimension of the second waveguide along y is about a few tens of millimeters, the bundle should contain some tens of PGFs. More specifically, the PGFs in the bundle should be disposed forming a layer: only one PGF along x and z, and tens of parallel PGFs along y, as Figure 3.16 shows. In the coupling, the propagation direction (z) in the Ti and air cores of the second and third waveguides should be as close as possible to the axes of the parallel PGFs (say, the propagation direction of neutrons in them). The neutron beam coupling facilitates the confinement of the neutrons in the air cores of the HCCs (a few thousands) of each PGF in the bundle forming the first waveguide, entering successively into the Ti and air cores of the second and third waveguides and to propagate confined along them, for distances about 1 m, possibly.

One could also entertain other possibilities: for instance, the opposite one in which confined neutrons propagating firstly in the second and third waveguides would enter into the air cores of the HCCs of the first waveguide. In the present proposal CW3, there would be confinement along x and, partially, also along y.

3.5.2.4   Comments on the Quantum-Mechanical Analysis

The quantum-mechanical description of the confined slow neutrons by means of propagation modes can be employed, by making use of certain analogies with theoretical formulations in classical optics and associated results. The smallness of the ratio (neutron characteristic wavelength)/(transverse dimension of the fiber) gives rise to computational challenges and different techniques must be used in order to estimate the effectiveness of the waveguide.

A slow neutron, as it propagates confined along a core, has always a non-vanishing probability for penetrating a distance d into the corresponding clad (which has a finite potential, in general). Such a probability decreases as the energy of the neutron decreases compared to the potential of the clad. The probability for such a penetration is exponentially small and, hence, negligible, for almost all thermal neutrons, for typical solid materials in the clad (for instance, those containing silica), provided that the clad thickness be about or larger than $d \geq 0.5$ μm. One could attempt to disregard that probability, as a zeroth-order approximation upon analyzing (admittedly, in an eventually crude way) various general features of confined slow neutron propagation. Just by passing, we remark that ultracold neutrons experience total internal reflection in the clad and, so, the probability for their penetration into the clad (even if not vanishing) becomes rather small: then, the comments below for an infinitely repulsive clad (Figure 3.17) could possibly be less crude, as a zero-order approximation of the description of the propagation phenomena.

Consequently, one simplified, but nevertheless interesting, approach is that in which the slow neutron is subject to an infinite repulsive potential in the clad. Then, the wavefunction $\varphi(z, x)$ of the thermal neutron vanishes in the whole clad and at its boundary (Dirichlet boundary conditions). The qualitative physical picture is that the incoming neutron wave, as it enters into one end-face of a waveguide and starts to propagate, gives rise, if certain conditions are met, to confined propagation modes which propagate along the waveguide (without entering into the clad, which is excluded). The generation of propagation modes out of the incoming neutron wave

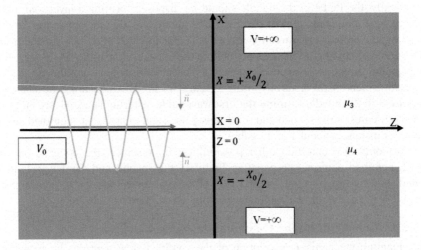

FIGURE 3.17 Confined neutron propagation along a very long planar film waveguide, with an infinitely repulsive cladding.

even in this simplified case and its physical picture are rather difficult to formulate and analyze quantitatively and, so, we shall not discuss them here. Just to complement the slow neutron propagation of modes for fibers in Subsection 3.4.1.1, we shall summarize here those in the ideal case of a very long film waveguide, with infinitely repulsive potential in the clad and constant potential V_0 in the core: see Figure 3.17 The two waveguide walls are located at $x = -x_0/2$ and $x = x_0/2$, for any z. The propagation modes have the following structure:

$$\varphi(z,x) = e^{izK_{z,pm}} \frac{2m}{\hbar^2} \frac{1}{2} \left\{ \mu_3 \exp\left[-i(x - x_0/2)\sqrt{\frac{2m(E-V_0)}{\hbar^2} - K_{z,pm}^2} \right] \right.$$

$$\left. + \mu_4 \exp\left[i(x + x_0/2)\sqrt{\frac{2m(E-V_0)}{\hbar^2} - K_{z,pm}^2} \right] \right\}$$

(3.22)

In Eq. (3.22) μ_3 and μ_4 are constant. $K_{z,pm}$ is the component of the wavevector along the axis (z) of the waveguide. We have to impose Dirichlet boundary conditions in the two-waveguide walls, for any z:
For: $x = -x_0/2$, $\varphi(z, -x_0/2) = 0$, so that:

$$\mu_4 + \mu_3 \exp\left(ix_0 \sqrt{\frac{2m(E-V_0)}{\hbar^2} - K_{z,pm}^2} \right) = 0.$$

(3.23)

For: $x = x_0/2$, $\varphi(z, x_0/2) = 0$ and, hence:

$$\mu_3 + \mu_4 \exp\left(ix_0 \sqrt{\frac{2m(E-V_0)}{\hbar^2} - K_{z,pm}^2} \right) = 0.$$

(3.24)

Eqs. (3.23) and (3.24) imply:

$$\exp\left(ix_0 \sqrt{\frac{2m(E-V_0)}{\hbar^2} - K_{z,pm}^2} \right) = \pm 1 \text{ for } \mu_3 = \mp\mu_4.$$

(3.25)

We shall discuss them in more detail.
For:

$$\exp\left(ix_0 \sqrt{\frac{2m(E-V_0)}{\hbar^2} - K_{z,pm}^2} \right) = +1 \Rightarrow$$

$$\sqrt{\frac{2m(E-V_0)}{\hbar^2} - K_{z,pm}^2} = \frac{2\pi n}{x_0}; \mu_3 = -\mu_4.$$

(3.26)

($n = 1,2,3,\ldots$, excluding the case $n = 0$). These are the odd modes. Then:

$$\varphi(z,x) = -e^{izK_{z,pm}} \frac{2m}{\hbar^2} \frac{\mu_+}{2} e^{i\pi n} (-2i) \sin\frac{2\pi nx}{x_0}.$$

(3.27)

For:

$$\exp\left(ix_0\sqrt{\frac{2m(E-V_0)}{\hbar^2}-K_{z,\mathrm{pm}}^2}\right)=-1\Rightarrow$$

(3.28)

$$\sqrt{\frac{2m(E-V_0)}{\hbar^2}-K_{z,\mathrm{pm}}^2}=\frac{\pi(2n+1)}{x_0};\,\mu_3=\mu_4.$$

($n=0,1,2,3,\ldots,$). These are the even modes. Then:

$$\varphi(z,x)=e^{izK_{z,\mathrm{pm}}}\frac{2m}{\hbar^2}\frac{\mu_+}{2}e^{i\pi n}(2i)\cos\frac{\pi x(2n+1)}{x_0}.$$

(3.29)

μ_3 plays in both cases the role of a normalization constant.

Recently [50], we have studied the somewhat less idealized case of a semi-infinite film waveguide ($0<z<+\infty$) with infinitely repulsive potential in the clad and an incoming neutron plane wave in $z<0$. Then, the waveguide has four walls. We have analyzed quantitatively the generation of propagation modes out of the incoming thermal neutron wave and its physical picture through a new formulation. Our aim is to find the wave function that solves the partial differential equation for $\varphi(z,x)$ following from Eq. (3.3), with Dirichlet boundary conditions and corresponding to the above incoming plane wave. We have been able to cast the determination of $\varphi(z,x)$ in terms of a suitable linear integral representation, involving four functions $\mu_i(x)$, $i=1,2,3,4$ associated to the four walls, the incoming thermal neutron wavefunction and a two-dimensional Green's function. By imposing Dirichlet boundary conditions in the four waveguide walls, one finds a system of four inhomogeneous linear integral equations of second kind, which determines uniquely the four functions μ_1, μ_2, μ_3, μ_4: $\mu_i=f_i(\mu_j,\mu_k,\mu_l);i\neq j,k,l$. This system of equations for μ_1, μ_2, μ_3, μ_4, is very useful and interesting for the following reasons: (i) it can be solved by successive iterations (thereby yielding an infinite series solution for $\varphi(z,x)$), (ii) one can set up simulations in order to compute numerically the low order iterations, (iii) the successive iterations allow for physically appealing interpretations of the propagation of the neutron wave along the core.

Detailed numerical computations have been carried out [50]. In certain cases, we find a strong decrease of the wavefunction as the neutron tries to enter into the waveguide (i.e., the formation of propagation modes is not completely achieved). However, in other cases we find that, due to interaction with the walls, propagation modes are starting to be formed. Estimates for a semi-infinite film waveguide ($0<z<+\infty$) with finitely repulsive potential in the clad and an incoming neutron wave in $z<0$ have also been undertaken. We omit details [50].

3.5.3   CONJECTURE: POSSIBLE APPLICATIONS TO BORON NEUTRON CAPTURE THERAPY (BNCT)

3.5.3.1   A Short Introductory Discussion on BNCT

An area with important applications is BNCT. Accounts, successively updated, of the status of BNCT are given in Refs. [7, 8] and references therein. We also address to

Chapter 6 in this book for a very wide and updated presentation of it. The short summary below of certain important features of BNCT is intended both as a motivation for our conjecture below and as an introduction to the very comprehensive account in Chapter 6. BNCT continues to be very attractive and active in nuclear medicine for promising (albeit non-exclusive or non-unique) therapies of certain malignant tumors.

Standard (normal size) tumors (for example, breast cancer) are clinically categorized as: T1: tumor < 2 cm. T2: tumor is larger than 2 cm, but no larger than 5 cm. T3: tumor is larger than 5 cm. T4: tumor has any size, but it has spread beyond the tissue (we do not treat this data in detail for obvious reasons). On the other hand, there are also small tumors, with smaller sizes (of order about 1 mm or smaller), which could be relevant for our purposes, as we shall comment later.

BNCT is a therapeutic modality for treating certain malignant tumors, like primary brain tumors (glioblastomas) and head and neck cancer. In short, in order to start the treatment, certain tumor- localizing drug containing the (non-radioactive) isotope B_5^{10} is injected into the patient. After decades, the drugs employed continue to be the boron-containing compounds BSH (with chemical formula $Na_2B_{12}H_{11}SH$ and known under several related names, like sulfhydryl borane or disodium undecahydro-mercapto-close-dodecacarbonyl) and BPA (an amino acid: a dihydroxyboryl derivative of phenylalanine, with chemical [8] formula $C_9H_{12}BNO_4$ also named p-boronphenylalanine), both of them specific for BNCT [7]. At a subsequent stage, the patient is irradiated with a slow neutron beam (epithermal, thermal), focused toward the location of the malignant tumor, to which the above drug has been delivered previously. The basic BNCT (nuclear) reactions are:

$$n + B_5^{10} \rightarrow He_2^4 + Li_3^7 \quad \text{and} \quad n + B_5^{10} \rightarrow He_2^4 + Li_3^7 + \gamma. \tag{3.30}$$

2.8 MeV and 2.3 MeV are respectively liberated in those two reactions, in the form of kinetic energies of the final He_2^4 and Li_3^7 particles. The second one of those reactions is shown in Figure 3.18. The photon (γ) in the second reaction has energy 0.48 MeV. The final He_2^4 and Li_3^7 particles are able to produce damage to tumor cells (namely to their DNA macromolecules) of malignant cells in small domains of sizes 8 μm and 5 μm, respectively. The basic aim of the therapy at the macromolecular level is that one single hit of one slow neutron could produce irreversible damage to both helices of one double-helix DNA macromolecule of a malignant cell (which has diameter dimensions between 1.8 nm and 2.3 nm), so that the possibility for its repair be practically forbidden. Damage to only one chain does not prevent that repair. Recall that typical linear scales along each strand in DNA and distances between both strands are of order 1 to 2 nm.

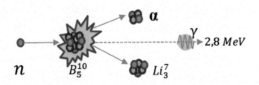

FIGURE 3.18 Basic BNCT reaction.

A very important aspect is to control, as much as possible, the delivery of neutrons to malignant cells and, so, to avoid that normal tissue could receive the neutron radiation. For a thermal neutron beam F to be effective in standard BNCT, its flux F should be not smaller than 5×10^8 (neutrons/(cm$^2 \cdot$ s)) (say, about a few dozen grays). The latter order of magnitude will be an important reference throughout this contribution. Standard BNCT treatments for standard (normal size) tumors last for a time interval between 30 minutes and 1 hour. A typical full BNCT treatment requires a total of about 10^{14} neutrons. These figures are consistent with one another (to within an order of magnitude). In fact, for one (standard) tumor having surface about 16 cm^2 (exposed to an incoming flux 10^9 (neutrons/(cm$^2 \cdot$ s)) and a treatment lasting 40 minutes, one has roughly: 10^9 (neutrons / cm$^2 \cdot$ s) $\times 2.4 \times 10^3 s \times 16$ cm$^2 = 4 \times 10^{13}$ neutrons. For a previous conjecture, based upon Ref. [29], see Ref. [51].

3.5.3.2   TCW1, TCW2, and BNCT

As stated above, a thermal neutron flux about 10^9 (neutrons/(cm$^2 \cdot$ s)) allows to treat a normal or large tumor of volume, say, about 4^3 cm^3. Then, one could argue, at least qualitatively (through a simple proportionality), that a small tumor with volume 1 mm^3 (requiring a smaller total number of neutrons) could possibly be treated, during approximately a similar time duration, by a flux about $(4 \times 10)^{-3} \times 10^9$ (neutrons/(cm$^2 \cdot$ s)), say, about 10^4 (neutrons/(cm$^2 \cdot$ s)).

A basic aim of our proposals is to control, as much as possible, the delivery of neutrons to malignant cells and, so, to avoid that normal tissue could receive the neutron radiation and it is the basic aim of this investigation. For a thermal neutron beam F to be effective in standard BNCT, for normal-sized (10 cm^3) tumors its flux F should be not smaller than 5×10^8 (neutrons/(cm$^2 \cdot$ s)).

Let us consider the accelerator-based neutron source, constructed by the private Japanese company Sumitomo and interesting for our discussion here, namely, the 30 MeV H$^-$ Cyclotron [52]. It employs a proton beam equivalent to 1 mA and produces a flux 10^9 (neutrons/(cm$^2 \cdot$ s)) (stable during about 30 minutes), which is specifically designed for BNCT.

Let the latter flux enter into the first waveguide in the above proposal TCW1 (see Figure 3.13 and consider confinement along the transverse direction x), for definiteness. Then, plausibly, the slow neutron flux at the exit of the first waveguide, just at the entrance of the second waveguide, could be about 10^5 to, perhaps, 10^6 (neutrons/(cm$^2 \cdot$ s)). If losses upon entering into the second waveguides are moderately small, as one would expect since the main flux reduction would occur between air and the entrance of the first waveguide, then, the outgoing flux at the exit of the second waveguide, having air core, should not suffer further appreciable attenuation, say, could be about 10^4 to, perhaps, 10^5 (neutrons/(cm$^2 \cdot$ s)).

Let us now turn very briefly to TCW2 (see Figure 3.15 and consider lateral confinement in both transverse directions x and y). A direct application of the θ_{cr}^2 reduction factor in the confined flux would suggest an outgoing flux at the exit of the second waveguide (having air core) would be about 10^4 (neutrons/(cm$^2 \cdot$ s)). One could also speculate that, also for TCW2, the coupling of the incoming neutrons to the propagation modes in the Ti core of the first waveguide could be enhanced adequately (perhaps, through some lateral illumination): a discussion of this aspect lies outside our scope here. Anyway, one could entertain the possibility that the outgoing flux at the exit of the second waveguide in TCW2 could be useful, possibly, for BNCT of small tumors, where smaller fluxes could suffice, as discussed above.

3.6 CONCLUSIONS AND DISCUSSIONS

We have outlined various general properties of the behavior of slow neutron in material media. Based upon them, we have reviewed the confined propagation of slow neutrons along waveguides of small transverse cross section. We have reviewed previous theoretical proposals in thin films, in single fibers and in multi fibers. We have reviewed two important experiments employing thin films and polycapillary glass fibers.

Based upon those experiments and on some of those previous proposals, we have formulated, based upon [49] some new proposals regarding possibilities for improved focalization (at and a bit below 1 μm) and propagation along longer lengths (up to about 1 m) with relatively small attenuation, and discuss possible applications. Specifically, our proposals here are based upon earlier experiments by Feng et al. displaying slow neutrons confined propagation in thin films of short length (not exceeding 2 cm) [44] and extending previous theoretical proposals (in which Ti and Si play a key role) in three-dimensional fiber waveguides [19, 20]. We present some proposals which could enable, possibly, confined propagation along distances about 1 m. (possibly with relatively small attenuation) and their focalization at the micron scales and somewhat below. Our proposals, named TCW1 and TCW2, consider two suitably coupled waveguides. TCW1 is formulated in two dimensions and is based upon experiments with thin films: it describes confinement in x direction and propagation in z direction. Moreover, TCW2 is formulated in three dimensions and is based upon previous proposals for thin fibers: it describes confinement in x and y directions and propagation in z direction. In both TCW1 and TCW2, the first waveguide (with a length less than 1 cm) has Ti in its core, in order to ensure proper confinement, and the second waveguide has air in its core, so as to ensure small attenuation. The claddings in both could well contain Si. For an incoming slow neutron flux about 10^9 (neutrons/(cm$^2 \cdot$ s)) entering the first waveguide in TCW2, our estimates would suggest that the outgoing one (at the exit of the second waveguide) could possibly be not smaller than about 10^4 (neutrons/(cm$^2 \cdot$ s)). We entertain the possibility that the latter outgoing flux, even if small, could possibly be useful for BNCT of small tumors (with sizes of order about 1 mm^3 or smaller).

To conclude, we would like to emphasize the possible, a priori, interest of pursuing the exploration of shorter scales by means of slow neutrons (like in the CW3 proposal) and that, in the specific case of BNCT, improvements (whatever the device employed) in focalizing neutrons could enable to concentrate them onto small malignant tissues and to reduce their delivery to normal tissues. We emphasize the general interest and possible applications of exploring short scales by means of slow neutrons, through the devices considered here and suitable combinations thereof.

ACKNOWLEDGMENTS

R. F. Álvarez-Estrada acknowledges Mineco (Ministry of Economy and Competitiveness, Spain) for Projects FIS2015-65078-C2-1-P and PGC2018-094684-B-C21 (partially funded by FEDER), for financial support. We are also grateful to Mr. David Fernández for kind help with the files and Mrs. Belén Ramos for kind help with the figures.

REFERENCES

1. J. Byrne, *Neutrons, Nuclei, and Matter: An Exploration of the Physics of Slow Neutrons* (IoP Publishing, Bristol, 1995).
2. G. E. Bacon, *Neutron Diffraction.* 3rd ed. (Clarendon Press, Oxford, 1975).
3. V. F. Sears, *Neutron Optics* (Oxford University Press, Oxford, 1989).
4. D. F. R. Mildner, Neutron Optics, Chapter 36 in *Handbook of Optics*, Volume III, 2nd ed., The Optical Society of America (Editors M. Bass, J. M. Enoch, E. W. Van Stryland, and W. L. Wolfe) (McGraw -Hill, New York, 2001).
5. A. Erko, M. Idir, T. Krist, and A. G. Michette, Introduction, pp. 1–5 in *Modern Developments in X-Ray and Neutron Optics* (Editors A. Erko, M. Idir, T. Krist, and A. G. Michette) (Springer, Berlin, 2010).
6. F. Ott, Focusing Optics for Neutrons, pp. 113–135 in *Modern Developments in X-Ray and Neutron Optics* (Editors A. Erko, M. Idir, T. Krist, and A. G. Michette) (Springer, Berlin, 2010).
7. W. Sauerwein, R. Moss, and A. Wittig, *Research and Development in NCT* (Monduzzi Editore, Bologna, 2002).
8. W. Sauerwein, P. M. Bet, and A. Wittig, Drugs for BNCT: BSH and BPA, pp. 117–160 in *Neutron Capture Therapy. Principles and Applications* (Editors W. Sauerwein, A. Wittig, R. Moss, and Y. Nakagawa) (Springer, Berlin, 2012). See also: P. Torres-Sanchez, I. Porras, F. Arias de Saavedra, M. P. Sabariego, and J. Praena, On the upper limit for the energy of epithermal neutrons for Boron Neutron Capture Therapy. *Radiat. Phys. Chem.*, 156, 240–244 (2019).
9. Y. Kasesaz et al., BNCT project at Tehran Research Reactor: current and prospective plans. *Prog. Nucl. Energy*, 91, 107–115 (2016).
10. J. Finney and U. Steigenberger, Neutrons for the Future. *Phys. World*, 27–32, December (1997).
11. H. Maier-Leibnitz and T. Springer, The use of neutron optical devices on beam-hole experiments. *J. Nuclear Energy A/B17*, 217–225 (1963).
12. B. Alefeld, J. Crist, D. Kukla, R. Scherm, and W. Schmatz, Berichte der Kernforschungsanlage [Reports of the Nuclear Research Plant], Julich [July] Ju 1-194-NP (1965).
13. B. Jacrot, Instrumentation for neutron inelastic scattering research, p. 225, *Proc Symp. Instrumentation for Neutron Inelastic Scattering Research.* (IAEA, Vienna, 1970).
14. O. Schaerpf and D. Eichler, A neutron guide tube with circular cross section: principles of construction and measured properties. *J. Phys. E: Sci. Instrum.*, 6, 774–780 (1973).
15. D. Marx, Microguides for neutrons. *Nucl. Instr. and Meth.*, 94, 533–536 (1971).
16. M. Rossbach, O. Scharpf, W. Kaiser, W. Graf, A. Schirmer, W. Faber J. Duppich, and R. Zeisler, The use of supermirror neutron guides to enhance cold neutron fluence rates. *Nucl. Instr. and Meth. in Phys. Res. B*, 35, 181–190 (1988).
17. A. Schebetov, A. Kovalev, B. Peskov, N. Pleshanov, V. Pusenkov, P. Schubert-Bischoff, G. Shmelev, Z. Soroko, V. Syromyatnikov, V. Ulianov, and A. Zaitsev, Multi-channel neutron guides of PNPI: results of neutron and X-ray refractometry tests. *Nucl. Instr. and Meth.in Phys. Res. A*, 432, 214–226 (1999).

18. R. E. De Wames and S. K. Sinha, Possibility of guided-neutron-wave propagation in thin films. *Phys. Rev. B*, 7, 917–921 (1973).
19. R. F. Alvarez-Estrada and M. L. Calvo, Neutron fibers: a possible application of neutron optics. *J. Phys. D: Appl. Phys.*, 17, 475–502 (1984). Ibid: *J. Phys.*, 45 (Suppl.3), C3-243-C4-248 (1984).
20. M. L. Calvo and R. F. Alvarez-Estrada, Neutron fibers (II): some improving alternatives and analysis of bending losses, a possible application of neutron optics. *J. Phys. D: Appl. Phys.*, 19, 957–973 (1986).
21. A. W. Snyder and J. Love, *Optical Waveguide Theory* (Springer, Berlin,1983).
22. A. Messiah, *Quantum Mechanics*, (Vol.1), (North-Holland Publishing, Amsterdam, 1960).
23. M. A. Kumakhov and F. F. Komarov, Multiple reflection from surface X-ray optics. *Phys. Rep.*, 191, 289–350 (1990).
24. M. L. Calvo, Neutron fibers: a three-dimensional analysis of bending losses. *J. Phys. D: Appl. Phys.*, 33, 1666–1673 (2000).
25. M. Abramowitz and I. A. Stegun. Eds., *Handbook of Mathematical Functions* (Dover, New York, 1965).
26. R. F. Alvarez-Estrada and M. L. Calvo, Neutron waveguides and applications, pp. 331–385 in *Optical Waveguides. From Theory to Applied Technologies* (Editors M. L. Calvo and V. Lakshminarayanan) (CRC Press, Boca Raton, 2007).
27. B. Rohwedder, Interference effects in capillary neutron guides. *Phys. Rev A.*, 65, 043619-1 to 043619-7 (2002).
28. M. A. Kumakhov and V. A. Sharov, A neutron lens. *Nature*, 357, 390–391 (1992).
29. H. Chen, R. G. Downing, D. F. R. Mildner, W. M. Gibson, M. A. Kumakhov, I. Yu. Ponomarev, and M. V. Gubarev, Guiding and focusing neutron beams using capillary optics. *Nature*, 357, 391–393 (1992).
30. D. F. R. Mildner and H. Chen, The neutron transmission through a cylindrical guide tube. *J. Appl. Cryst.*, 27, 316–325 (1993).
31. D. F. R. Mildner, V. A. Sharov, and H. H. Chen-Mayer, Neutron transport through tapered channels. *J. Appl. Cryst.*, 30, 932–942 (1997).
32. D. F. R. Mildner, H. H. Chen-Mayer, and W. M. Gibson, Focusing neutrons with tapered capillary optics. *J. Appl. Phys.*, 92, 6911–6917 (2002).
33. W. M. Gibson et al., Polycapillary focusing optic for small-sample neutron crystallography. *J. Appl. Cryst.*, 35, 667–683 (2002).
34. W. M. Gibson et al., Convergent-beam neutron crystallography. *J. Appl. Cryst.*, 37, 778–785 (2004).
35. A. A. Bjeoumikhov, N. Langhoff, R. Wedell, V. I. Beloglasov, N. Lebed'ev, and N. Skibina, New generation of polycapillary lenses: manufacture and applications. *X-Ray Spectrometry*, 32, 172–178 (2003).
36. M. A. Kumakhov, Neutron capillary optics: status and perspectives. *Nucl. Instr. and Meth. in Phys. Res. A*, 529, 69–72 (2004).
37. R. F. Alvarez-Estrada, and M. L. Calvo, Neutron fibers: confined propagation and focusing of neutrons in the micron range and residual stress analysis. *Materials Science Forum (Scientific.Net, Materials Science and Engineering), Trans Tech Publications*, 772, 51–56 (2014).
38. C. Bergemann, H. Keymeulen, and J. F. van der Veen, Focusing X-ray beams to nanometer dimensions. *Phys. Rev. Lett.*, 91, 204801-1 to 204801-4 (2003).
39. S. Iijima, Helical microtubules of graphitic carbon. *Nature*, 354, 56–58 (1991).
40. M. S. Dresselhaus and H. Dai, Carbon nanotubes: continued innovations and challenges, pp. 237–239 in *Advances in Carbon Nanotubes* (Guest Editors M. S. Dresselhaus and H. Dai). Materials Research Society (MRS) Bulletin 29, Number 4, April (2004).

41. J. Liu, S. Fan, and H. Dai, Recent advances in methods of forming carbon nanotubes, pp. 244–250 in *Advances in Carbon nanotubes* (Guest Editors M. S. Dresselhaus and H. Dai). Materials Research Society (MRS) Bulletin 29, Number 4, April (2004).

42. W. A. de Heer, Nanotubes and the pursuit of applications, pp. 281–295 in *Advances in Carbon Nanotubes* (Guest Editors M. S. Dresselhaus and H. Dai). Materials Research Society (MRS) Bulletin 29, Number 4, April (2004).

43. G. F. Calvo and R. F. Alvarez-Estrada, Confined propagation of thermal neutrons using nanotubes. *Nanotechnology*, 15, 1870–1876 (2004).

44. Y. P. Feng, C. F. Marjkrzak, S. K. Sinha, D. G. Wiesler, H. Zhang, and H. W. Deckman, Direct observation of neutron-guided waves in a thin-film waveguide. *Phys. Rev. B*, 49(15), 10814–10817 (1994).

45. S. P. Pogossian, A. Menelle, H. Le Gall, J. M. Desvignes, and M. Artinian, Experimental observation of guided polarized neutrons in magnetic-thin-film waveguides. *Phys. Rev. B*, 53, 14359–14363 (1996).

46. A. Menelle, S. P. Pogossian, H. Le Gall, J. M. Desvignes, and J. Ben Youssef, Observation of magnetic films neutron waveguides. *Physica B*, 234–236, 510–512 (1997).

47. S. P. Pogossian, A. Menelle, H. Le Gall, J. Ben Youssef, and J. M. Desvignes, Observation of neutron guided waves from the open end of a thin film waveguide interferometry. *J. Appl. Phys.*, 83, 1159–1162 (1997).

48. S. P. Pogossian and H. Le Gall, Neutrons and X-ray guided wave optics in nanostructures and multilayers. *Recent Res. Devel. Optics*, 2, 597–618 (2002).

49. R. F. Alvarez-Estrada, I. Molina de la Peña, and M. L. Calvo, Focalizing slow neutrons beams at and below micrón ranges: discussion on BNCT. *Phosphorus Sulfur Silicon Relat. Elem.*, 193(2), 64–73 (2018).

50. I. Molina de la Peña, M. L. Calvo, and R. F. Alvarez-Estrada, Focalizing slow neutrons beams at and below micron ranges: discussion on BNCT (II). *Phosphorus Sulfur Silicon Relat. Elem.*, 194(10), 956–966 (2019).

51. R. F. Alvarez-Estrada and M. L. Calvo, Neutron fibers and possible applications to NCT. *Appl. Radiat. Isot.*, 61, 841–844 (2004).

52. T. Mitsumoto, K. Fujita, T. Ogasawara, H. Tsutsui, S. Yajima, A. Maruhashi, Y. Sakurai, and H. Tanaka, BNCT system using 30 Mev H-cyclotron pp. 430–432 in *Proc. of Cyclotrons and their Applications (International Conference CYCLOTRON 2010*, Editors You-Jin et al.) (Lanzhou, China, 2010).

Section II

Neutron Optics-Based
Technologies and Applications

4 Neutron Imaging

Markus Strobl

CONTENTS

4.1 NEUTRON IMAGING INTRODUCTION

Neutron imaging nowadays has developed from a pure non-destructive testing tool for industrial and technological applications into a distinguished research instrument [1, 2]. In particular, the introduction of digital neutron imaging detectors, nearly completely replacing analog image recording such as on film material today, has enabled outstanding progress. The dynamic range and digital data especially of CCD and CMOS cameras allows for superior data treatment and quantification as well as advanced methods adding a third dimension to the imaging data through time-resolved and 3D tomography studies. Today even 4D and more dimensions can be addressed in time-resolved tomography [3] and novel contrast modalities [4–6].

The main applications of neutron imaging today cover a wide range of fields and disciplines prominently including engineering materials science, energy devices and materials, industrial R&D, geology, biology/agriculture and cultural heritage to name but a few (Figure 4.1) [1–10]. Particular fields of study are often defined by

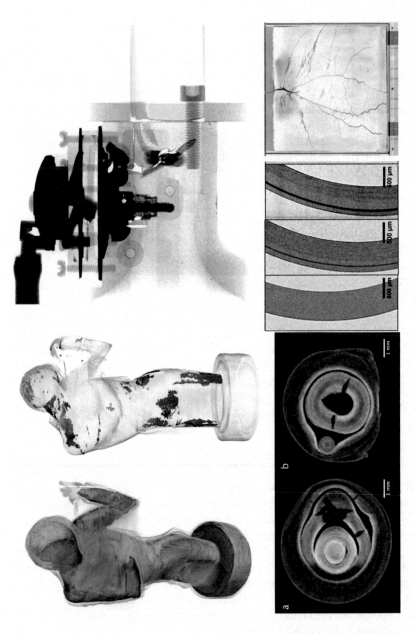

FIGURE 4.1 Five illustrative examples of neutron imaging: inner corrosion in a statuette with wooden core and metallic surface [7], liquid distribution in a carburetor (courtesy PSI, NIAG), the inner structure of a pearl [8], hydrogen uptake in a nuclear fuel gladding [9], and root behavior in soil [10].

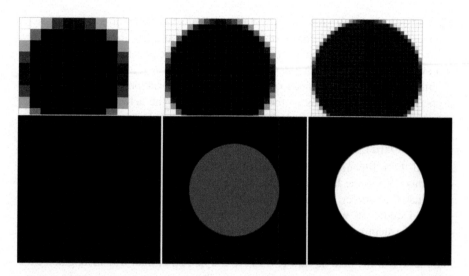

FIGURE 4.2 Spatial resolution (top) and contrast (bottom) as basic requirements for any kind of imaging.

the complementarity to X-ray imaging, and neutrons are predominantly used where X-rays cannot penetrate, like bulk metal materials and components and/or where X-rays struggle to provide contrast. The latter involves a large number of studies of water content and water content evolution, e.g., in porous media or devices such as fuel cells; hydrogen, especially in metals or particular hydrogen storage applications, etc.; other hydrogen containing materials and liquids; and lately with respect to energy research also batteries.

The fundamental requirements for imaging are principally twofold (Figure 4.2) and not different for neutrons than for other imaging modalities and can be expressed as (i) spatial resolution and (ii) contrast.

Both imply requirements to neutron optics and neutron optical devices. While this might appear obvious for the spatial resolution requirement, it will be seen that in fact the requirements of the latter are even more and more complex.

4.1.1 Spatial Resolution

The ability to achieve spatial resolution has two major aspects. First, the beam geometry and neutron optics, and second the spatial resolution capabilities of the detector system utilized. These will be discussed separately below. However, let us first consider resolution more generally and in the context of dimensions. The most basic image resolution is spatial resolution in two dimensions of the projection of an object collapsed from 3D to 2D. Assuming an object not stable over time, it is even a reduction from 4D to 2D. However, imaging has developed to nowadays enable 3D studies with neutrons, as tomography, where a volume can be reconstructed from a number of 2D projection images at different angles, and as time-resolved radiography, where the temporal evolution of a projection at a fixed angle is observed. Recently, even full

4D studies have been reported, providing access to the observations of processes, changes in materials and compositions in 3D over time [3]. Resolutions of a few seconds paired with 3D spatial resolutions of the order of a few 100 μm could be achieved with neutrons to date in such 4D studies [11].

Other novel methods enable to observe multiple material properties or non-scalar properties, thus also adding to the dimensionality of the achieved images, which hence go well beyond 4D already today [12, 13].

4.1.1.1 Spatial Resolution: Imaging Detectors

Around the millennium, neutron-sensitive film methods were replaced by digital detectors. While the latter in the beginning provided worse spatial resolution, the advantages of direct digital data handling for analyses, tomography, and time-resolved studies as well as their superior dynamic range implied a seminal development for neutron imaging [14]. Modern digital detectors are able to outperform film also in spatial resolution capabilities, though with digital detectors resolution and the size of the field of view (FOV) are coupled other than with analog detection techniques. Apart from some amorphous Si flat panel detectors, detector systems combining neutron-sensitive scintillators, visible light optics, and digital cameras are dominating in state-of-the-art neutron imaging applications. A huge advantage of these detectors, which we refer to as scintillator/camera detectors, is their flexibility in tailoring the detector parameters to the needs of a measurement. Nowadays FOV dimensions range from a few millimeters to several tenths of meters [15, 16] and utilized resolutions from a few micrometers [17, 18] to the millimeter range (Figure 4.3). Special scintillators are tuned to either high resolution or highest efficiency for best spatial or temporal resolution respectively. Most spread scintillators for cold and thermal neutrons are LiF/ZnS or GADOX (gadolinium oxysulfide) and thicknesses range from a few hundreds to a few micrometers. Those scintillators are easily exchangeable in the course of an experiment and cover the beam facing side of a light tight box, in which the scintillator light is directed via a mirror to the light optics of a digital camera system, operating outside the direct neutron beam direction. Cameras utilized are mainly scientific CCD and CMOS cameras, which are cooled in order to keep electronic noise low. The light optics utilized are commercial products and differ in sophistication depending on the resolution requirements. The best currently routinely achieved resolution is 4 μm and entails the use of a custom made Gd-157 enriched GADOX scintillator [18]. Better resolutions down to 1 μm have so far only been reported in proof of principle experiments beyond practical application.

Figure 4.3 represents on the one hand the most common detector solutions utilized in neutron imaging today and puts them in a context of spatial and temporal resolutions provided. The bottom panel of Figure 4.3 on the other hand illustrates a state-of-the-art detector suite with respect to resolution and FOV, exemplified by the detectors available at the neutron imaging facilities of the Neutron Imaging and Applied Materials Group (NIAG) at the Paul Scherrer Institut (PSI) in Switzerland.

Currently there is a significant effort ongoing for novel neutron imaging detectors that would be capable of time-of-flight (ToF) imaging at modern pulsed spallation neutron sources. The most prominent example of these developments is a

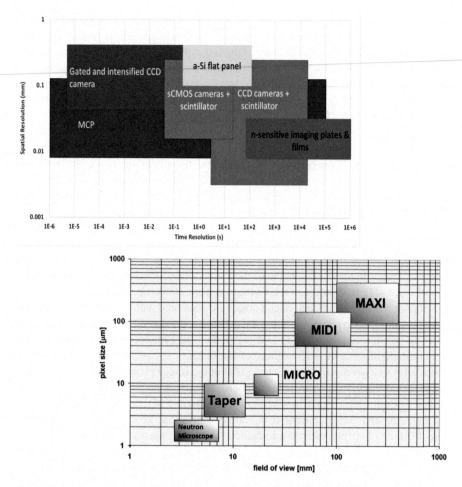

FIGURE 4.3 Top: Common state-of-the-art detector solutions for neutron imaging. Bottom: State-of-the-art detector suite exemplified at the neutron imaging instruments at the Paul Scherrer Institut (PSI) in Switzerland.

detector combining B-10 doped micro-channel plates (MCP) with medipix/timepix chips [19]. In contrast to scintillator/camera detectors, such detectors provide sub-millisecond time resolution down to microseconds and below. Often this comes at the expense of spatial resolution and limitations in FOV and/or count rate capabilities. In the specific case of the so-called MCP detectors, the availability of the new generation of timepix3 chips is expected to overcome some limitations in resolution and count rate limitations. Also, the current limit in FOV to about 28×28 mm^2 is meant to be overcome at least in one dimension. The high time resolution enabled by such detectors does not only enable wavelength resolved image detection for ToF but can also, e.g., increase the efficiency of formerly stroboscopically measured repetitive processes, where time resolutions can reach microseconds due to the accumulation of exposures of coinciding process phases [20, 21]. In contrast to earlier approaches, the fast detectors enable to measure all phases simultaneously, i.e., separately within each period, for extended exposure times of numerous process periods, instead of the conventional single phase for each extended exposure. This enables also complex multi-phase exposures like demonstrated, e.g., in Ref. [18]. Another advantage of such new technologies of counting detectors for imaging applications is basically an elimination of read-out noise. Therefore, the detectors are also well suited for low count rate applications [12].

Finally films or imaging plates are still in use for some limited applications. In particular, films are sometimes preferred by industry in certified inspection procedures. Also, they still provide for 2D inspections the advantage of a combination of high resolution on very large FOVs, where digital detectors are limited by the number of pixels available. However, ever larger pixel arrays become available with 2k × 2k being standard nowadays and 4k × 4k being available already.

4.1.1.2 Spatial Resolution: Beam Geometry

Due to the absence of efficient lenses for neutrons, the principle of neutron imaging today still rests on the basic scheme of the Camera Obscura utilizing a pinhole geometry (Figure 4.4). In contrast to the camera obscura, however, neutrons provide transmission images and hence source and object are separated with the object moved to the detector side. In such setup we define the collimation ratio L/D where L is the distance between pinhole and sample and D is the pinhole size (diameter of a round pinhole). This ratio defines the local divergence of the beam in any point of the sample, which contributes to a geometric blur of such point in the image plane of $d = l/(L/D)$ according to the collimation and the distance l between sample and detector.

For optimum efficiency this geometric resolution will match the resolution of the detector and those combined match the resolution need of the measurement. The distance l between sample and detector, however, has to be minimized in order to enable best efficiency, as this is the only parameter that influences the resolution without impacting the available flux density. An excessive sample to detector distance will increase the requirement on L/D, where an increase implies a loss in flux density by the square of the collimation improvement, which is due to the two-dimensionality of the geometry. This applies equally to an increase in distance L and

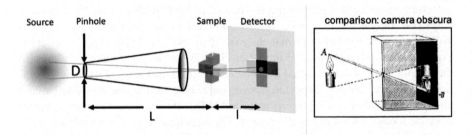

FIGURE 4.4 Schematic of the pinhole imaging geometry of neutron imaging (left) and the analogy to the camera obscura principle (right); (round dots in image planes denote geometrical image blur).

decrease of pinhole D. In order to adapt the geometric resolution to the requirements of measurements and detectors, a pinhole exchanger is standard at state-of-the-art installations for imaging and typically several measurement positions are available along the flight path of the beam downstream of the pinhole.

Theoretically for a fixed sample to detector distance an increase of the resolution by a factor of 2 implies a loss of flux by a factor of 4. In addition, the decreased pixel size implies another factor 4 in flux deficiency for achieving comparable statistics. Together, this is an impact of a factor of 16, i.e., more than an order of magnitude on the exposure time, which does not yet take into account the potential need for a less efficient higher resolution scintillator, etc. This implies that improved flux conditions are limited in their impact on improved resolution conditions, despite their key role. In the field of temporal resolution, however, increased brightness translates one to one into improved resolution, when not considering technological limitations.

Typical pinhole sizes D range from a few millimeters to a few centimeters and distances L are several meters. Consequently, L/D ratios range from around 100 up to several 1000s, where the latter are mainly used for phase effects, rather than real space image resolution. Due to the fact that spatial resolution scales with the size of the FOV for digital detectors, samples for high resolution are typically small and can be placed accordingly close to the detector and vice versa.

However, the large range of resolutions and sizes of FOV also implies that conventional imaging instruments are optimized for a large range of beam sizes, implying that homogeneous beams with sizes up to several $100 \, cm^2$ have to be considered and require a large divergence of the beam at the pinhole, with a homogeneous distribution. In order to avoid excessive illumination, the beams are cut down to the required beam size of a specific measurement by multiple beam scrapers and variable slits.

So far, only limited tests of alternative optical setups involving focusing neutron optics have been explored, which shall be discussed in a section dealing with alternative optics solutions.

4.1.2 CONTRAST

Contrast is generally provided by the measureable interaction of the object with the beam. An image forms when these interactions vary spatially. The most utilized and conventional interaction is the attenuation of the beam. Advanced neutron imaging techniques utilize a number of interactions to achieve different contrast and information about the samples studied. However, let us start with attenuation contrast, which will also establish the motivations for using neutrons for imaging.

Here we shall also note, that the focus of this chapter is on imaging with thermal and cold neutrons, even though also fast and epithermal neutrons are partially used for imaging. These can provide yet additional contrast mechanisms such as resonance absorption that provide, e.g., chemical sensitivity.

4.1.2.1 Attenuation Contrast

While attenuation contrast imaging is best known in the context of X-rays, e.g., through medical imaging, neutrons provide some unique features, which indicate the potential of their complementary use for imaging in science and non-destructive testing [1, 2]. The attenuation of neutron beams as utilized for imaging is based on the absorption and scattering of the beam, where the latter has to be such as to remove neutrons from the transmitted beam, i.e., by scattering to significant angles away from the forward direction. The attenuation can be described by the Beer Lambert law which for imaging reads

$$I(x,z) = I_0(x,z)e^{-\int \mu(x,y,z)dy} \tag{4.1}$$

and where the linear attenuation coefficient $\mu = N\sigma_{tot} = N(\sigma_a + \sigma_s)$ and correspondingly is based on the total microscopic cross section σ_{tot}, which is the sum of the microscopic absorption σ_a and scattering cross section σ_s. With N being the particle density the linear attenuation coefficient has the unit of a length, generally given in cm^{-1}. The linear attenuation coefficient $\mu(x,y,z)$ here represents the sample and is projected along the beam direction y (compare integral) to form an image $I(x,z)$ across the incident neutron beam cross section $I_0(x,z)$.

Accordingly, the image contrast is based on the linear attenuation coefficient, constituted by density and microscopic cross sections as well as macroscopic material thickness. While this is equally valid for X-rays, the cross sections of different materials differ significantly for neutrons and X-rays. The reason is the obvious difference of interactions, where neutrons are subject to strong interaction with the nucleus while X-rays interact via the electromagnetic force with the charge mainly of the electron cloud. From these differences some unique features of the neutron can be established for imaging, such as thermal neutron cross sections of many metallic structural materials are smaller than those for X-rays (with energies for comparable imaging, i.e., around 150 keV) enabling the neutrons to probe relevant thicknesses of centimeters. In contrast, neutrons are highly sensitive, meaning displaying a large cross section, for some important light elements, which appear transparent to X-rays, in particular hydrogen. This enables them to detect small local quantities of hydrogen and hydrogenous materials such as, especially also, water and organic constituents in bulk matrices of structural or other materials (Figure 4.5). As the strong interaction does not feature a systematic relation with the atomic number, as is the case of X-rays due to the increasing number of electrons, neutrons are often more sensitive providing contrast between elements of similar atomic number. Finally, the strong interaction with the core makes neutrons uniquely sensitive to different isotopes, which also enables contrast variation for imaging when required.

Radiographic attenuation contrast images in general represent the transmission as

$$T(x,z) = \frac{I(x,z)}{I_0(x,z)} = e^{-\int \mu(x,y,z)\partial y}, \tag{4.2}$$

while tomographic reconstructions of cross sections or volumes are represented in terms of the linear attenuation coefficient $\mu(x,y,z)$ $[cm^{-1}]$. Conventionally,

FIGURE 4.5 Examples of high penetration power (top, examples from TUM and GKSS) and additional sensitivity to hydrogenous materials in contrast to X-rays (bottom, X-ray left sides, examples from PSI).

attenuation contrast images with neutrons are produced utilizing a wide thermal neutron spectrum ΔE as available from the thermal and cold moderators.

As the cross sections of materials are indeed energy dependent, the measured linear attenuation coefficients are more correctly and completely characterized as $\mu(x,y,z,\Delta E)$ in which the actual spectral distribution of intensity in ΔE has to be considered. Because in some cases the variations of the attenuation can display significant variations, beam hardening due to early depletion of strongly attenuated energy ranges can occur upon transmission. In contrast to the meaning in X-rays, beam hardening does not necessarily imply a spectral shift toward higher energies. Such artifacts can be corrected partially or they can be avoided by the use of monochromatic neutrons, however, with the corresponding penalty in available flux densities. Nevertheless, a considerate choice of energy, for advanced energy dispersive techniques rather referred to by wavelength, can on the other hand improve overall transmission.

Significant and highly systematic variations in the cross section in the thermal energy range can be attributed in particular to the coherent elastic scattering cross section in crystalline materials. These can be correlated to the Bragg scattering from the crystal lattice planes and give rise to a number of exploitations in wavelength-resolved attenuation contrast neutron imaging referred to as Bragg edge imaging or more general as diffraction contrast imaging [4].

4.1.2.2 Diffraction Contrast

Diffraction contrast refers to the utilization of contrast variations appearing in the wavelength-resolved transmission spectrum in neutron imaging [4]. This, however, requires the addition of wavelength-resolving capabilities to neutron imaging and implies severely reduced neutron flux densities. In return, it provides access to spatially resolved crystal lattice information in neutron imaging. This is due to the fact that for many relevant materials, in particular structural engineering materials, the scattering cross section is dominant over the absorption cross section in the contribution to the attenuation coefficient and in addition the coherent cross section is most significant. This means, as can be seen in Figure 4.6, that Bragg scattering characterizes the features of the attenuation spectrum. For powder-like crystals with randomly oriented crystal grains, characteristic transmission patterns containing so-called Bragg edges can be extracted for each pixel of multiple wavelength dispersive radiographic images containing detailed crystal structure information. The Bragg edges in particular are found at wavelengths matching the crystal lattice distances d_{hkl} for specific hkl lattice plane families. The reason being that due to the Bragg equation the longest wavelength that can be diffracted at a crystal lattice distance d_{hkl} is diffracted at θ being 90 degrees where

$$\lambda = 2d_{hkl}\sin\theta = 2d_{hkl}. \tag{4.3}$$

Because beyond this wavelength no diffraction can take place at this lattice plane family anymore, the cross section displays a discontinuity toward a significantly lower cross section dominated by larger d values for different lattice families, or, beyond the largest d of a specific crystal lattice by absorption only, assuming that other scattering effects, like inelastic and incoherent scattering are negligible, as is in general the case for relevant materials (compare Figure 4.6).

The linear attenuation is now described in full wavelength dependence based on the elastic coherent cross section of a crystalline material, which can in the simplest form for an isotropic sample be written as

$$\sigma_{\text{coh}}^{\text{el}}(\lambda) = \frac{\lambda^2}{4V_0} \sum_{2d_{hkl}>\lambda}^{2d_{\max}} |F_{hkl}|^2 d_{hkl}. \tag{4.4}$$

(Note that in recent literature the summation has wrongly been displayed mostly over the complementary part of lattice distance parameter, due to a single original mistake in an early publication.) Corresponding to the wavelength-dependent diffraction contribution to the attenuation, for such crystalline systems, e.g., local phase distributions and transformations of phases can be measured through the height of specific Bragg

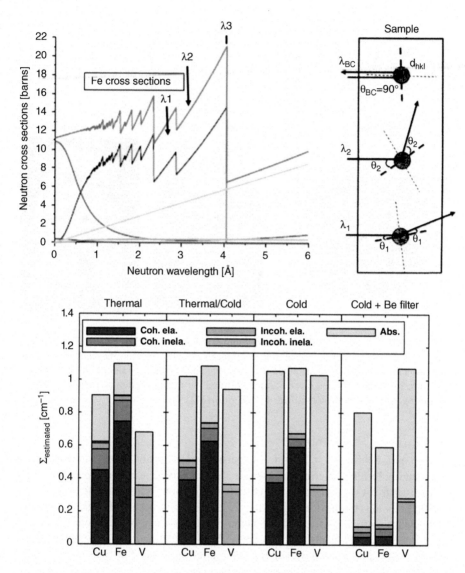

FIGURE 4.6 A schematic illustrating the origin of Bragg edges from the transmission of isotropic powder-like polycrystalline materials (left) and a comparison of the different contributions to the attenuation by such materials for different utilized neutron spectra [22].

edges (Figure 4.7 left). Lattice parameters and strains along the beam direction can be deduced from exact pixel-wise evaluation of Bragg edge positions [4].

When the crystal grains of a material reach a size in which they can be resolved spatially, then the grain morphology becomes visible and can be reconstructed in 3D including the indexing of the individual crystal grains based on corresponding tomography data [13] (Figure 4.7 right).

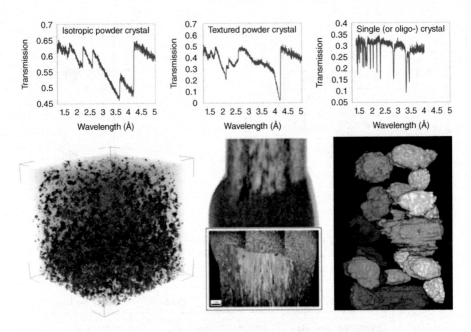

FIGURE 4.7 On top, typical spectra depending on basic crystalline features [4], the bottom line illustrates some corresponding examples of applications: a 3D reconstruction of martensite/austenite phase distribution in an isotropic polycrystalline sample, a visualization of texture in a welded rolled metal sample [23] and the reconstruction of individual grains composing an iron sample consisting of large grains [13].

4.1.2.3 Dark-Field Contrast

Neutron dark-field contrast [24] refers in general to the capability to visualize in neutron imaging locally differing small angle scattering behavior of a sample due to its microstructure, which is beyond direct spatial resolution. While this can be achieved in different ways, the most efficient and widespread approach is the use of spatially modulated beams. In particular, the beam modulation can be realized in conventional imaging instruments. For a beam with a well-defined spatial modulation, in general in one dimension only, small angle scattering induced by the sample locally will reduce the modulation contrast, i.e., the visibility $V = (I_{max} - I_{min})/(I_{max} + I_{min})$, locally in the image. The loss of visibility can be related to the small angle scattering function $S(q)$, where q is the modulus of the scattering vector, by [25]

$$\frac{V_s(\xi_{GI})}{V_0(\xi_{GI})} = \int_{-\infty}^{\infty} dq\, S(q)\cos(\xi_{GI}q) = G(\xi_{GI}), \qquad (4.5)$$

and hence represents the Fourier back transformation of the scattering function into real space, in which ξ being the real space parameter of the correlation length and G is the projected real space correlation function of the scattering structure containing the corresponding structural information [5, 25]. Taking into account multiple scattering, in principle a requirement for this approach because the scattering is here measured superimposed to the direct transmitted beam, the relative visibility in modulated beam dark-field contrast imaging reads [25]

$$\frac{V_s\left(\xi_{GI}\right)}{V_0\left(\xi_{GI}\right)} = e^{\Sigma t(G(\xi_{GI})-1)}. \tag{4.6}$$

It features analogy to the Beer-Lambert law and is hence suited for equivalent tomographic reconstruction processes giving access in principle to small angle scattering with 3D spatial resolution. Σ represents here the total macroscopic small angle scattering cross section and t the thickness, here representing the integral along the path through the sample. The probed correlation length parameter ξ is expressed through [25]

$$\xi = L_s\lambda/p \tag{4.7}$$

where L_s is the sample to detector distance (note in some setups the definition has to be adapted) and p being the modulation period. From this it is obvious that for quantitative local small angle scattering analyses, the relative visibility has to be probed as a function of ξ [25, 26]. This can again be achieved by a wavelength dispersive approach or through the variation of the geometric parameters L_s and/or p. Different approaches can be realized depending on specific setups and realizations of modulated beam imaging experiments.

Different approaches have been explored for this relatively recent technique [24, 27, 28]. First of all, Talbot Lau grating interferometers are used with different geometries and grating periods. Modulation of the beam through spin-echo has been explored with polarized neutrons and magnetic field regions; so-called far field interferometers utilizing Moiré fringes have been investigated. Finally, also simple absorbing structures projected on the sample and detector can be utilized.

The quantitative approach has been applied so far to a number of reference samples, but also to observe rheological processes and structural evolutions, e.g., in cohesive powder under pressure. Qualitative measurements have been successful in visualizing structural flaws in engineering materials, but in particular of magnetic domains in bulk materials (Figure 4.8).

4.1.2.4 Phase Contrast

Phase contrast with neutrons has been prominently communicated and has been realized through very different approaches either as direct phase contrast [29], differential phase contrast [30, 31], or propagation-based inline phase contrast [32]. The first case can only be tackled by split beam perfect crystal interferometers, not providing flux conditions required for routine measurements. Differential phase contrast has faced similar issues when using perfect crystal double crystal diffractometers, but became an efficient approach with the introduction of the Talbot Lau grating interferometers. Finally, propagation-based phase contrast refers to the intensity variations that can be

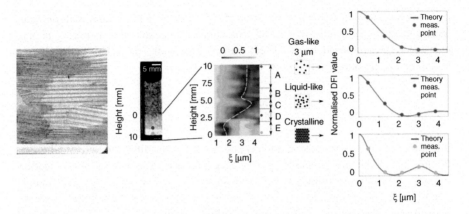

FIGURE 4.8 Typical examples of dark-field contrast neutron imaging, a visualization of magnetic domains in a transformer steel as well as a quantitative analyses of quasi crystalline particle sedimentation [21, 26].

detected in well collimated beams at a certain distance from the sample as described by transport of intensity equations accounting for phase effects. In general, it can be summarized that neutron phase contrast detects the refraction contributions of neutrons at interfaces, which very often lead to image artifacts and in contrast to X-rays have not lead to important application cases. In particular the efficient measurements with gratings enable quantitative reconstructions of the refractive index distribution of samples rather than the attenuation coefficient in conventional imaging. While for X-rays that means that low contrast and biological tissue can be resolved, in the context of neutron attenuation contrast behavior this feature appears of minor relevance and applicability.

However, it should be considered that in particular the dark-field contrast enabling the visualization of magnetic domains is in fact differential phase contrast affecting the two spin states oppositely, and therefore rather manifests as dark-field contrast than as differential phase contrast.

4.1.2.5 Polarization Contrast

The magnetic moment of neutrons implies their sensitivity to magnetic fields and structures based on corresponding interactions. The investigations of magnetic phenomena and structures in bulk materials are thus one of the main applications in neutron science and in particular neutron scattering. Therefore, this interaction can also be exploited in neutron imaging for spatially resolved studies in the macroscopic domain [6]. Contrary to examples with respect to dark-field contrast, as outlined above [21], polarization contrast imaging uses a polarized neutron beam coupled with polarization analyses. This way, local depolarization and spin precession are measured in order to observe and quantify magnetic fields and structures. This technique of polarized neutron imaging builds on the basic principle of Larmor precession of the neutron polarization vector in the presence of a magnetic field.

In a magnetic field described by the vector $\boldsymbol{B}$, the polarization vector $\boldsymbol{P}$ precesses around the field vector according to

$$\frac{d\boldsymbol{P}}{dt} = \gamma \boldsymbol{P} \times \boldsymbol{B}. \tag{4.8}$$

The accumulated spin phase φ of the polarization vector of a polarized neutron beam traversing a magnetic field B can be written by the path integral:

$$\varphi = \omega_L t = \frac{\gamma}{v} \int_{path} B ds = \frac{\gamma m \lambda}{h} \int_{path} B ds \tag{4.9}$$

with $\gamma = -4\pi\mu/h = 1.832 \ 10^8$ rad/s/T the gyromagnetic factor of the neutron, and h the Planck constant. v represents the velocity of the neutron with the wavelength λ and m is its mass. Measurements of such polarization vector rotation can be realized with a polarized incident beam and a polarization analyzer subsequent to the sample in front of the detector. When assuming the simplest case of a sample with a magnetic field-oriented perpendicular to the polarization vector of the incident beam the resulting image superimposing attenuation by the sample and transmission of the polarization analyzer is described by

$$I(x,y) = I_0(x,y)\exp\left(-\int\sum(x,y,s)ds\right)(1+\cos\varphi(x,y))/2. \tag{4.10}$$

Normalizing by the attenuation yields

$$I(x,y) = I_0(x,y)(1+\cos\varphi)/2. \tag{4.11}$$

and, thus, enables a visualization of the field, however, with an intrinsic modulus of 2π in φ. Therefore, a quantification is possible only for well-designed experiments, mostly simple cases offering sufficient a priori knowledge or with significant efforts of field modeling and signal modeling and simulation. Nevertheless, the visualization and detection of magnetic field variations in particular in the bulk of materials, where it is not accessible with other techniques, have shown valuable insights. In particular, simple implementations, not striving for field reconstruction but detecting, e.g., the paramagnetic to ferromagnetic phase transition at the Curie temperature by simply probing depolarization of the beam, turned out to be of outstanding importance in numerous applications. In addition, significantly, more sophisticated approaches have been developed in the meantime and the 3D reconstruction of magnetic vector fields has become feasible. Figure 4.9 illustrates some examples of different sophistication applied in different contexts.

4.2 BEAM REQUIREMENTS FOR NEUTRON IMAGING

The main beam requirement for neutron imaging might simply be expressed as a well characterized homogeneous beam of sufficient collimation and cross section. However, such simple statement contains a lot, in particular, when in practice it has

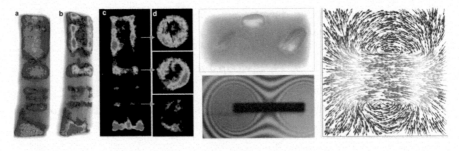

FIGURE 4.9 Left: Curie temperature mapping through depolarization imaging, providing 3D maps of the already ferromagnetic parts of a sample at a certain temperature [6]; mid top: 3D reconstruction of magnetic field areas trapped in a lead superconductor close to the critical temperature [33]; mid bottom: straightforward polarized neutron radiography of the field of a dipole magnet [34] and right-hand side: a result of the first 3D magnetic field reconstruction from a polarimetric neutron tomography (color indicates field strength) [12].

to be applied in a single instrument for sample sizes ranging from half meters to half centimeters and for spatial resolutions of fractions of a millimeter to a few micrometer as well as considered time resolutions, spectral resolution, etc. Also we will see that both, spatial resolution and contrast modality chosen have a strong influence on neutron optical requirements and devices utilized.

Based on the fact that neutron beam ports are limited and a large variety of techniques probing various condensed matter characteristics and phenomena require access, neutron imaging instruments generally have to be able to provide flexibility to serve a large user community with very different requirements. The first requirement shaping a neutron imaging instrument more than any other is the inspection of large scale objects. This is an inherent requirement due to the outstanding transmission characteristics for neutrons, which allow large objects to be inspected, yet with high sensitivity for certain materials and features. Standard maximum beam size requirements for stand-alone neutron imaging instruments range around 30×30 cm^2 (Figure 4.10) [35]. This is ± 10 cm each direction. What can be achieved strongly depends on source parameters and geometrical boundary conditions including beam extraction system

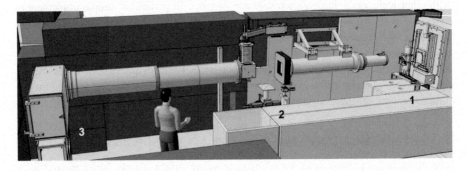

FIGURE 4.10 Typical end station of a conventional imaging instrument illustrated by one of the first state-of-the-art installations NEUTRA at Paul Scherrer Institut [35].

and space. The beam area shall be well defined and fast neutron flux as well as intrinsic gamma background should be kept as low as reasonably possible. The illuminated area, however, has also to be flexible, i.e., smaller beam sizes shall be easily achievable in order to limit irradiation and background. Together, this requires a beam with a large and smooth, ideally flat, divergence at the pinhole position of the instrument.

Within this large beam area, deviations of the beam characteristics shall be at a minimum. This includes beam collimation, spectrum, and flux. Spectral deviations lead to different attenuation cross sections of the same material in different areas of the beam and hinder correct correlation and quantification. In most cases a correction is not possible due to generally unknown content and homogeneity of a sample. Spatial flux variations on the other hand are corrected by normalization with the incident beam. However, this correction is limited in cases of abrupt changes, which can cause artifacts. Deviations toward low flux can lead to, e.g., premature beam starvation and issues with counting statistics, which cannot be corrected either. Very often, the beam size is considered limited at 50% peak flux, however, a recommended definition at about 80% is certainly more useful in this respect.

For flexible but efficient application the collimation ratio of the incident beam needs to be tunable. This implies the possibility in a pinhole setup for variation of the distance from the pinhole, which however also impacts the available beam size, or a variation of the pinhole size. Also asymmetric configurations can be considered, which allow larger collimation in one than in the other direction. L/D ratios from about 100 to 1,000 are most relevant, but also ratios as small as 50 and as large as 10,000 are considered for some exceptional cases. This translates into a local divergence between 1.1 and 5×10^{-3} degrees, while in the pinhole divergence values around 2 degrees are favorable.

For advanced wavelength-resolved techniques corresponding variable monochromatic beams or ToF encoded wavelength resolution imply additional requirements on variable wavelength resolutions, depending on the specific application and technique. Necessary resolutions cover the whole spectrum of requirements of scattering instruments from 10% down to sub-percent. However, in particular here the demands for still substantial beam sizes and homogenous spectral conditions over the FOV constitute high and unparalleled exigencies. On the other hand, requirements about cross talk, contamination with secondary wavelengths are often more relaxed, due to the direct transmitted beam measurements and their analyses.

Methods such as grating interferometry and polarized neutron imaging in addition require neutron optical devices such as absorption and phase grating with partially micrometer precision as well as neutron spin filter and spin manipulation components, respectively. In particular for the polarization devices such as for the monochromatization the difficulty and distinction of requirements for imaging in contrast to scattering techniques with regards to such optical devices lies in the fact that in imaging the local variations are resolved, while in scattering only the overall average contributes and local variations, e.g., of wavelength or polarization remain undetected and without relevance.

Accordingly, state-of-the-art neutron imaging instruments [35–43] are highly flexible installations with batteries of optional neutron optical devices to add on or remove from the basic optical setup and geometry (Figure 4.11).

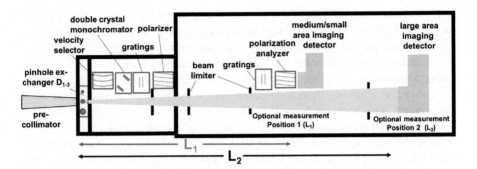

FIGURE 4.11 Schematic of the layout of an imaging beamline with basic installations beyond the pinhole, including various options of optical devices to be added or removed regularly.

4.3 OPTICS SOLUTIONS FOR NEUTRON IMAGING

The main distinction of different cases of basic optics for neutron imaging is first the decision whether the instrument features a neutron guide or not. Other main characteristics are the source geometry and time structure, i.e., a pulsed or continuous source. Neutron imaging instruments at pulsed sources are a relatively new development with the first dedicated beamlines coming online only within the current decade.

4.3.1 BASIC OPTICS SOLUTIONS

The initial decision for or against a beam guide is in general driven by the available moderator and space for an imaging beamline and the default option working well for both thermal and cold neutrons is the direct beam solution without a neutron guide.

4.3.1.1 Direct Beam Geometry

The direct beam geometry avoids neutron guide optics and thus directly views the neutron source, i.e., a thermal or cold moderator. The main choices to be made with respect to optics are the location of the collimating pinhole and the length of the instrument. In general, utilized moderators have an extended emitting surface of the order of 10×10 cm^2. The length of the instrument is often constraint by available space and shielding needs, but also by the implications of the geometry of the direct beam extraction. When, e.g., assuming a fixed maximum length L from pinhole to sample and detector position, the impact of the choice for the distance of the pinhole from the source becomes obvious as depicted in Figure 4.12. The distance of the pinhole from the source l_0 defines together with the source size the maximum beam size that can be achieved for a given distance L [44]. At the same time this distance l_0 determines the source area seen from a point in the sample. Outside the central fully illuminated area the intensity drops slowly in the penumbra region which is only relatively larger for larger l_0. The latter however creates a smaller homogenously illuminated central area useful for imaging with the same efficiency. For smaller target sizes of the FOV for imaging a longer distance l_0 can hence be chosen providing the same flux density. The larger source size seen is balanced by the longer distance in terms of available flux density. Inhomogeneities of the moderator source distribution

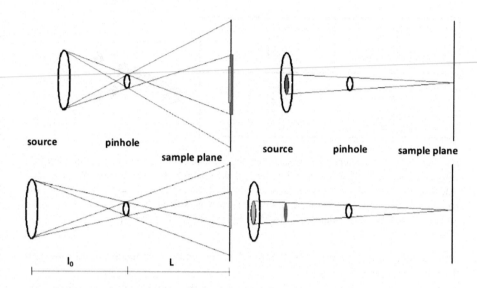

FIGURE 4.12 Illustration of basic considerations for a direct imaging geometry from source to detector [44].

are imaged on the detector through the pinhole. A smaller source size requires a smaller l_0 and/or a longer distance L to achieve equivalent beam sizes.

In reality, the smallest source to pinhole distance is mostly limited by the biological shield of the source and corresponding limitations for moving parts such as a pinhole exchanger in this area. Therefore, sometimes solutions with a double pinhole geometry are realized. An inner fixed pinhole in the biological shield would be fixed for a minimum L/D and maximum beam size. It is followed downstream, outside the inner monolith by a pinhole selector wheel, enabling the choice of a number of variable effective pinhole sizes (sometimes also varying geometries). Sometimes this selector is combined with the function of a secondary shutter, with a position without pinhole in the selector wheel.

A main concern then remains the penumbra, unutilized flux transported into the instrument. Therefore, the pinhole is ideally not only a local slit but consists of extended collimators [45]. This also improves the general background situation in the instrument. Care has to be taken of sufficient attenuation at the narrowest part of such device which is therefore best amended with a rig of higher absorbing material than the channels, which are of borated steel or similar material. The pinhole defining ring consists rather of Gadolinium or Cadmium. Figure 4.13 illustrates some possible designs of collimator pinholes.

The term "beam adjusted collimator" [45] refers to collimators which do not display round cross sections throughout, but account for the change of the beam geometry from a rectangular source through a round pinhole onto a generally square FOV. Under the assumption that the pinhole should be round for isotropic resolution despite the squared detector and image pixels, this leads to a further optimization of transporting only utilized beam through the instrument to the sample and detector position. However, such optimization might have limited effect with respect to

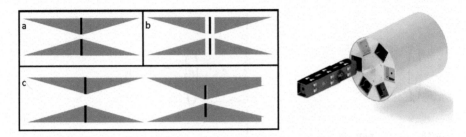

FIGURE 4.13 Some schematic models for direct beam geometry pinhole collimation [45] with (a) a compact pinhole collimator with implemented high absorption edge, (b) a separated realization of the individual parts, and (c) a typical double pinhole with an exchangeable second part for increased collimation; on the right-hand side is the realization of a pinhole exchanger from the instrument ANTARES at FRM2 at TU Munich [39].

significantly smaller beam sizes than the designed maximum used regularly in such flexible instrument.

As such instruments have a direct view into the active area of a source they receive also a significant radiation background from non-moderated neutrons and gamma radiation streaming through the pinhole. This can affect the image quality and contaminates recorded images with a large number of white spots. Filters like 1-cm thick layers of sapphire, lead, and beryllium are implemented and can often be added and removed remotely.

Some beam profiles of a typical instrument (ICON at Paul Scherrer Institut, Figure 4.14 [36]) are shown in Figure 4.15. These illustrate the potential of the direct

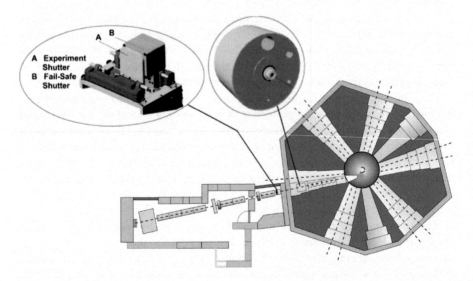

FIGURE 4.14 Example of the basic layout of a direct geometry instrument: ICON at SINQ source of PSI [36] highlighting the pinhole selector placed in the source monolith following the heavy shutter units (circles) as well as outer secondary experiment and fail-safe shutter.

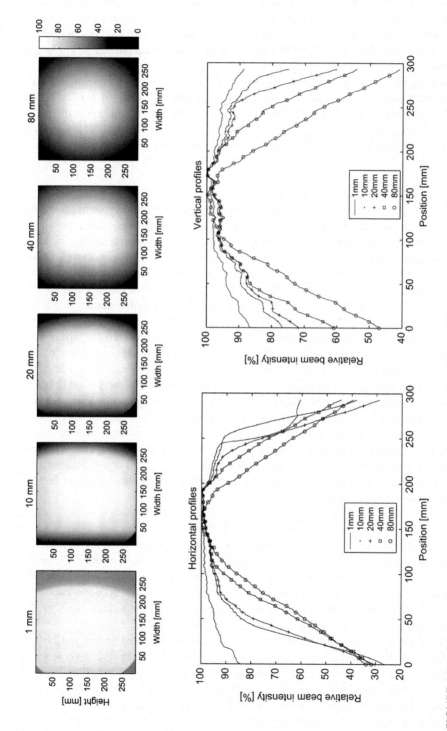

FIGURE 4.15 Measurements of the maximum beam cross sections at ICON at PSI in dependence on the chosen pinhole collimation [36].

beam geometry in providing very large homogeneous beam cross sections for imaging large objects with varying collimation. Note that the graphs are normalized to the peak as 100% and that the flux increases with the square of the chosen pinhole size. It also illustrates how the profile gets more peaked with increasing flux of larger pinholes due to divergence losses at the sides.

Examples of such direct beam geometry state-of-the-art instruments are NEUTRA and ICON at PSI [36], ANTARES at FRM2, TUM [39], BT2 at NIST [37], DINGO at ANSTO [43], NRAD at KFKI (Hungary), the neutron imaging beamline at KAERI, (Korea) as well as RADEN at J-PARC [41], a pulsed source instrument. Such instruments are in use for both cold and thermal neutron spectra.

4.3.1.2   Neutron Guide Solutions

There are various reasons for choosing or having a neutron guide system providing the neutron beam to the instrument. A significant advantage of a guide is certainly the opportunity to suppress the background in particular of gamma and fast neutron radiation from the source. The shielding requirements for the end station relax. Other advantages are significant simplifications for the collimation system and its accessibility and replaceability. Basically, all critical and movable components of the instrument move further away from the source and its critical systems. This however comes at a price. The transported divergence is limited, which implies smaller maximum beam sizes. Furthermore, over the large range required the divergence is not a smooth function anymore. Through the pinhole the inhomogeneity of the whole optical system is imaged on the detector (Figure 4.16) [46, 47], and features get more pronounced with higher resolution at smaller pinhole sizes. In addition, the guide influences the spectral homogeneity and, e.g., a bent guide best suppressing the background leads to an asymmetric spectral distribution of the beam. All these negative effects stand against the beam homogeneity requirement of neutron imaging, and it has to be considered carefully if the requirements are still fulfilled for a specific instrument and what countermeasures can be taken. It is an advantage when the guide can be designed specifically for the instrument, but also budgetary factors might play a role in optimization.

4.3.1.2.1   *Neutron Guide Solution at Continuous Source*

At a continuous neutron source the reason to choose a neutron guide for an imaging instrument can be spatial considerations and available instrument positions. It might also be the convenience of relaxed effort and requirements at an existing guide end position or relaxed requirements for large FOVs. Also scanning procedures can solve the issue of a small beam size. It might also be the desire for a clean beam or the possibilities to engineer the beam conditions through the guide design.

In general, some parameters for a guide system are fixed initially, like the distance of the guide entrance from the source, which defines the divergence a guide picks from the source. At existing guides in a neutron guide hall and due to other instruments fed by a guide at a continuous source also the upstream guide geometry might be fixed to a conventional rectangular cross section with a fixed m-value throughout. Because the m value limits the divergence at the guide exit, the

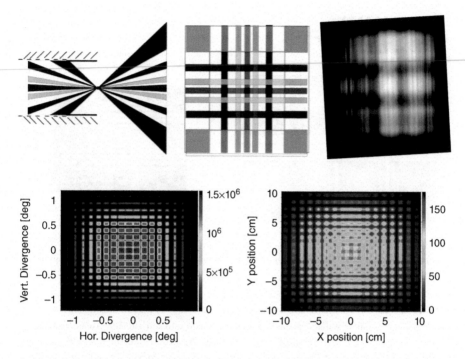

FIGURE 4.16 Illustration of the inhomogeneities introduced by a guide (here straight constant cross section) [46] (top left) compared to a measurement at a specific guide of the CG-1 beamline at HFIR of ORNL [47]; a certain asymmetry in the measurement displays the effect of a bent guide; in the bottom line a simulation of divergence and intensity profile at a straight guide [46].

minimum L/D for an instrument is predefined (Figure 4.17 top) through the guide and wavelength dependent through

$$(L/D)_{\min} = 1/\tan(2\gamma_c(\lambda)). \tag{4.12}$$

In some cases, imaging is performed directly at the guide exit in order to utilize the high flux, at low spatial resolution conditions; however, the guide end position is normally the place to position a pinhole for a subsequent imaging end station analog to those in direct geometry. The pinhole behind a generally bent guide, being out of line of sight of the source, reduces to a simple slit or pinhole in an absorber mask of, e.g., a few mm boron carbide or the like. It can easily be adapted for special requirements and exchanged when a specific size or shape is not foreseen in a remote-controlled exchanger easily accessible. Increasing the collimation ratio leads to increased detection of the artifacts of the guide due to the discontinuous divergence profile (Figure 4.17 bottom) imaging the guide performance. A countermeasure to weaken or eliminate the distinct structured inhomogeneities is the use of a diffuser just before the pinhole. Carbon or Al_2O_3 powder are common solutions with minimum beam attenuation, which has to be balanced for any filter used.

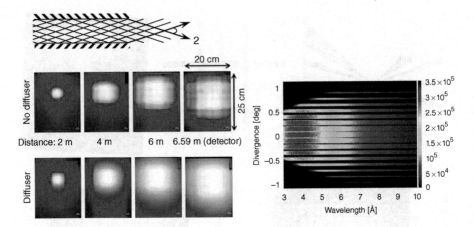

FIGURE 4.17 Illustration of the divergence limitation of a guide [44] (top) and the effect of the countermeasure of a diffuser [40] as well as the simulated effects of a straight guide on the wavelength distribution [46].

Figure 4.17 (bottom) shows the effect of a diffuser at several positions downstream the pinhole, i.e., also with regards to the beam area and collimation ratio increase.

An example of a relatively short instrument for neutron imaging at a neutron guide is depicted in Figure 4.18 (top). It is the beamline BOA at SINQ [48] which is fed by a straight guide and a multi-channel (polarizing) bender. The bender efficiently avoids line of sight to the source and all neutrons undergo at least one reflection. The result is a beam with limited divergence in the horizontal direction of bending. The beam is, however, already polarized for corresponding applications. The length of this instrument is that of a direct geometry instrument, however, with lower background and smaller maximum beam size.

Figure 4.18 (bottom) illustrates the more general case of a neutron imaging station at the end of a long cold neutron guide (>50 m), where the cold neutron imaging beamline is at the end of the neutron guide NG-6 at NIST [38]. In such case, a mild curvature of the guide is sufficient to take the instrument out of direct line of sight and the long distance from the source further reduces the background. The used m-coating of the guide defines the maximum divergence available.

Also, the use of an elliptic guide element after a conventional neutron guide has been reported for focusing the beam into the pinhole and thus increasing the flux in the pinhole and subsequently the achieved beam size [49].

Illustrated in the image of Figure 4.18 (bottom) (like in Figure 4.10) is also the typical flight tube of an imaging instrument. It is either evacuated or flooded with He in order to avoid losses from air scattering along the significant distance of the flight path L. It includes scrapers to keep the beam hitting the walls and thin windows of fractions of mm in general of Al. In most cases it is modular in order to accommodate other optical devices at different measurement positions along the path in the end station.

Typical examples of instruments installed at neutron guides of continuous sources are the cold neutron imaging instrument at the neutron guide NG-6 at NIST and the CG-1D beamline of the HFIR of ORNL.

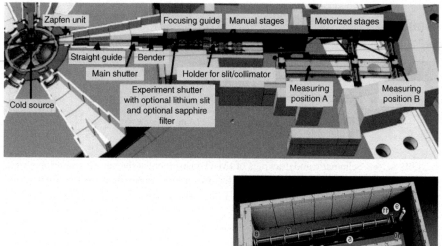

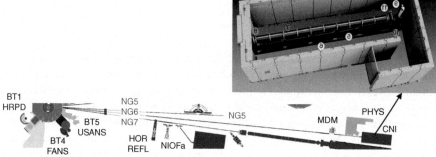

FIGURE 4.18 Top: BOA beamline [48] with short guide segment including a polarizing bender, however, otherwise rendering the instrument very similar to direct geometry instruments (compare Figures 4.10 and 4.11); Bottom: NG-6 cold neutron imaging instrument at NIST at the end of a long typical guide hall neutron guide (with two other upstream monochromator instruments) [38].

4.3.1.2.2   Neutron Guide Solution at Pulsed Source

While most optical considerations and principles remain the same at the modern pulsed spallation sources, the motivation for the decision for or against a neutron beam guide for a neutron imaging instrument is different. It is the choice of moderator and time structure that imply the need for a guide. On the other hand, the specific concept and layout of pulsed spallation sources enable highly individual and advanced optimized guide concepts for individual instruments such as for neutron imaging.

Because a key aspect also of imaging at spallation sources is the ToF wavelength resolution, the length of the instrument is determined by the source pulse width and the targeted (maximum) wavelength resolution. Also, the required wavelength bandwidth can play a role. The ToF imaging instrument IMAT at ISIS TS2 (Figure 4.19) [42] is the first dedicated imaging instrument at a state-of-the-art neutron source featuring a neutron guide to transport the neutrons to the instrument. The guide is straight and hence does not avoid line of sight, despite the particularly cumbersome background produced by such source. The instrument instead draws on a T0 chopper

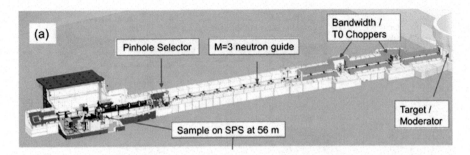

FIGURE 4.19 Schematic and drawings of IMAT instrument at the pulsed spallation source of ISIS target station TS2. The instrument features a long guide (42 m) with an integrated chopper system; the end station does not only facilitate imaging, but also diffraction experiments [42].

in order to remove the most severe background from the prompt pulse. The guide takes the instrument starting with the pinhole some 50 m away from the source. This length, however, also contributes to limit the background from the source. The motivation for such length is to achieve at this distance a sub-percent wavelength resolution suited for strain mapping and diffraction, while at the same time preserving the high flux density. Additional neutron optical devices required in such beamline are the choppers, which here are limited to T0 and bandwidth choppers avoiding contamination and pulse overlap.

At IMAT an $m = 3$ coated guide with a constant square cross section 95×95 mm^2 leads up from the shutter in the source monolith to the pinhole position at 46 m from the source. This guide is straight and about 42 m long. Another initial 2 m piece with a cross section of 100×100 mm^2 sits in the heavy shutter itself. The flight distance L from pinhole to sample position is 10 m and with that a maximum nominal beam size of 20×20 cm^2 can be achieved. A single wavelength frame has a width of about 6 Å and with pulse suppression a double bandwidth of about 12 Å can be achieved at 10 Hz. The wavelength resolution based on the moderator burst time and the instrument length is around 0.7% $\Delta\lambda/\lambda$ and even better for wavelengths shorter than 3 Å. The guide is interrupted for the choppers. Simulations of the beam profile in Figure 4.20 taken with low resolution illustrate the spectral effects of the guide system. Longer wavelengths contribute to the full beam size significantly more than the short wavelengths. The spectrally homogeneous beam size appears limited to rather about 10×10 cm^2, where about also an 80% intensity criterion would identify the beam size limit.

Another dedicated imaging instrument, ODIN [50], currently under construction at the long pulse spallation neutron source of ESS, will feature an individually optimized guide and chopper system. This system in particular has seen a number of optimizations [51], based on the changing design of the ESS moderator. The system and boundary conditions are unique in many ways. First, ESS is a long pulse source, which implies that high wavelength resolution cannot be achieved through flight path length alone. Second, ESS has moved from 3D moderator to a 2D moderator. This

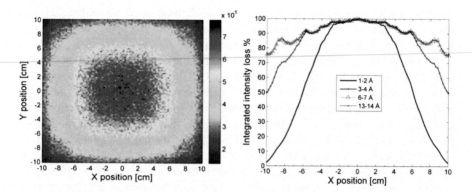

FIGURE 4.20 Simulations of the IMAT beam at the detector (left) and the wavelength-dependent beam profiles [49].

means one dimension of the moderator face has been significantly decreased. This has an impact in general on the variable pinhole approach and for a guide system, which requires now different geometries for the two dimensions. Hence, the instrument design features two novelties: (i) a complex optimized guide design (Figure 4.21)

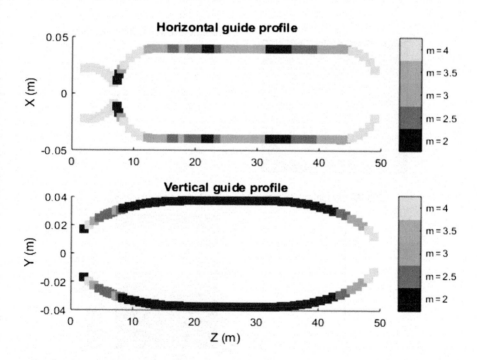

FIGURE 4.21 Guide specification of ODIN neutron guide for ESS with significantly different vertical and horizontal profiles due to the ESS flat moderator design (pancake), as well as a chopper system with pulse shaping choppers at around 7 m. The color code indicates the different m-coatings used.

and (ii) a pulse shaping chopper system for variable wavelength resolution [52]. The chopper system shall be discussed later for advanced imaging optics devices, while here the focus should be on the primary beam optics.

The basic idea for the guide system is to diverge the beam from a small beam area and converge it into the pinhole at the entrance to the end station. This is straightforward possible for the vertical direction, where the moderator with a few cm height matches the size of the largest pinholes considered. In the horizontal direction however, the beam is extended on the moderator side, like a conventional moderator, but a limited beam width is required at the position of the pulse shaping choppers at about 7 m, just outside the source monolith. This limitation is necessary to enable sufficient chopping accuracy, which would suffer from an extended beam. Hence, for the horizontal direction the beam is first focused from the large scale of the moderator to a small cross section for beam chopping and only from there the elliptic "point to point" geometry picks up the beam for transport to the pinhole of the imaging geometry. Correspondingly the m-coating values change throughout the guide system and relax from maxima of $m = 4$ in the most inclined sections to $m = 2$ in the flattest sections.

This configuration has been simulated to provide a FOV of about 15×15 cm^2 at a distance $L = 10$ m from the pinhole when applying a criterion of 75% maximum flux (Figure 4.22). A high level instrument requirement is to achieve an area of at

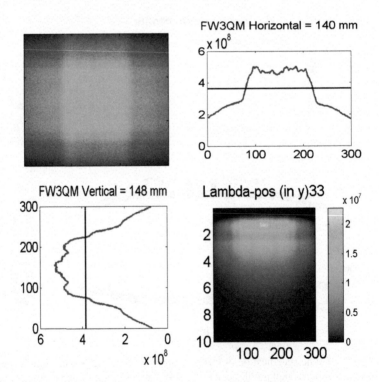

FIGURE 4.22 Simulated beam profiles and spectral distribution from the complex ODIN neutron guide at an $L = 10$ m measurement position and L/D of 300.

least 20×20 cm^2 at the outmost sample position at $L = 14$ m, which should hence be achieved. The divergence in the pinhole is approximately ±0.8 degree. Moreover, good spectral homogeneity appears to be possible with this optimized guide system (Figure 4.22). Nevertheless, the typical horizontal and vertical structures of a discontinuous divergence observed through the pinhole remain also in this guide system. Interruptions at the chopper positions contribute additionally to the inhomogeneities. Also, here the use of a diffuser is foreseen to minimize such disturbances of a flat beam profile. However, the wavelength-dependent transmission of the diffuser can lead to significant diminishing of flux at certain wavelengths as shown in a measurement of a diffuser transmission spectrum (Figure 4.23) in ToF at the ESS test beamline at HZB, Berlin [52].

In addition, the arrangement of ESS moderators enables bi-spectral extraction [53], which is for imaging an attractive opportunity. It provides more flux for white beam applications and a choice between a thermal part with higher penetration in many cases and cold with higher sensitivity. Also, a wider range of wavelengths to choose the range of interest from for diffractive, dark field, or polarized neutron imaging enables to better tailor experiments to the needs of a study. In order to view the neighboring cold and thermal moderators the guide axis is pointing on the thermal part, while in the extraction section a number of additional supermirrors deflect cold neutrons into the thermal beam (Figure 4.24). This enables efficient and homogeneous mixing of the two spectra parts with efficiencies ranging for both spectra parts around 80% with a minimum of around 2/3 in the cross over a range between 2 and 3 Å. However, this is so far only simulated and the commissioning of the instrument expected in 2022 will still have to prove the success and efficiency of the system and approach.

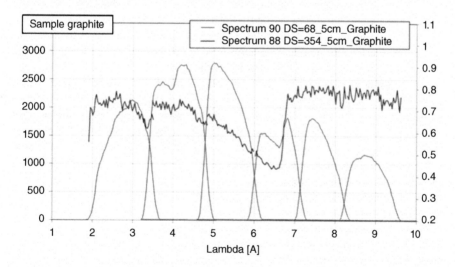

FIGURE 4.23 Illustration of the downside of a diffuser element as measured at the ESS test beamline at HZB [52] with a test system of the ODIN wavelength multiplication chopper system; The attenuation of the diffuser ranges around 80% but in the steep Bragg edge at around 6.5 Å it reaches a minimum of 50%.

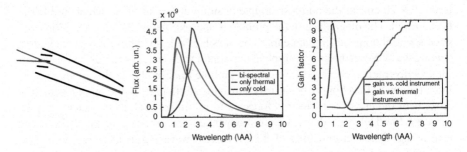

FIGURE 4.24 The bi-spectral extraction system of the ODIN guide system in a schematic drawing (left) as well as the extracted spectrum as compared to the two separate thermal and cold spectra (mid) at the right the gains as compared to a pure thermal or cold moderator view, respectively (right).

4.3.2 OPTICAL DEVICES FOR ADVANCED NEUTRON IMAGING

Advanced neutron imaging methods pose additional requirements to the optical system of an instrument. First of all, most advanced techniques require wavelength resolution. Therefore, standard neutron optical devices for wavelength resolution need to be implemented in neutron imaging beamlines.

4.3.2.1 Wavelength Resolution

Different methods and experimental aims require different wavelength resolutions ranging from around 10% like in SANS instruments to sub-percent resolution like in high resolution diffractometers. However, the specific requirements of neutron imaging for large divergence and large cross section of a beam, but in particular the spatial homogeneity pose some issues to the standard devices, which do not become apparent in the spatially integral scattering measurements. Based on beam size considerations, these neutron optical elements have to be integrated close to the pinhole, in general as soon as possible downstream of the aperture.

At a continuous beam for the requirement of around 10% monochromatic beam resolution a velocity selector is the device of choice [36]. However, the transmission of a velocity selector is divergence dependent and in an imaging beamline this dependence leads to systematic spectral inhomogeneity over the FOV. Depending on the geometry of the selector with the rotation axis above, below, or at the side of the beam the mean wavelength varies over the FOV either vertically or horizontally, with a distribution even displaying a radial dependence due to the corresponding optical conditions (Figure 4.25) [54]. While this implies that the achieved wavelength resolution, which nominally concerns the full integrated beam is locally improved accordingly, different contrast conditions have to be taken into account across the beam and can only in some measurements be accounted and corrected for, but not in all. Another limitation implied by selectors is the relatively high minimum wavelength in standard devices.

A similar problem occurs with monochromators, which are used at continuous sources for wavelength resolution requirements in the order of a few percent [55]. Typically, PG crystals with around 1 degree of mosaicity are used in nondispersive

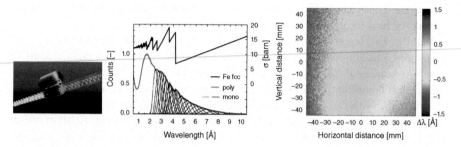

FIGURE 4.25 Functional scheme of a velocity selector as well as the impact on spectral resolution compared to the Bragg edge profile of fcc iron [36]; on the right-hand side the spectral inhomogeneity introduced by a selector in the example from the ICON beamline at PSI [54].

double crystal arrangements. The latter is required to remain in the axis of the extended downstream instrument and enable a significant divergence to pass to the sample and imaging detector. While the crystal arrangement will somewhat limit the divergence, in particular in the direction perpendicular to the reflection plane dependent on the mosaicity, in the reflection plane again a strong wavelength gradient will result (Figure 4.26). And again the nominal resolution is integral, while the local resolution is enhanced. This directional dependence has to be considered and chosen carefully. As in general the rotation axis for tomographic scans is vertical, in a horizontal double crystal geometry the wavelength will average over all tomographic slices, while in a vertical geometry each tomographic slice will represent a slightly different wavelength. The gradient also has to be considered when choosing sample geometry and orientation.

The mosaicity has to be selected to enable the envisaged resolution, which however is wavelength dependent. It has to be considered that mosaicities larger than the beam divergence do relax the wavelength resolution, but do not increase the flux density, but the perpendicular beam size. Finally, the crystal monochromator arrangements also contaminate longer wavelengths with shorter wavelength second-order

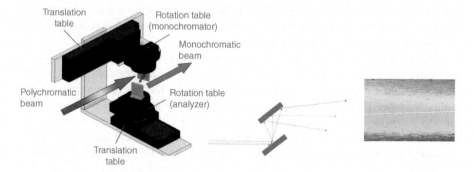

FIGURE 4.26 Drawing of the arrangement of a horizontal double crystal monochromator for neutron imaging, a schematic illustrating the principle leading to the wavelength gradient and an illustration of the gradient on the detector.

reflections. These can be filtered either by an additional crystal or through preconditioning the beam, e.g., with a velocity selector.

Finally, wavelength resolution of a large variety of resolutions is also possible in imaging utilizing a ToF approach. This was naturally first and mainly applied at pulsed sources [4, 41, 42], while only chopper systems at continuous sources can provide full flexibility for tailoring measurements to the needs of a particular investigation [56], both in terms of wavelength resolution and bandwidth. This tailoring can have a decisive impact on the efficiency of a measurement. At short pulse sources the pulse burst function from the moderator has to be taken into account as it is asymmetric and has a tail. Otherwise a chopper system is mainly required to avoid overlap or to suppress pulses in order to increase the bandwidth [41, 42]. A ToF approach is in general only efficient for wavelength dispersive but not for monochromatic measurements, where hence at same source brightness continuous sources are superior [57]. At same source brightness even the ToF approach at a continuous source can be superior when the bandwidth and resolution requirements can be better matched than at the pulsed source [58]. However, advanced pulsed sources nowadays have superior brightness, so that efficiency advantages have to be judged carefully. A special case is the long pulse source of ESS. It is expected to not only feature a time averaged flux outperforming continuous sources but also a time structure that enables tailoring the resolution [59]. Instruments at ESS are hence expected to outperform all existing instruments. However, ESS instruments will overall be more complex in order to achieve this flexibility. Figure 4.27 depicts the complex ToF chopper system functionalities of the imaging instrument ODIN under construction at ESS. The system features a pair of optical blind pulse shaping choppers with variable distance between them. This enables to increase from the initial resolution given by 3 ms burst time and 60 m distance, i.e., about 10% at 2 Å and increasing toward longer wavelengths [50], to 1% and better through complex wavelength frame multiplication [52]. The system also accommodates onefold pulse suppression for the respective resolutions.

While a chopper functioning as pulse source provides in contrast to a short pulse source a (tunable) symmetric pulse shape, it has been shown that also here the geometry of the pulse chopping is imaged as an asymmetry of the ToF wavelength, which requires correction (Figure 4.28) [58]. As these measurements are wavelength dispersive, that means always a series of wavelengths are measured, correction is more straightforward as for monochromatic measurements.

The installation of a ToF system, in particular a versatile one, at a standard direct geometry neutron imaging instrument is complex and has not been attempted so far. A realization at a guide instrument appears easier, but has not been attempted yet either, but other ToF instruments have been utilized for imaging. Cost and complexity are the downside of such systems, but they also come with a limitation to beam size. A limited beam size eases some of the chopper system's technical specifications but is mainly caused by available technology for efficient imaging with ToF resolution.

4.3.2.2 Polarization
Aiming at polarized neutron studies in neutron imaging implies the need for all kinds of polarized neutron devices including neutron spin filters, guide fields, spin

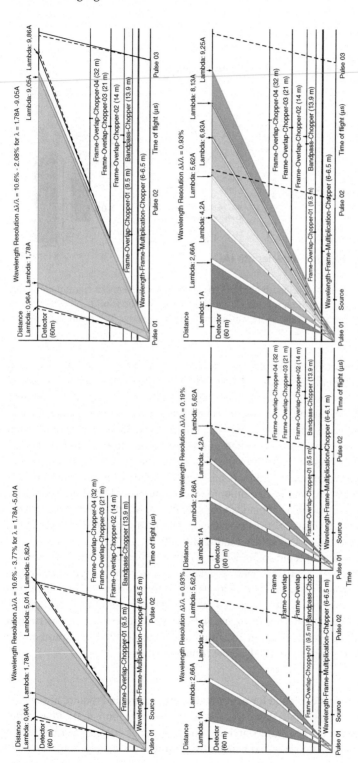

FIGURE 4.27 Different ToF options achievable with the complex ODIN wavelength frame multiplication chopper system: (top left) natural bandwidth and resolution defined by the time structure and instrument length, (top right) same resolution but double bandwidth due to pulse suppression; (bottom left) wavelength frame multiplication mode to tune wavelength to 1%, (bottom mid) same setting as before but reduced distance of pulse shaping optical blind chopper pair tuning the resolution to about 0.2% and (bottom right) tuned resolution with wavelength band multiplication at double the bandwidth through pulse suppression.

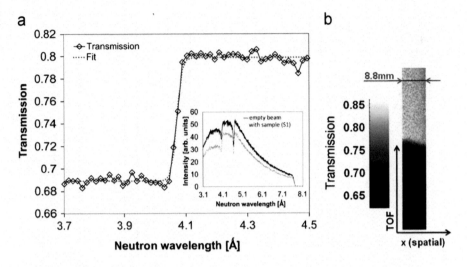

FIGURE 4.28 Measured wavelength gradient from a chopper system at a continuous source in Bragg edge transmission imaging measurements at the former BioRef reflectometer at HZB [56]; the image at the left shows the variation of the Bragg edge ToF position across the small beam [58].

flippers, and turners as well as magnetic field shielding. When limiting the beam size to a few centimeters in both dimensions, equipment generally used for polarized neutron experiments can be utilized. The main specific requirements of imaging again concern the homogeneity across the beam because polarization is investigated locally. A specific issue for neutron imaging consists in the need for a polarization analyzer, which has to be placed between sample and detector and thus extends the sample to detector distance l, which in the pinhole geometry implies loss of spatial resolution. Therefore, compact analyzers are sought which at the same time do not excessively attenuate the beam in an inhomogeneous manner. Therefore, solid state polarizing benders are often used but also He spin filters, which however require the corresponding infrastructure. Setups can display significant complexity depending on a specific measurement as displayed in Figure 4.29.

The most sophisticated setup so far, which is routinely used, is at the ToF imaging instrument RADEN at J-PARC. It enables polarimetric neutron imaging, i.e., probing all three orthogonal polarization directions for all three orthogonal incident spin directions as proposed in Ref. 60. It enabled the first 3D reconstruction of a magnetic vector field [12]. Other than seen in Figure 4.30, large partitions of the instrument setup, i.e., around the sample position and spin turners are shielded with mu-metal to optimize the field conditions and reduce the influence of stray fields.

4.3.2.3 Gratings

Grating interferometers exploited for neutron imaging feature and require absorption and phase gratings with periods down to a few micrometers (Figure 4.31). Phase gratings are generally made of Si wafers and absorption gratings consist of Gd, often on quartz glass wafers. Beam sizes and hence also grating sizes mostly range between

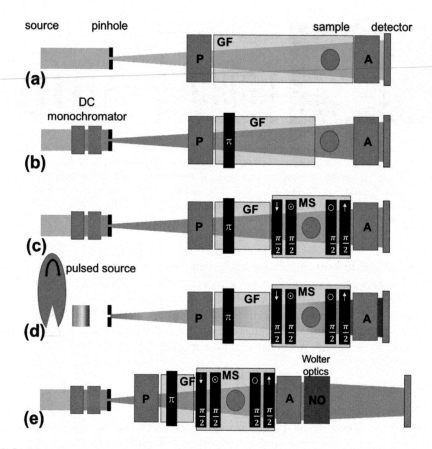

FIGURE 4.29 Schematic of polarized neutron imaging setups with increasing complexity [6]; P refers to polarizer, A to analyzer, GF to guide field, π and $\pi/2$ for corresponding spin flippers/turners, MS for magnetic shielding, and NO for a neutron optical lens.

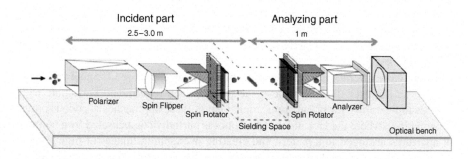

FIGURE 4.30 Currently most advanced polarimetric imaging setup (without the magnetic field shielding) at RADEN instrument at JPARC [41].

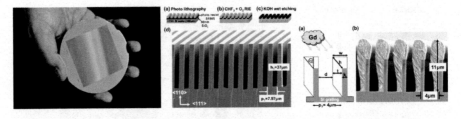

FIGURE 4.31 Photograph of a typical grating of a Talbot Lau interferometer as well as microscopies of the structures of a phase and absorption grating [61] with some details of manufacture. [Reproduced from C. Grünzweig, F. Pfeiffer, O. Bunk, T. Donath, G. Kühne, G. Frei, M. Dierolf, and C. David, Design, fabrication, and characterization of diffraction gratings for neutron phase contrast imaging, Rev. Sci. Inst., 79, 053703 (2008), with the permission of AIP Publishing.]

5×5 cm^2 and 10×10 cm^2. Talbot Lau grating interferometers are generally optimized for a specific wavelength, but as wavelength dispersive measurements and also the utilization at ToF sources becomes of interest, designs are investigated, which overcome too strict wavelength requirements, in order to measure over a larger range with good visibility contrast. Grating production is today mainly done commercially according to the specifications required for a specific study or setup.

4.3.3 ALTERNATIVE APPROACHES AND OUTLOOK

Today measurements of smaller samples become more and more dominant and high resolution or special techniques gain importance. However, the design of instruments is to some extent governed by the ability to illuminate large samples, while smaller samples can in principle be imaged at less specialized beam optics. A number of techniques, e.g., ToF imaging have been pioneered at scattering instruments and an optimized beamline for imaging smaller size samples only could enable more compact instruments and maybe even more efficient optics.

In addition, highest resolution imaging does not necessarily require ever increased collimation values but rather use the higher flux of a more divergent beam, because the typically very small samples can be placed exceptionally close to the detector. For example, an L/D value of 250 enables a resolution of 500 μm at 25 cm sample to detector distance, 50 μm for 2.5 cm and 5 μm for 2.5 mm. The latter is about the best currently achieved in standard operation for samples of up to 5 mm in size.

In the past, several alternative approaches have been explored for neutron imaging. These however did not achieve the performance required in particular with regards to flexibility for broad application. One of these approaches is coded source imaging [62], where the lack of lens like optics is partially compensated by a number of apertures arranged in a pattern and creating overlapping images of high resolution, defined by the single aperture opening, on the detector. These images can be deconvoluted in order to provide a well-focused single image recorded with higher flux and hence shorter exposure and/or increased resolution than in the standard

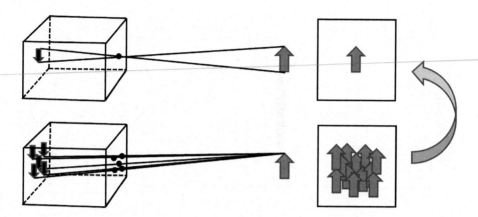

FIGURE 4.32 Sketches of pinhole camera vs a multi-pinhole coded aperture and representations of acquired images; reconstruction algorithms enable to recover the top image from the bottom one with well-defined and known coded apertures; this concept can be transferred to neutron imaging in pinhole geometry [62].

approach (Figure 4.32). However, it has been found that this technique is well suited for sharp structures but fails for smooth structures, which finally disqualified widespread application.

In addition, compound refractive lenses have been tested for neutron imaging (Figure 4.33) [63]. However, this approach suffers from small apertures and significant material that has to be penetrated and weakens the beam. Nevertheless, the ability to move the sample away from the detector without sacrificing flux or resolution might have still potential for methods, e.g., with polarized neutrons, where extended sample detector distances are implied and the FOV is limited by other devices, while samples are small in general anyways. In addition, the latter technique requires monochromatic neutrons, which otherwise is another severe drawback of this method with serious chromatic aberration.

Also, the use of an expanding guide to decrease divergence has been considered and/or the use of collimators, e.g., MCP. Both approaches are limited in the flexibility of different collimation settings and/or sample sizes.

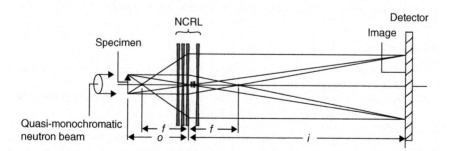

FIGURE 4.33 Principle setup and concept of imaging with neutron compact refractive lenses [63].

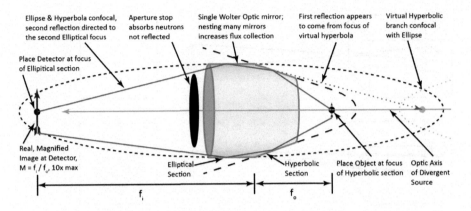

Ellipse & Hyperbola confocal, second reflection directed to the second Elliptical focus

Aperture stop absorbs neutrons not reflected

Single Wolter Optic mirror; nesting many mirrors increases flux collection

First reflection appears to come from focus of virtual hyperbola

Virtual Hyperbolic branch confocal with Ellipse

Place Detector at focus of Ellipitical section

Real, Magnified Image at Detector, M = f$_i$ / f$_o$, 10x max

Elliptical Section

Hyperbolic Section

Place Object at focus of Hyperbolic section

Optic Axis of Divergent Source

f$_i$

f$_o$

FIGURE 4.34 Principle of Wolter optics as considered as neutron lenses for neutron imaging [65].

A more recent proposal for lens-like optics are reflective optics of so-called Wolter mirrors as proposed by Mildner, M.V. Gubarev [64] in analogy to X-ray optics. The system is in principle achromatic and consists of a constant and a convergent cone, which in ideal case are covered with supermirrors on the inside. These reflect divergent neutrons from a point in the object placed in the focus of the convergent part onto one point on the detector in the focal point of the straight cone considered the central part of an ellipse (Figure 4.34). For the system to work efficiently a beam of significant divergence on the one hand and a very exact construction of nested cones on the other hand is required. So far, a number of considerations and a few prove of principle measurements have been reported. Again, usability appears to mainly concern cases where an extended sample to detector distance is required, like in polarized neutron imaging. While for a medium resolution range advantages might be achievable, a limitation appears to be small sample sizes due to the curvature of the focal plane on the detector. Improvements for the resolution limit in contrast are not expected, where additional flux is the most required, because the accuracy of the geometry of the device required for this appears currently not feasible. However, technology improves and novel ideas are generated constantly and progress of many devices and techniques for neutron imaging have to be expected.

REFERENCES

1. M. Strobl, N. Kardjilov, A. Hilger, I. Manke, and J. Banhart, Topical review: Advances in neutron radiography and tomography, *J. Phys. D*, 42, 243001 (2009).
2. N. Kardjilov, I. Manke, A. Hilger, M. Strobl, and J. Banhart, Neutron imaging in material science, *Mater. Today*, 14(6), 248–256 (2011).
3. M. Zarebanadkouki, A. Carminati, A. Kaestner, D. Mannes, M. Morgano, S. Peetermans, E. Lehmann, and P. Trtik, On-the-fly neutron tomography of water transport into lupine roots, *Phys. Proc.*, 69, 292–298 (2015).
4. R. Woracek, J. Santisteban, A. Fedrigo, and M. Strobl, Diffraction in neutron imaging—a review, *Nucl. Inst. Meth. A*, 878, 141–158 (2018).

5. M. Strobl, R. P. Harti, C. Grünzweig, R. Woracek, and J. Plomp, Small angle scattering in neutron imaging—A review, *J. Imaging*, 3, 64 (2017).

6. M. Strobl, H. Heimonen, S. Schmidt, M. Sales, N. Kardjilov, A. Hilger, I. Manke, T. Shinohara, and J. Valsecchi, Topical review: Polarisation measurements in neutron imaging, *J. Phys. D*, 52, 12 (2019).

7. D. Mannes, E. Lehmann, A. Masalles, K. Schmidt-Ott, and A. Przychowski, et al., The study of cultural heritage relevant objects by means of neutron imaging techniques, *INSIGHT*, 56(3), 137–141 (2014).

8. D. Mannes, C. Hanser, M. Krzemnicki, R. Harti, I. Jerjen, and E. Lehmann, Gemmological investigations on pearls and emeralds using neutron imaging, *Phys. Proc.*, 88, 134–139 (2017).

9. W. Gong, P. Trtik, A.W. Colldeweih, L.I. Duarte, M. Grosse, E. Lehmann, J. Bertsch J. Nucl. Mater. (2019) In press; doi.org/10.1016/j.jnucmat.2019.151757.

10. M. Zarebanadkouk, X. Kim Yangmin, and A. Carminati, Where do roots take up water? Neutron radiography of water flow into the roots of transpiring plants growing in soil, *New Phytol.*, 199, 1034–1044 (2013).

11. C. Tötzke, N. Kardjilov, I. Manke, and S. E. Oswald, Capturing 3D water flow in rooted soil by ultra-fast neutron tomography, *Sci. Rep.*, 7, 6192 (2017).

12. M. Sales, M. Strobl, T. Shinohara, A. Tremsin, L. Theil Kuhn, W. Lionheart, N. Desai, A. Bjorholm Dahl, and S. Schmidt, Three dimensional polarimetric neutron tomography of magnetic fields, *Sci. Rep.*, 8, 2214 (2018).

13. A. Cereser et al., Time-of-flight three dimensional neutron diffraction in transmission mode for mapping crystal grain structures, *Sci. Rep.*, 7, 9561 (2017).

14. H. Pleinert, E. Lehmann, and S. Körner, Design of a new CCD-camera neutron radiography detector, *Nucl. Instr. Meth. A*, 399(2–3), 382–390 (1997).

15. C. Grünzweig, D. Mannes, A. Kaestner, F. Schmid, P. Vontobel, J. Hovind, S. Hartmann, S. Peetermans, and E. Lehmann, Progress in industrial applications using modern neutron imaging techniques, *Phys. Proc.*, 43, 231–242 (2013).

16. P. Trtik, C. Scheuerlein, P. Alknes, M. Meyer, F. Schmid, and E. H. Lehmann, Neutron microtomography of MgB2 superconducting multifilament wire, *Phys. Proc.*, 88, 95–99 (2017).

17. S. H. Williams, A. Hilger, N. Kardjilov, I. Manke, M. Strobl, P. A. Douissard, T. Martin, H. Riesemeier, and J. Banhart, Detection system for microimaging with neutrons, *JINST*, 7, P02014 (2012).

18. P. Trtik and E. H. Lehmann, Progress in high-resolution neutron imaging at the Paul Scherrer Institut—the neutron microscope project, *J. Phys.: Conf. Ser.*, 746(1), (2016).

19. A. S. Tremsin, J. B. McPhate, A. Steuwer, W. Kockelmann, A. M Paradowska, J. F. Kelleher, J. V. Vallerga, O. H. W. Siegmund, and W. B. Feller, High-resolution strain mapping through time-of-flight neutron transmission diffraction with a microchannel plate neutron counting detector, *Strain*, 48, 296–305 (2012).

20. A. Tremsin, N. Kardjilov, M. Strobl et al., Imaging of dynamic magnetic fields with spin-polarized neutron beams, *New J. Phys.*, 17, 043047 (2015).

21. R. P. Harti, M. Strobl, R. Schäfer, N. Kardjilov, A. S. Tremsin, and C. Grünzweig, Dynamic volume magnetic domain wall imaging in grain oriented electrical steel at power frequencies with accumulative high-frame rate neutron dark-field imaging, *Sci. Rep.*, 8, 15754 (2018).

22. M. Raventos, E. H. Lehmann, M. Boin, M. Morgano, J. Hovind, R. Harti, J. Valsecchi, A. Kaestner, C. Carminati, P. Boillat, P. Trtik, F. Schmid, M. Siegwart, D. Mannes, M. Strobl, and C. Gruenzweig, A Monte Carlo approach for scattering correction towards quantitative neutron imaging of polycrystals, *J. Appl. Cryst.*, 51 (2018).

23. E. H. Lehmann, S. Peetermans, L. Josic, H. Leber, and H. vanSwygenhoven, Energy-selective neutron imaging with high spatial resolution and its impact on the study of crystalline-structured materials, *Nucl. Instr. Meth. A*, 735, 102–109 (2014).

24. M. Strobl, C. Grünzweig, A. Hilger, I. Manke, N. Kardjilov, C. David, and F. Pfeiffer, Neutron dark-field tomography, *Phys. Rev. Lett.*, 101, 123902 (2008).

25. M. Strobl, General solution for quantitative dark-field contrast imaging with grating interferometers, *Sci. Rep.*, 4, 7243 (2014).

26. R. P. Harti, M. Strobl, B. Betz, K. Jefimovs, M. Kagias, and C. Gruenzweig, Sub-pixel correlation length neutron imaging: Spatially resolved scattering information of microstructures on a macroscopic scale, *Sci. Rep.*, 7, 44588 (2017).

27. M. Strobl, M. Sales, J. Plomp, W. G. Bouwman, A. S. Tremsin, A. Kaestner, C. Pappas, and K. Habicht, Quantitative neutron dark-field imaging through spin-echo interferometry, *Sci. Rep.*, 5, 16576 (2015).

28. D. S. Hussey, H. Miao, G. Yuan, D. Pushin, D. Sarenac, M. G. Huber, D. L. Jacobson, J. M. LaManna, and H. Wen, Demonstration of a white beam far-field neutron interferometer for spatially resolved small angle neutron scattering, arXiv:1606.03054 [physics. ins-det] (2016).

29. H. Rauch and S. A. Werner, *Neutron Interferometry: Lessons in Experimental Quantum Mechanics, Wave-Particle Duality, and Entanglement*, 2nd ed., Vol. 12. Oxford, UK: Oxford University Press, 2015.

30. W. Treimer, M. Strobl, A. Hilger, C. Seifert, and U. Feye-Treimer, Refraction as imaging signal for computerized (neutron) tomography, *Appl. Phys. Lett.*, 83(2), 398–401 (2003).

31. F. Pfeiffer, C. Grünzweig, O. Bunk, G. Frei, E. Lehmann, and C. David, *Phys. Rev. Lett.*, 96, 215505 (2006).

32. B. E. Allman, P. J. McMahon, K. A. Nugent, D. Paganin, D. L. Jacobson, M. Arif, and S. A. Werner, Phase radiography with neutrons, *Nature*, 408, 158 (2000).

33. N. Kardjilov, I. Manke, M. Strobl, A. Hilger, W. Treimer, M. Meissner, T. Krist, and J. Banhart, Three-dimensional imaging of magnetic fields with polarized neutrons, *Nat. Phys.*, 4, 399–403 (2008).

34. M. Strobl, N. Kardjilov, A. Hilger, E. Jericha, G. Badurek, and I. Manke, Imaging with polarized neutrons, *Physica B*, 404, 2611 (2009).

35. E. H. Lehmann, P. Vontobel, and L. Wiezel, Properties of the radiography facility NEUTRA at SINQ and its potential for use as European reference facility, *Nondestruct. Test. Eva.*, 16(2–6), 191–202 (2001).

36. A. P. Kaestner, S. Hartmann, G. Kuehne, G. Frei, C. Gruenzweig, L. Josic, F. Schmid, and E. H. Lehmann, The ICON beamline—A facility for cold neutron imaging at SINQ, *Nucl. Instr. Meth. A*, 659, 387–393 (2011).

37. D. S. Hussey, D. L. Jacobson, M. Arif, K. J. Coakley, and D. F. Vecchia, In situ fuel cell water metrology at the NIST neutron imaging facility, *J. Fuel Cell Sci. Tech.*, 7, 021024 (2010).

38. D. S. Hussey, C. Brocker, J. C. Cook, D. L. Jacobson, T. R. Gentile, W. C. Chen, E. Baltic, D. V. Baxter, J. Doskow, and M. Arif, A new cold neutron imaging instrument at NIST, *Phys. Proc.*, 69, 48–54 (2015).

39. M. Schulz, B. Schillinger, E. Calzada, D. Bausenwein, P. Schmakat, T. Reimann, and P. Böni, The new neutron imaging beam line ANTARES at FRM II, *Restaurierung und Archäologie*, 8 (2015).

40. L. Santodonato, H. Z. Bilheux, B. Bailey, J. Bilheux, P.T. Nguyen, A. S. Tremsin, D. L. Selby, and L. Walker, "The CG-1D neutron imaging beamline at the Oak Ridge National Laboratory high flux isotope reactor", *Phys. Proc.*, 69, 104–108 (2015).

41. Y. Kiyanagi, T. Shinohara, T. Kai, T. Kamiyama, H. Sato, K. Kino, K. Aizawa, M. Arai, M. Harada, K. Sakai, K. Oikawa, M. Ooi, F. Maekawa, H. Iikura, T. Sakai, M. Matsubayashi, M. Segawa, and M. Kureta, Present status of research on pulsed neutron imaging in Japan, *Phys. Proc.*, 42, 92–99 (2013).

42. W. Kockelmann et al., Time-of-flight neutron imaging on IMAT@ISIS: A new user facility for materials science, *J. Imaging*, 4, 47 (2018).

43. U. Garbe, T. Randall, and C. Hughes, The new neutron radiography/tomography/imaging station DINGO at OPAL, *Nucl. Instr. Meth. Phys. Res. A*, 651, 42–46 (2011).

44. B. Schillinger et al., The design of the neutron radiography and tomography facility at the new research facility reactor FRM-II at Technical University Munich, *Appl. Radiat. Isotopes*, 61(4), 653–657 (2004).

45. F. Grünauer et al., Optimization of the beam geometry for the cold neutron tomography facility at the new neutron source in Munich, *Appl. Radiat. Isotopes*, 61(4), 479–485 (2004).

46. Y. Wang, G. Wei, H. Wang, Y. Liu, L. He, K. Sun, S. Han, and D. Chen, A study on inhomogeneous neutron intensity distribution origin from neutron guide transportation, *Phys. Proc.*, 88, 354–360 (2017).

47. L. Crow, L. Robertson, H. Z. Bilheux, M. Fleenor, E. B. Iverson, X. Tong, D. Stoica, and W. T. Lee, The CG1 instrument development test station at the high flux isotope reactor, *Nucl. Instr. Meth. Phys. Res. A*, 634, S71–S74 (2011).

48. M. Morgano, S. Peetermans, E. H. Lehmann, T. Panzner, and U. Filges, Neutron imaging options at the BOA beamline at Paul Scherrer Institut, *Nucl. Instr. Meth. A*, 754, 46–56 (2014).

49. G. Burca, W. Kockelmann, J. A. James, and M. E. Fitzpatrick, Modelling of an imaging beamline at the ISIS pulsed neutron source, *J. Instr.*, 8, 10001 (2013).

50. M. Strobl, The scope of the imaging instrument project ODIN at ESS, *Phys. Proc.*, 69, 18–26 (2015).

51. A. Hilger, N. Kardjilov, I. Manke, C. Zendler, K. Lieutenant, K. Habicht, J. Banhart, and M. Strobl, Neutron guide optimisation for a time-of-flight neutron imaging instrument at the European spallation source, *Optics Exp.*, 23(1), 301–311 (2015).

52. M. Strobl, M. Bulat, and K. Habicht, The wavelength frame multiplication chopper system for an ESS test-beamline and corresponding implications for ESS instruments, *Nucl. Instr. Meth. A*, 705, 74–84 (2013).

53. C. Zendler, K. Lieutenant, D. Nekrassov, L. D. Cussen, and M. Strobl, Bi-spectral beam extraction in combination with a focusing feeder, *Nucl. Instr. and Meth. A*, 704, 68–75 (2013).

54. S. Peetermans, F. Grazzi, F. Salvemini, and E. Lehmann, Spectral characterization of a velocity selector type monochromator for energy-selective neutron imaging, *Phys. Proc.*, 43, 121–127 (2013).

55. R. Woracek, D. Penumadu, N. Kardjilov, A. Hilger, M. Strobl, R. C. Wimpory, I. Manke, and J. Banhart, Neutron Bragg-edge-imaging for strain mapping under *in situ* tensile loading, *J. Appl. Phys.*, 109, 093506 (2011).

56. M. Strobl, R. Steitz, M. Kreuzer, M. Rose, H. Herrlich, F. Mezei, R. Dahint, and M. Grunze, BioRef—A versatile time-of-flight reflectometer for soft matter applications at Helmholtz-Zentrum Berlin für Materialien und Energie, Berlin, *Rev. Sci. Instr.*, 82, 055101 (2011).

57. M. Strobl, Future prospects of imaging at spallation neutron sources, *Nucl. Instr. and Meth. A*, 604, 646–652 (2009).

58. M. Strobl, R. Woracek, N. Kardjilov, A. Hilger, D. Penumadu, R. Wimpory, A. Tremsin, T. Wilpert, C. Schulz, and I. Manke, Time-of-flight neutron imaging for spatially resolved strain investigations based on Bragg edge transmission at a reactor source, *Nucl. Instr. Meth. A*, 651, 149–155 (2012).

59. S. Peggs, et al. ESS Conceptual Design Report. Research Report. European Spallation Source, Lund, Sweden (2012).

60. M. Strobl, N. Kardjilov, A. Hilger, E. Jericha, G. Badurek, and I. Manke, Imaging with polarized neutrons, *Physica B*, 404, 2611–2614 (2009).

61. C. Grünzweig, F. Pfeiffer, O. Bunk, T. Donath, G. Kühne, G. Frei, M. Dierolf, and C. David, Design, fabrication, and characterization of diffraction gratings for neutron phase contrast imaging, *Rev. Sci. Inst.*, 79, 053703 (2008).

62. A. L. Damato, Berthold K. P. Horn, and Richard C. Lanza, Coded Source Imaging for Neutrons and X-Rays, 2006 IEEE Nuclear Science Symposium Conference Record, N6-11 (2006).

63. H. R. Beguiristain, I. S. Anderson, and C. D. Dewhurst, A simple neutron microscope using a compound refractive lens, *Appl. Phys. Lett.*, 81, 4290 (2002).

64. D. F. R. Mildner and M.V. Gubarev, Wolter optics for neutron focusing, *Nucl. Instr. Meth. A*, 634, S7–S11 (2011).

65. D. S. Hussey, H. Wen, H. Wu, T. R. Gentile, W. Chen, D. L. Jacobson, J. M. LaManna, and B. Khaykovich, Demonstration of focusing Wolter mirrors for neutron phase and magnetic imaging, *J. Imaging*, 4, 50 (2018).

5 Neutron Spin Optics: Concepts, Verification, and Prospects

Nikolay Pleshanov

CONTENTS

5.1 INTRODUCTION: NEUTRON SPIN OPTICS AS EXPANSION FROM 1D TO 3D

Polarized neutrons are used to obtain direct, detailed, and reliable information about the magnetic state of the sample and about the features of the dynamical processes that define fundamental physical properties of high-tech materials. The increase both in the throughput and in the polarizing efficiency of neutron polarizers and analyzers facilitates the solution of the research tasks.

Neutron coatings used in polarizers and analyzers select spins in one of the opposite directions [1–3]. The possibility of expansion from 1D (spin selection) to 3D (spin rotations) by using coatings that rotate neutron spins seems to be quite promising. To be of practical interest, the neutron spins should be rotated about one axis, the angle of rotation should weakly depend on the momentum transfer and the reflectivities should be close to 1.

The appropriate solutions were found [4, 5] and new possibilities for spin manipulations were dubbed as neutron spin optics (NSO) [6]. NSO is based on polarization effects related to superposition of the opposite spin components of the waves reflected from one or more interfaces in magnetic layered structures, in particular, on a spatial separation of the spin components ("spin splitting").

A new direction in polarized neutron instrumentation is based on quantum aspects of interaction of neutrons with magnetically anisotropic layers. NSO opens new possibilities for spin manipulations and may lead to numerous innovations in neutron techniques. Among the advantages of anticipated NSO devices over the existing spin manipulation options [7] are

- compactness (miniaturization is practically unlimited);
- zero-field option (no external fields are required, guide fields are optional);
- multifunctionality (spin manipulations, beam spectrum and divergence can be handled at the same time); and
- new possibilities and new solutions, e.g., hyperpolarization.

5.2 NEUTRON SPIN SELECTION WITH POLARIZING COATINGS

It follows from Chapter 1 (see Section 1.6) that neutrons with the spin up (+) and down (−) are differently reflected from magnetic coatings. When one of the reflectivities significantly exceeds the other (usually $R_- \ll R_+$), such coatings may be used for designing neutron optical polarizers and analyzers (Figure 5.1).

In order to minimize the spin-down neutron reflectivity (R_-), a material with the magnetic potential V_m equal to the nuclear potential V_n should be chosen. To achieve a reasonably high reflectance (R_+), the spin-up neutron potential ($V_+ = V_n + V_m$) must be sufficiently large. Most suitable material is CoFe alloy.

Neutron polarizers on the basis of mirrors made of bulk magnetic materials (Figure 5.2a) required very strong magnetizing fields produced by heavy and cumbersome magnets, considerable roughness, and curvature of the mirrors worsening the reflectance. Therefore, transition to easily magnetized thin (about 100 nm) ferromagnetic films on glass substrates with roughness below 1 nm and almost perfect flatness (Figure 5.2b) was an important achievement. In order to suppress total reflection of neutrons with the undesired spin from the glass substrate, an absorbing $Ti_{85}Gd_{15}$ layer weakly reflecting neutrons ("antireflective underlayer") was used (Figure 5.2c) [8]. The integral polarizing efficiency of the guide assembled from such thin-film mirrors reached a record high level 0.97 [9].

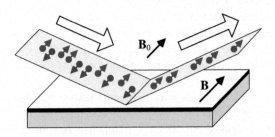

FIGURE 5.1 The principle of operation of a polarizing neutron coating: the probabilities of reflection of neutrons with the opposite spins are essentially different.

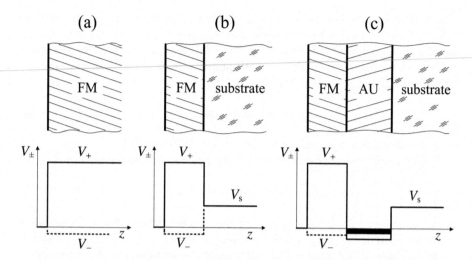

FIGURE 5.2 Polarizing mirrors on the basis of bulk ferromagnetic (FM) (a), thin ferromagnetic film on a substrate (b), and on an antireflective underlayer (AU) (c).

The neutron beam can also be monochromatized by reflection. To this end, one uses periodic structures made of alternating materials with potentials V_1 and V_2, quite different from each other [10]. Each pair of neighboring layers forms a bilayer, its thickness being equal to the multilayer period. Write the Bragg equation for the first-order reflection as

$$\lambda_B = 2D, \tag{5.1}$$

where D is the optical thickness of the bilayer, λ_B is the first-order Bragg peak position at the axis of the "perpendicular" wavelength $\lambda_\perp = \lambda/\sin\theta$, θ is the glancing angle. Reflection of neutrons from the monochromator is maximum with quarter-wavelength ($\lambda_B/4$) layers. Taking account of Bragg Eq. (5.1) and refraction in the layers, one finds the quarter-wavelength thicknesses of layers made of materials 1 and 2:

$$d_{1,2} = \frac{\lambda_B}{4\sqrt{1-(\lambda_B/\lambda_{c1,2})^2}}, \tag{5.2}$$

where λ_{c1} and λ_{c2} are characteristic wavelengths of the materials. Designate the respective critical momentum transfers as q_{c1} and q_{c2}.

It follows from the dynamic theory of reflection from a multilayer with N bilayers with quarter-wavelength layers and perfect interfaces that the peak reflectivity (at $q = q_B$) is [11]

$$R_{dyn}(q_B) = \tanh^2\left[N\frac{q_{c2}^2 - q_{c1}^2}{2q_B^2}\right]. \tag{5.3}$$

The peak reflectivity approaches 1, when N grows. The higher the optical contrast $\left| q_{c2}^2 - q_{c1}^2 \right|$ and the thicker the layers (the smaller q_B), the fewer bilayers are required to achieve the designed reflection level. The following expression of the kinematic theory of reflection from a multilayer with quarter-wavelength layers is often used to estimate the Bragg peak width [11]:

$$\Delta\lambda_B = \lambda_B / N. \tag{5.4}$$

Note that the peak is narrower with more bilayers in the multilayer.

The total reflection edge is defined either by the average potential of the multilayer

$$\bar{V} = (V_1 d_1 + V_2 d_2) / (d_1 + d_2) \tag{5.5}$$

or by the substrate potential V_s (by the greater potential). Note that the potential of a silicon substrate is approximately two times lower than that of a glass substrate. Reflection from the substrate can be suppressed by an antireflective absorbing under-layer. To reduce the portion of neutrons beyond the Bragg peak, one can also shift the peak along the q-axis from the total reflection edge by diminishing the thickness of the bilayers. To suppress higher Bragg reflection orders, it is normally sufficient to reflect the beam in succession from two multilayer monochromators.

When one of the materials in the multilayer (assume that it is material 2) is magnetic, its potential is different for neutrons with the spin up (V_2^+) and down (V_2^-) (Figure 5.3). Provided that $V_1 = V_2^-$, the Bragg reflection of spin-down neutrons is minimum and the reflected monochromatic beam is polarized. The efficiency of the first polarizing monochromator Fe/Ge was about 0.9 [12]. The polarizing efficiency of the up-to-date monochromators exceeds 0.99 [13].

To increase the reflected beam intensity, supermirror coatings made as numerous bilayers with different thickness are used. Each bilayer consists of quarter-wavelength layers. An increase in the angular range where neutrons are efficiently reflected is achieved by a coherent enhancement of reflection from one or another group of bilayers with close thicknesses, i.e., due to the quasi-Bragg reflection. A gradual variation of thicknesses of the layers from bilayer to bilayer makes it possible to efficiently reflect neutrons from the supermirror at all glancing angles up to an angle in m times exceeding that which corresponds to the total reflection edge of one of the best neutron reflectors, nickel.

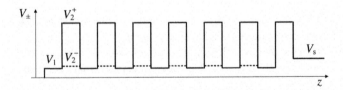

FIGURE 5.3 The polarizing monochromator is made as a periodic structure of alternating nonmagnetic and magnetic layers with potentials, respectively, V_1 and $V_2^\pm$ (spin-up and spin-down neutron potentials are different) on a substrate with a potential V_s.

The first supermirror was produced in 1977 [14]. Since then the supermirrors became important elements of neutron optical devices used in all leading neutron centers for transportation, formation, polarization, and polarization analysis of neutron beams.

Polarizing supermirrors are built as a sequence of alternating nonmagnetic and magnetic layers tuned to minimize neutron optical contrast for one of the spin components. An increase of m for polarizing supermirrors means not only an increase in luminosity of instruments, but also a higher degree of beam polarization, the possibility to polarize beams with a larger divergence and a wider spectral range, as well as the possibility to analyze the polarization of such beams. The increase in the degree of the beam polarization not only facilitates the polarized neutron data analysis, but also gives new experimental possibilities, e.g., measurement of a weaker signal of spin-flip scattering.

To achieve a maximum polarization of the beam, the crossed geometry of two supermirror polarizers, one with vertical channels and the other with horizontal channels, was proposed and tested [15]. A coil was used to produce a field adiabatically guiding the neutron spins between the polarizers. The beam polarization was found to be as high as $99.72 \pm 0.10\%$. However, the intensity was reduced due to a double, horizontal and vertical, collimation of the beam. To avoid such a loss in the intensity, one needs to use one polarizer with essentially improved supermirror coatings.

The reflectivity R_- at working angles can be as low as 0.7–0.8% for the supermirrors, e.g., Fe/Si supermirrors on glass (SwissNeutronics, Switzerland) and CoFe/TiZr supermirrors on the TiZrGd underlayer (PNPI NRC KI, Russia) (Figure 5.4). To achieve the beam polarization as high as 99.7% with one polarizer, R_- should be reduced to 10^{-3},

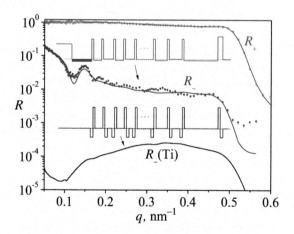

FIGURE 5.4 The experimental (points) and fitting (curves) spin-up (R_+) and spin-down (R_-) neutron reflectivities of a CoFe/TiZr supermirror with the TiZrGd underlayer as functions of the momentum transfer q [17]. The curve R_-(Ti) is the spin-down neutron reflectivity calculated for a supermirror with barriers and Ti (1.2 nm) interlayers at the interfaces, and other factors enhancing the reflection excluded. In the inserts: spin-down neutron potential profiles used in fitting R_- and in calculating R_-(Ti). (Modified from Ref. [16].)

i.e., practically by one order of magnitude. Such a method of suppressing reflection of neutrons with the undesired spin has been suggested and verified [16].

First of all, a satisfactory and consistent fitting of R_+ and R_- was achieved (Figure 5.4) [17]. The difference between structural and magnetic roughness was the last straw that broke the camel's back. The model also included interfacial roughness growing from layer to layer. The main factors that contribute into the reflectivity R_- for a CoFe/TiZr supermirror are the numerous regions with zero magnetization at each interface and the surface oxide layer, both leading to formation of the potential barriers for spin-down neutrons. Therefore, to improve the polarizing efficiency of supermirrors, one has to suppress neutron reflection from the barriers.

The theory of layers suppressing reflection from the barriers is presented in Ref. [16]. An *antibarrier* layer adjoins a layer, appearing as a barrier of height V_b and thickness b, and is designed as a layer with a negative potential V_a and thickness

$$a = \frac{2}{|q_a|}\left[\arccos\frac{|q_a|}{\sqrt{|q_a|^2 + q_b^2}} - \arctan\frac{2|q_a|q_b\exp(-bq_b)}{|q_a|^2 + q_b^2 + \left(|q_a|^2 - q_b^2\right)\exp(-bq_b)}\right], \quad (5.6)$$

where we used the critical momentum transfers for the materials of the barrier and the well:

$$q_{a,b} = \sqrt{8mV_{a,b}} / \hbar. \quad (5.7)$$

Suppression of reflection from the barriers is efficient, when $bq_b < 1$. In most practical cases $bq_b \ll 1$; then Eq. (5.6) coincides with the zero average potential condition:

$$bV_b + aV_a = 0. \quad (5.8)$$

Calculations in Figure 5.4 demonstrate the potential of the method: Ti interlayers (with the negative potential) at each interface reduce reflection from the barriers at the interfaces by two orders of magnitude (cf. curves R_- and $R_-(Ti)$). Note that other factors enhancing the reflection, except the barriers at the interfaces, are excluded in calculation of $R_-(Ti)$.

Calculations of R_- for a supermirror with Ti interlayers at the interfaces are represented in Figure 5.5; only one of the main factors enhancing reflection of neutrons with the undesirable spin is included in each calculation, namely: surface oxide layer (curve 1), antireflective underlayer (2), mismatch of the potentials of magnetic and nonmagnetic layers (3). It is to be noted that reflection from the oxide layer may be reduced by depositing a Ti layer on top: due to oxidation, a Ti/TiO_2 bilayer is formed with a well/barrier potential profile. The efficiency of the bilayer in suppressing neutron reflection at low q was experimentally demonstrated [18]. The composition and the thickness of the antireflective underlayer were optimized in Ref. [19] so that it does not worsen the polarizing efficiency of CoFe/TiZr supermirrors. It was also shown in the paper that one can reduce reflection from the underlayer to any reasonable value by decreasing the content of Gd and increasing the thickness of the underlayer. Reflection due to a difference in the potentials of magnetic and nonmagnetic layers is minimized by matching the composition of materials of the layers.

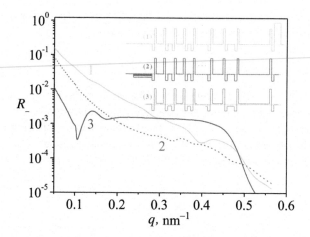

FIGURE 5.5 The spin-down neutron reflectivities R_- calculated as functions of q for a supermirror with Ti (1.2 nm) interlayers at the interfaces. In the insert: spin-down neutron potential profiles used in calculation of R_- and including surface oxide layer (1), antireflective underlayer (2), a 5 neV mismatch of the potentials of magnetic and nonmagnetic layers (3). (Modified from Ref. [16].)

In the first experiments on the use of Ti nanolayers the reflectivities R_- have been obtained for CoFe/Ti/TiZr/Ti supermirrors first without [16] and then with [20] the underlayer (Figure 5.6). The underlayer suppresses R_- at small q; comparatively high level of R_- at large q is related to presence of unwanted oxygen in the working chamber at the beginning of the given sputtering process. The measurement of R_-, which is less than R_+ almost by three orders of magnitude, could be made only owing to high polarization of the beam. Additional measurements with the coatings, named *superpolarizing* coatings, led to the following conclusion [21]: in spite of the fact that CoFe/Ti/TiZr/Ti supermirror contained about 200 magnetic layers, the portion of

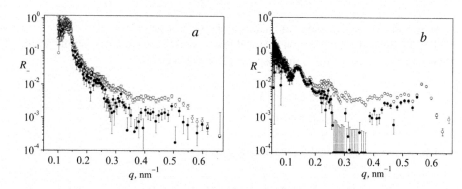

FIGURE 5.6 The spin-down neutron reflectivities R_-, not corrected (unfilled circles) and corrected (filled circles) for the incident beam polarization and the flipper efficiency, for CoFe/Ti/TiZr/Ti supermirrors (a) without and (b) with the TiZrGd underlayer as functions of q. The thickness of Ti interlayers is 1.5 nm. (Data from Ref. [16].)

neutrons reflected with the spin flip is less than 10^{-3}. New sputtering machines with four targets are needed to develop the advanced coatings.

5.3   NEUTRON POLARIZATION VECTOR

NSO exploits 3D nature of polarization. Therefore, consider first the theoretical background concerning the concept of the neutron polarization vector.

The neutron spin is represented by a quantum mechanical operator $\hat{\mathbf{s}}$, which acts in the spin subspace and transforms the spin components without noticing the spatial coordinates. In practice, the Pauli operator is used, which is proportional to the spin operator:

$$\hat{\sigma} = 2\hat{\mathbf{s}} / \hbar. \tag{5.9}$$

Its measurement yields the polarization vector

$$\mathbf{P} = \langle \hat{\sigma} \rangle \equiv \langle \Psi | \hat{\sigma} | \Psi \rangle / \langle \Psi | \Psi \rangle. \tag{5.10}$$

Since the spatial and spin subspaces are orthogonal, the quantization axis Z can be chosen independently of the axes in the coordinate representation. Choosing the quantization axis Z and the orths $\mathbf{e}_X$, $\mathbf{e}_Y$, $\mathbf{e}_Z$ of a Cartesian coordinate system, the Pauli operator can be written with the respective spin projection operators as

$$\hat{\sigma} = \hat{\sigma}_X \mathbf{e}_X + \hat{\sigma}_Y \mathbf{e}_Y + \hat{\sigma}_Z \mathbf{e}_Z. \tag{5.11}$$

The operators of the spin projection onto the respective axes are represented by the Pauli matrices:

$$\hat{\sigma}_X = \begin{pmatrix} 0 & 1 \\ 1 & 0 \end{pmatrix}_Z, \ \hat{\sigma}_Y = \begin{pmatrix} 0 & -i \\ i & 0 \end{pmatrix}_Z, \ \hat{\sigma}_Z = \begin{pmatrix} 1 & 0 \\ 0 & -1 \end{pmatrix}_Z. \tag{5.12}$$

Here and further the quantization axis is shown as a subindex.

The length of the polarization vector $\mathbf{P}$ is assumed to be equal to 1, so the vector $\mathbf{P}$ visualizes a vector of the spin subspace, a normalized spinor

$$|A\rangle = \begin{pmatrix} a_\uparrow \exp(i\alpha_\uparrow) \\ a_\downarrow \exp(i\alpha_\downarrow) \end{pmatrix}_Z, \quad a_{\uparrow,\downarrow} \geq 0, \quad a_\uparrow^2 + a_\downarrow^2 = 1, \tag{5.13}$$

which can always be brought into canonical form:

$$|A\rangle = \exp\left(i \frac{\alpha_\uparrow + \alpha_\downarrow}{2} \right) \begin{pmatrix} \cos(\gamma/2)\exp(-i\varphi/2) \\ \sin(\gamma/2)\exp(i\varphi/2) \end{pmatrix}_Z, \tag{5.14}$$

$$\varphi = \alpha_\downarrow - \alpha_\uparrow, \ a_\uparrow = \cos(\gamma/2), \ a_\downarrow = \sin(\gamma/2). \tag{5.15}$$

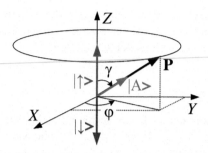

FIGURE 5.7 A geometrical visualization of the spinor: the angles γ and φ give the orientation the polarization vector **P**. The spinor can also be interpreted as a superposition of the orthogonal states, |↑> (spin-up) and |↓> (spin-down); the resultant spin along the polarization vector **P** visualizes the state |A>.

It follows from

$$\mathbf{P} = \left(\langle A | \sigma_X | A \rangle, \langle A | \sigma_Y | A \rangle, \langle A | \sigma_Z | A \rangle \right)_Z = \left(\sin\gamma\cos\varphi, \sin\gamma\sin\varphi, \cos\gamma \right)_Z \quad (5.16)$$

that γ is the angle between **P** and the quantization axis Z, φ is the angle between the axis X and the projection of the vector **P** onto the plane XY (Figure 5.7).

The spinor |A> can also be interpreted as a superposition of the orthogonal states, |↑> (spin-up) and |↓> (spin-down), and visualized by two spins, along and opposite to the quantization axis Z (Figure 5.7). A change in the ratio of the probability amplitudes of these states, $a_\uparrow / a_\downarrow$, brings about a change in the angle γ. An increase in the phase difference between these states is equivalent to an anti-clockwise rotation of **P** (neutron spin) about the quantization axis Z. When $a_\uparrow = a_\downarrow$, the vector **P** rotates in the plane XY.

In a uniform field **B**∥Z the components P_X and P_y, formally obtained from Eq. (5.10) in the conventional approach, vanish after integration over the entire configuration space. The polarization vector **P** averages out due to precession of the spin in the field **B**. An adequate approach is to introduce the spin operator at a point [22]:

$$\hat{\sigma}(\mathbf{r}) \equiv \hat{\sigma}\delta(\mathbf{r}' - \mathbf{r}) = \delta(\mathbf{r}' - \mathbf{r})\hat{\sigma}. \quad (5.17)$$

Its measurement yields the neutron polarization vector at a point **r** (at any time t):

$$\mathbf{P}(\mathbf{r},t) = \left\langle \Psi(\mathbf{r}',t) \middle| \hat{\sigma}(\mathbf{r}) \middle| \Psi(\mathbf{r}',t) \right\rangle / \left\langle \Psi(\mathbf{r}',t) \middle| \delta(\mathbf{r}' - \mathbf{r}) \middle| \Psi(\mathbf{r}',t) \right\rangle. \quad (5.18)$$

Now, introducing a trajectory by a velocity **v** and an initial point $\mathbf{r}_0$, one finds the equation of motion of the polarization vector along the trajectory:

$$\mathbf{P}(t) = \frac{\left\langle \Psi(\mathbf{r},t) \middle| \hat{\sigma}(\mathbf{r}_0 + \mathbf{v}t) \middle| \Psi(\mathbf{r},t) \right\rangle}{\left\langle \Psi(\mathbf{r},t) \middle| \delta(\mathbf{r} - \mathbf{r}_0 - \mathbf{v}t) \middle| \Psi(\mathbf{r},t) \right\rangle} = \frac{\left\langle \Psi(t) \middle| \hat{\sigma} \middle| \Psi(t) \right\rangle}{\left\langle \Psi(t) \middle| \Psi(t) \right\rangle}. \quad (5.19)$$

In this manner, the behavior of **P** was analyzed in Ref. [22] by using the theorem on superposition of two spinors [23] and plane wave solutions for neutrons that traverse regions with magnetic fields without being scattered.

5.4   REFLECTION OPERATOR

Assume that the z-axis is perpendicular to the reflecting surface; z_0 and z_s are the top and bottom coordinates of a layered structure. The wave functions of incident, reflected and transmitted neutrons are bound as follows with the reflection operator $\hat{r}$ and the transmission operator $\hat{t}$:

$$|\psi_r(z_0)\rangle \equiv \hat{r}|\psi_i(z_0)\rangle \text{ and } |\psi_t(z_s)\rangle \equiv \hat{t}|\psi_i(z_0)\rangle. \tag{5.20}$$

The plane wave solutions above and below the layered structure are

$$|\Psi(z)\rangle = \begin{cases} \exp[i\hat{k}_{0,z}(z-z_0)]|\psi_i(z_0)\rangle + \exp[-i\hat{k}_{0,z}(z-z_0)]\hat{r}|\psi_i(z_0)\rangle & (z \le z_0), \\ \exp[i\hat{k}_{s,z}(z-z_s)]\hat{t}|\psi_i(z_0)\rangle & (z \ge z_s). \end{cases} \tag{5.21}$$

The value $|\psi_i(z_0)\rangle$ of the spinor at the topmost interface of the layered structure is defined by orientation of the incident neutron spin on the sample surface. Substituting the incident and reflected wave functions, $|\psi_i(z_0)\rangle$ and $|\psi_r(z_0)\rangle$, into Eq. (5.19), one obtains $\mathbf{P}_0$ and $\mathbf{P}$, the polarization vectors of the incident and reflected beams at the mirror surface ($\mathbf{v} = 0$). In particular,

$$\mathbf{P}_0(t) = \frac{\langle \psi_i(z_0)|\hat{\sigma}|\psi_i(z_0)\rangle}{\langle \psi_i(z_0)|\psi_i(z_0)\rangle}. \tag{5.22}$$

Reflection of neutrons from magnetic layers is completely described by the reflection operator $\hat{r}$, which is represented by a reflection matrix. The reflection matrix elements are defined by the parameters of the layers and the external media. They are independent of the incident neutron spin orientation, but, sure, they depend on the representation basis and on the momentum transfer

$$q = (4\pi \sin \theta) / \lambda. \tag{5.23}$$

Usually, the quantization axis is chosen along a physically selected direction. When it is along the guide field $\mathbf{B}_0$, the states with the opposite spins will be designated with the signs (+) and (–), and the reflection matrix can be written as

$$\mathbf{r} = \begin{pmatrix} r_{++} & r_{-+} \\ r_{+-} & r_{--} \end{pmatrix}_{\mathbf{B}_0}. \tag{5.24}$$

Note that the order of indices of the diagonal elements corresponds to the initial and final states; it is opposite to the standard numeration of the elements in a matrix, e.g., $r = \begin{pmatrix} r_{11} & r_{12} \\ r_{21} & r_{22} \end{pmatrix}$.

The diagonal elements of the reflection matrix describe reflection without reversal of the spin, the non-diagonal elements describe spin-flip reflection. The probabilities of non-spin-flip (NSF) and spin-flip (SF) reflections are given by the squares of the moduli of the respective reflection matrix elements:

$$R_{++} = |r_{++}|^2, \quad R_{-+} = \frac{k_0^+}{k_0^-}|r_{-+}|^2, \quad R_{+-} = \frac{k_0^-}{k_0^+}|r_{+-}|^2, \quad R_{--} = |r_{--}|^2, \quad (5.25)$$

where $k_0^\pm$ are the wave vector components normal to the sample surface. Account is taken of the changes in the wave vector components under reflection with the spin reversal.

5.5 SPIN ROTATION BY NEUTRON REFLECTION

Quantum mechanically, the neutron spin evolution is the result of superposition of the coherent spin states. Wherever the moduli of the opposite spin components do not change in time and space, the spin evolution may be described as precession. The quasi-classical Larmor precession picture and the quantum approach can be reconciled, when the spin-dependent scattering is negligible. Particularly, in a homogeneous magnetic medium both quasi-classical and quantum approaches lead to precession with the Larmor frequency, even in the essentially non-classical case when the energies of the opposite spin components are different. It is only when the neutron state is a superposition of non-orthogonal states with different energies, the spin behavior in a homogeneous magnetic medium may have no classical counterpart [23, 24].

For specular reflection of neutrons from a layered structure the Schrödinger equation is reduced to one-dimensional equation with plane wave solutions for the incident and reflected neutrons. As a result of reflection from a layered structure with magnetic moments not collinear to the incident neutron spin, the spin orientation changes. Thus, one can design magnetic structures that provide a required change in the spin orientation.

The idea to make neutron mirror spin turners including flippers was mentioned as early as 1994 [25]. Yet, at first glance, the idea to turn spins by reflection of neutrons seemed to be quite impractical. For example, an efficient flipping under total reflection from a $Co_{36}Fe_{64}$ (170 nm) film was observed [26] in a field 5.9 mT, but, quite disappointing and discouraging, the respective q-range is narrowed down to 1 point (Figure 5.8). To be of practical interest,

- the spin-up (+) and spin-down (−) reflectivities $R_\pm$ should be close to 1,
- the spins should rotate about one axis, and
- the rotation angles should be independent of q.

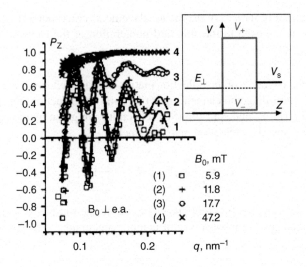

FIGURE 5.8 The measured (points) and fit (solid curves) polarization component P_z of neutrons reflected from a CoFe film with a thickness $d = 170$ nm and potentials $V_\pm$ on a glass substrate with a potential V_s in different fields is represented as a function of q. In the neutron experiment the applied field is perpendicular to the easy axis. In the inset: the spin-up (+) and spin-down (–) neutron components with a normal energy $E_\perp$ undergo total reflection at different interfaces of the magnetic film (spin splitting). (Modified from Ref. [26].)

Therefore, magnetically anisotropic layers with collinear fields are of special interest for us. Magnetic moments in layers with high coercitivity and high remanence remain to be collinear to the easy axis, independently of the orientation of a sufficiently weak external field. One can describe the spin evolution under reflection of neutrons from such a structure as rotation about the direction **b** of the fields in the reflecting layers only when the reflectivities of neutrons with the spin up (+) and down (–) the vector **b** are equal:

$$R_+ = R_-. \tag{5.26}$$

Otherwise, the angle between the spin and **b** changes and the spin evolution is not mere rotation about **b**. Note that *precession condition* (5.26) is in contrast to the requirement of maximum difference between R_+ and R_- for polarizing coatings.

To observe the precession under reflection, the spins of the incident neutrons should be inclined to the direction of the fields in the reflecting layers. By definition, the precession angle is positive for the anticlockwise rotation of the neutron spin. The sign is the same as in Larmor precessions of the neutron spin.

Precession condition (5.26) is satisfied under total reflection of neutrons: $R_+ = R_- = 1$. Total reflection of neutrons from a mirror was demonstrated by Fermi and Zinn [27] as early as 1946. The phase shift of the wave function under total reflection of a neutron was demonstrated by the author [28] in 1994.

Reflection of a neutron wave from a boundary of partition with a nonmagnetic medium is known to be described by the Fresnel coefficient

$$r = (1 - n_\perp) / (1 + n_\perp). \qquad (5.27)$$

The refractive index $n_\perp = \sqrt{1 - V / E_\perp}$ is defined for a neutron in the medium with a potential V, $E_\perp = E \sin^2\theta$ is the energy of the neutron with the velocity equal to the velocity component perpendicular to the boundary of partition, θ is the glancing angle. When the total reflection condition $E_\perp < V$ is fulfilled, the refractive index $n_\perp$ is a pure imaginary number and

$$r = (1 - i\sqrt{V / E_\perp - 1}) / (1 + i\sqrt{V / E_\perp - 1}) = z / z^* = \exp(i\delta). \qquad (5.28)$$

Therefore, the phase of the totally reflected wave is shifted with respect to the phase of the incident wave by an angle [28]

$$\delta = -2 \arccos \sqrt{E_\perp / V} = -2 \arccos(q / q_c). \qquad (5.29)$$

where q_c is the critical momentum transfer of material of the medium.

The potential of a medium with a magnetic field **B** is different for neutron states with the spin up (+) and down (−) the field:

$$V_\pm = V_n \pm |\mu_n B|, \qquad (5.30)$$

where V_n is the nuclear potential and $|\mu_n B|$ is the Zeeman energy. According to Eq. (5.29), the phase shifts of the opposite spin components under total reflection ($E_\perp \leq V_- < V_+$) of the neutron from the boundary with a semi-infinite magnetic medium differ, i.e., $\delta_+ \neq \delta_-$. It means that the neutron spin at the moment of reflection rotates about **B** by an angle

$$\varphi(q) = \delta_-(q) - \delta_+(q) = 2[\arccos(q / q_+) - \arccos(q / q_-)] \qquad (q \leq q_\pm), \qquad (5.31)$$

where $q_\pm$ are the critical momentum transfers of material of the magnetic medium for spin-up and spin-down neutrons. The sign of the spin rotation is the same as in Larmor precessions (anticlockwise) about **B**. However, the angle φ decreases with the neutron wavelength, whereas it is proportional to the neutron wavelength for the quasi-classical Larmor precession. Therefore, we shall talk about *quantum precession* of the spin.

A pronounced magnetic anisotropy with the related high remanence of an iron (120 nm) film was used in the experiments [29] on the direct observation of the spin rotation under total reflection of monochromatic neutrons from a boundary with a sufficiently thick magnetic layer by using 3D-analysis of neutron polarization at the neutron reflectometer NR-4M. To explain the experiments, define the X-axis parallel to the in-plane magnetization **M** in the sample and the Y-axis perpendicular to **M**. Under total reflection the neutron spin will rotate about the X-axis. In Figure 5.9 the polarization vectors of the incident and reflected beams are designated as P_0 and P, accordingly. When P_0 is made parallel to the Z-axis, the quantities P_Z / P_0 and P_Y / P_0 are proportional to $\cos\varphi$ and

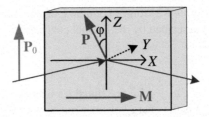

FIGURE 5.9   The reference frame XYZ is related to the in-plane magnetization **M** and the sample surface normal. Due to phase shifts under total neutron reflection the spin rotates by an angle φ about **M** (quantum precession).

$-\sin\varphi$. In Figure 5.10 the respective experimental points obtained by measurements at three glancing angles are represented by symbols □ and ■. The solid curves are the functions, $\cos(\varphi(q))$ and $-\sin(\varphi(q))$, calculated with the use of Eq. (5.31).

In addition, time-of-flight (TOF) measurements with the same sample were carried out [30] at the neutron reflectometer EROS (LLB, Saclay, France). The sample was magnetically saturated in a strong external field $\mathbf{B}_0$ parallel to the sample surface. Then it was turned about its surface normal by $\pi/2$ in zero field and adjusted to a glancing angle $\theta = 0.4°$. When a weak field $\mathbf{B}_0$ was introduced, the sample magnetization **M** remained perpendicular to it. The incident beam polarization $\mathbf{P}_0$ was parallel to $\mathbf{B}_0$. As a consequence, the reference frame XYZ as shown in Figure 5.9 can be used again. As before, the X-axis is parallel to **M** and the Z-axis is parallel to $\mathbf{P}_0$.

The projection of the reflected beam polarization **P** onto the guide field was measured by 1D-analysis of polarization. Polarizing CoFe/TiZr supermirrors were used to polarize the incident beam and analyze the reflected beam. The experimental

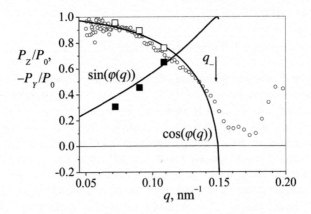

FIGURE 5.10   The experimental ratios P_Z/P_0 and $-P_Y/P_0$ are obtained for an Fe (120 nm) film with 1D (o) and 3D (□ and ■) polarization analysis as functions of q. To protect it from oxidation, the Fe film was coated with a Ti (20 nm) layer. Solid curves are the functions $\cos(\varphi(q))$ and $\sin(\varphi(q))$ calculated for $q < q_-$ (the edge of the total reflection q-range is marked with an arrow) by using Eq. (5.31).

points (o) are represented in Figure 5.10. The edge of the total reflection region is characterized by the critical momentum transfer q_- which is marked with an arrow.

Although Eq. (5.31) was derived for a boundary of partition of two semi-infinite media, the agreement between theory and experiment in the region of total reflection $q < q_-$ is quite satisfactory. A discrepancy observed points out to the sensitivity of the method to the surface state. The smaller q, the thinner the surface region of interaction with the neutron; the method is depth-resolving and can be used to study surface magnetism and magnetism of nanolayers.

The maximum of the quantum precession angle φ in this experiment is close to $\pi/2$ (Figure 5.10). According to Eq. (5.31), φ increases from 0 at $q = 0$ to a maximum $2\arccos(q_- / q_+)$ at $q = q_- < q_+$. For a given θ, the neutron with a shorter λ penetrates deeper into the non-classical under-barrier region with a field. It corresponds with a larger precession angle φ at larger q. Note that a decrease in q_- reduces the range of working glancing angles. The precession angle φ can be made $\pi/2$ (when $q_+ \geq \sqrt{2}\, q_-$), but it can never reach π.

Theoretically, once reflecting neutrons one can rotate the spin by $\pi/2$. However, the use of such spin manipulators is restricted by a rather strong dependence of the precession angle on the momentum transfer, i.e., on the neutron wavelength and the glancing angle. Methods of leveling the spin rotation angles are described in Section 5.6.

Even a more peculiar precession is induced by spatial separation of the opposite spin components ("spin splitting"), when the spin-up and spin-down neutron components are totally reflected from different boundaries of a magnetic layer with a thickness d and the potential eigenvalues $V_\pm$ (Figure 5.8). Neglecting reflection of spin-down neutrons from the first boundary, one finds the quantum precession angle as

$$\varphi = qd[1 - (q_- / q)^2]^{1/2} + \varphi_- - \varphi_+ \qquad (E_\perp \leq V_+, V_s), \qquad (5.32)$$

where V_s is the neutron potential of the substrate. The first term is related to different paths of the neutron waves representing the opposite spin states and the other two terms are the phase shifts caused by total reflection from (now different) boundaries. The spin-up neutron wave, totally reflected from a boundary with a sufficiently thick magnetic layer, decays exponentially in the layer, and only the spin-down component reaches the substrate, the respective phase shifts being

$$\varphi_+ = -2\arccos(q / q_+), \quad \varphi_- = -2\arccos(q' / q_-'), \qquad (5.33)$$

where account of the changes in the kinetic energy and in the barrier height for the spin-down neutrons within the magnetic layer is taken with

$$q' / q_-' = \sqrt{(E_\perp - V_-) / (V_s - V_-)}. \qquad (5.34)$$

Again the quantum precession angle is anticlockwise for most parameters, but it can also be made clockwise ("antiprecession"), e.g., when $V_- = 0$, $V_+ < V_s$ and $qd < \varphi_+ - \varphi_-$. When $V_+ = V_s$ and $V_- = 0$, the total reflection conditions for the opposite spin components are identical $\left(\varphi_{TR}^+ = \varphi_{TR}^-\right)$ and the precession angle $\varphi = qd$.

To summarize, the non-classical spin behavior is related to purely quantum features of the interaction, in which the probabilities to find one and the same particle in the opposite spins states may essentially differ in a region with magnetic fields. Such interaction is impossible for a classical particle, which can be found at any time in a certain point and with a certain angular momentum. For the first time the signature of the phase shifts of the wave function under total reflection of a massive particle, the neutron, has been observed by means of a thin semi-transparent film as an interference element [28]. A shift in the phases of the opposite spin states upon total reflection manifests itself in the polarization vector rotation ("quantum precession"). Direct evidences of the quantum precession under total reflection have been furnished with vector polarization analysis [29] and spin-echo analysis [31]. The spatial separation of the opposite spin components ("spin splitting" [32]) has been observed by measuring one of the polarization vector components [26, 33] and exploited in NSE spectrometers [34] at J-PARC (Tokai, Japan).

In conclusion, note that, when neutrons traverse a magnetic layer without being reflected at its boundaries, the neutron spin rotation under reflection obeys the Larmor precessions law, i.e., the angle φ is proportional to magnetic induction, total neutron path in the magnetic layer, and neutron wavelength λ. The latter means that $\varphi \sim q^{-1}$. Generally, one can speak about Larmor and quantum (non-Larmor) precessions under reflection.

5.6 LEVELING THE SPIN ROTATION ANGLE

Mirror spin turners are the basic NSO elements. Define the efficiency of a spin turner as the portion of neutrons reflected with the spin in the desired direction **u** to be achieved by turning the incident beam polarization $\mathbf{P_0}$ by a certain angle φ about a rotation axis **b**:

$$\varepsilon(\varphi) = (1 + P_u / P_0) / 2, \tag{5.35}$$

where P_u is the projection of the reflected beam polarization **P** onto **u** and

$$P_0 = (I_+ - I_-) / (I_+ + I_-) \tag{5.36}$$

($I_\pm$ are the intensities of neutrons incident in the states with the spin up (+) and down (−) the guide field.) Note that Eq. (5.35) is a generalization of the conventional definition of the efficiency of flippers for spin turners. In particular, with respect to a quantization axis Z, the efficiency of a mirror flipper is

$$f = \varepsilon(\pi) = (1 - P_Z / P_{0,Z}) / 2. \tag{5.37}$$

The flipper efficiency is unity, when $P_Z = -P_{0,Z}$.

To find an algorithm for building layered structures that reflect neutrons with designed precession angles is still a formidable task. Several approaches to building mirror spin turners with $R \sim 1$ and precession angles but slightly depending on q were suggested and substantiated by calculations [4, 5]. Some possibilities of different methods of stabilizing the precession angle can be perceived from the calculations of the efficiency of $\pi/2$-turners in Figure 5.11. All further calculations are made

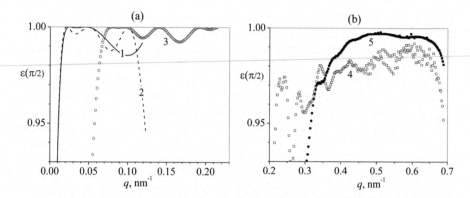

FIGURE 5.11 The efficiencies $\varepsilon(\pi/2)$ calculated as functions of q in the region of total reflection for the $\pi/2$-turners: glass coated with (1) Fe(20 nm)/Ti(20.5 nm)/TiO$_2$(3 nm); (2) Fe(20 nm)/optimized layer (40 nm, $V_n = 0$ neV, $V_m = 3$ neV)/Ti(5 nm)/TiO$_2$(3 nm); (3) NiMo(100 nm)/CoFe (5 nm)/optimized layer (57 nm, $V_n = -10$ neV, $V_m = 10$ neV)/Ti(5 nm)/TiO$_2$(3 nm); (4) NiMo/Ti ($m = 3$) (300 bilayers) supermirror, in which Ti layers in bilayers with numbers 1, 15, 77, 128 (counted from thicker bilayers) were substituted by Co layers, and a Co (7 nm) layer was added on top; (5) NiMo/X ($m = 3$) (300 bilayers) supermirror, where X stands for a material with optimized potentials $V_n = 0$, $V_m = 1.5$ neV, and with an additional Co (26.5 nm) layer on top. The points are obtained by convolution with a resolution $\Delta q/q = 0.03$. (Modified from Ref. [6].)

with the *generalized matrix* method [35, 36] (its later description is known as the *supermatrix* method [37]) on the assumption that the fields in magnetic layers, and hence the precession axis, are perpendicular to the incident neutron spins.

In the *biased total reflection* method the precession angle is leveled by using a layer with a negative potential $V < 0$ on top of the magnetic layer totally reflecting neutrons. Then $q/q_\pm$ in Eq. (5.31) should be replaced with

$$q' / q'_\pm = \sqrt{(E_\perp - V) / (V_\pm - V)}. \tag{5.38}$$

As a consequence, the range of precession angles in the total reflection region decreases. The precession is also affected by interference of the waves reflected from the two boundaries of the additional layer and, optimizing its thickness b, the dependence $\varphi(q)$ is additionally smoothened in the q-range, where reflection from the upper boundary is still insignificant. Curve 1 in Figure 5.11a is the efficiency of a $\pi/2$-turner on a glass substrate coated with Fe (20 nm), Ti (20.5 nm) with a negative potential and TiO$_2$ (3 nm) (effect of oxidation in air).

Another method of leveling the precession angle boils down to balancing Larmor and quantum precessions, the former decreasing and the latter increasing with q (*compensated precessions* method). Curve 2 in Figure 5.11a is the efficiency calculated for a $\pi/2$-turner on the basis of an Fe (20 nm) layer and a weakly reflecting magnetic layer with optimized thickness, nuclear V_n and magnetic V_m potentials. The magnetic layer is protected from oxidation by an antireflection bilayer Ti/TiO$_2$ (here TiO$_2$ is the native oxide layer) [16, 38] with a zero average potential for neutrons.

The upper limit of the working q-range is determined by the total reflection from the glass substrate with a low neutron potential. It can be increased in almost 2 times with a sufficiently thick (say, 100 nm) nonmagnetic NiMo layer on top of a glass substrate (see curve 3 in Figure 5.11a).

The method of *spatially distributed reflection* allows a multiple increase of the working q-range by using a sequence of bilayers with different thicknesses (as in a supermirror). Neutrons, reflected at different depths depending on q, spend the same time in magnetic layers so that their spins rotate by the same angle. This can be achieved either by incorporating magnetic layers (curve 4 in Figure 5.11b), or by using layers with low magnetization (curve 5), or else by including magnetic inter-layers at each interface of the supermirror (seems to be the most flexible, yet more detailed data on magnetism of such interlayers are needed for realistic estimations). These possibilities are illustrated in Figure 5.12.

The efficiencies in Figure 5.11b are good only at $q > 0.4$ nm^{-1}. Thick layers of the supermirrors function poorly. Removing these layers, we get efficiencies for $\pi/2$- and

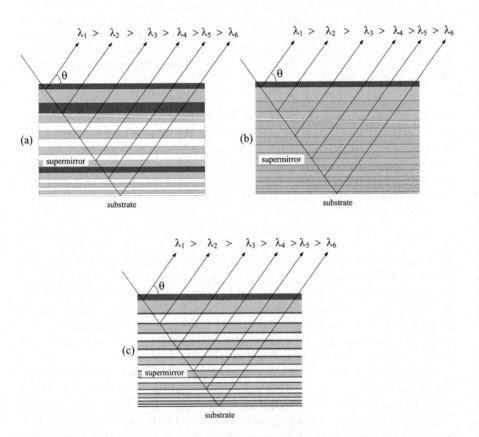

FIGURE 5.12 Building spin turners on the basis of supermirrors by incorporating magnetic layers (a), by using layers with low magnetization (b), or else by including magnetic interlayers at each interface (c).

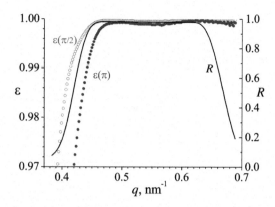

FIGURE 5.13 The reflectivity R and the efficiencies ε of spin $\pi/2$- and π- turners on the basis of the filters obtained from a NiMo/X ($m = 3$) supermirror by removing 50 thicker bilayers. In calculations the potentials of the X layers are $V_n = 0$ and $V_m = 1.5$ neV ($\pi/2$-turner), 3 neV (π-turner). A top Co layer is additionally deposited with a thickness 33 (66) nm, for the $\pi/2$ (π)-turners. The points are obtained by convolution with a resolution $\Delta q/q = 0.03$.

π-turners as high as 99.9% (Figure 5.13) in the working q-range, where neutron reflectivities for the truncated supermirrors (filters) are close to 1.

Record glancing angles and record efficiencies can be obtained with spin turners for monochromatic neutrons by using a magnetic coating on top of a nonmagnetic multilayer (multilayer-backed spin turners). This approach is described and experimentally verified in Section 5.8. It is based on the use of Larmor precession in a weakly reflecting magnetic layer backed by a strongly reflecting nonmagnetic multilayer. Then the rotation of neutron spins can be combined with monochromatization of the beam.

5.7 DESIGN OF MIRROR SPIN TURNERS

Performance of a mirror spin turner on the basis of magnetically anisotropic reflecting layers, in which the in-plane fields are along a unit vector **b**, is illustrated in Figure 5.14. The Z-axis is chosen along the incident beam polarization vector $\mathbf{P}_0$, the Y-axis is in the ($\mathbf{P}_0$,**b**) plane, and the X-axis is perpendicular to this plane. Represent the vector $\mathbf{P}_0 = P_0 (0, 0, 1)$ as the sum of the component $\mathbf{P}_{0\parallel}$ parallel to the precession axis **b** and the precessing component $\mathbf{P}_{0\perp}$. By definition,

$$\mathbf{P}_{0\parallel} = P_0 \cos\chi(0, \sin\chi, \cos\chi) = P_0 (0, \sin\chi\cos\chi, \cos^2\chi), \tag{5.39}$$

$$\mathbf{P}_{0\perp} = \mathbf{P}_0 - \mathbf{P}_{0\parallel} = P_0 (0, -\sin\chi\cos\chi, \sin^2\chi), \tag{5.40}$$

where χ is the angle between $\mathbf{P}_0$ and **b**.

Assume that the reflected beam polarization vector $\mathbf{P}$ is a result of precession of $\mathbf{P}_0$ about **b** by a designed value φ. As a consequence, $P = P_0$. The precession plane is

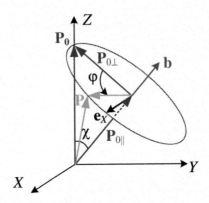

FIGURE 5.14 Performance of a mirror spin turner. Neutron spins rotate under reflection about the unit vector **b** parallel to the fields in a magnetically collinear layered structure.

perpendicular to the vector **b** lying in the (Y,Z) plane. The vector **P** may be written as the sum of $\mathbf{P}_{0\parallel}$ and the vector component in the precession plane, which, in its turn, can be decomposed in two mutually perpendicular components, along the vector $\mathbf{P}_{0\perp}$ and along the orth $\mathbf{e}_X$ (Figure 5.14):

$$\mathbf{P} = \mathbf{P}_{0\parallel} + \mathbf{e}_X P_{0\perp}\sin\varphi + \mathbf{P}_{0\perp}\cos\varphi$$
$$= P_0(\sin\chi\sin\varphi,\ \sin\chi\cos\chi(1-\cos\varphi),\ \cos^2\chi + \cos\varphi\sin^2\chi), \qquad (5.41)$$

where it is taken into account that the precessing component length $P_{0\perp} = P_0\sin\chi$.

Eq. (5.41) can be used to design mirror spin turners. Particularly, the condition $P_z/P_0 = \cos^2\chi + \cos\varphi\sin^2\chi = -1$ is equivalent to $\cos\varphi = 1 - 2/\sin^2\chi$, which is satisfied only with $\chi = \pm\pi/2$, $\varphi = \pm\pi$. It will be used in Section 5.7 to design a mirror flipper. The condition $P_x/P_0 = \sin\chi\sin\varphi = \pm1$ for a $\pi/2$ $(Z\to X)$ spin turner is satisfied with $\chi = \pm\pi/2$, $\varphi = \mp\pi/2$, and the condition $P_y/P_0 = \sin\chi\cos\chi\,(1-\cos\varphi) = \sin(2\chi)(1-\cos\varphi)/2 = \pm1$ for a $\pi/2$ $(Z\to Y)$ spin turner is satisfied with $\chi = \pm\pi/4$, $\varphi = \pm\pi$. Thus, one can turn spins from the initial Z-direction to $-Z$, $\pm X$, and $\pm Y$ directions (Figure 5.15).

Because of a beam divergence and a wavelength spread, the momentum transfer and the precession angle vary near the designed values q_0 and φ. The resultant polarization vector can be found by averaging the polarization vectors weighted by the reflected intensities for the corresponding q. When the precession angle deviates from φ by $\Delta\varphi$, the polarization vector $\mathbf{P}'$ can be found by replacing φ in Eq. (5.41) with $\varphi + \Delta\varphi$. According to Eq. (5.35), the respective efficiency of the spin turner is

$$\varepsilon(\varphi + \Delta\varphi) = (1 + P''/P_0)/2 = 1 - \sin^2\chi\,\sin^2(\Delta\varphi/2), \qquad (5.42)$$

where $P'' = \mathbf{P}'\mathbf{P}/P$ is the projection of the vector $\mathbf{P}'$ onto the designed direction $\mathbf{P}/P$.

For a mirror spin turner based on the Larmor precession, the precession angle is inversely proportional to q, hence

$$\Delta\varphi = -\varphi\Delta q/q, \qquad (5.43)$$

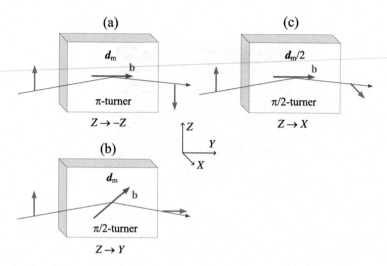

FIGURE 5.15 In a mirror flipper (a), the thickness of the magnetic layers (d_m) is chosen so that $\varphi = \pi$, and the magnetization is perpendicular to the incident neutron spins. Rotating the mirror flipper about its surface normal by $\pi/4$, we obtain a spin $\pi/2$-turner (b) that leaves spins in the mirror plane ($Z\rightarrow Y$). With magnetic layers thinner by half, one obtains a spin $\pi/2$-turner (c) that turns spins to the direction perpendicular to the mirror surface ($Z\rightarrow X$).

where $\Delta q = q - q_0$, and q_0 is the momentum transfer, at which the precession angle is equal to the designed value φ. E.g., $\chi = \pi/2$, $\varphi = \pi$ for a mirror flipper, hence $\varepsilon(\varphi + \Delta\varphi) = \cos^2(\Delta\varphi/2)$, which is consistent with the probability of the projection of the spin oriented along $\mathbf{P}'$ onto the designed direction $\mathbf{P}$. Hence, the NSF portion of neutrons as a function of q is

$$1 - f = \sin^2(\Delta\varphi / 2) = \sin^2[\pi(q - q_0) / 2q]. \tag{5.44}$$

5.8 EXPERIMENTAL VERIFICATION

The potential of the innovative optics is still to be revealed by studying polarization effects under neutron reflection from real structures, structural and magnetic factors that influence the functionality of magnetically anisotropic layers. The probing of the basic principles and elaboration of the basic elements for NSO, investigation of their performance and the effect of structural and magnetic imperfections are required. The basic elements for innovative neutron devices include mirror spin turners ($\pi/2$- and π-turners, in the first place).

A multilayer-backed spin turner for monochromatic beams is the simplest for realization. It is designed as a magnetic layer weakly reflecting neutrons on top of a nonmagnetic multilayer (Figure 5.16). At the Bragg peak (large q) the layers of Co, Fe, and their alloys weakly reflect neutrons and may be used to build mirror spin turners. Spins inclined to the magnetic induction vector $\mathbf{B}$ in the magnetic layer undergo Larmor precession twice, before and after reflection of neutrons from the multilayer. The precession angle depends on q, so the spread of precession angles is

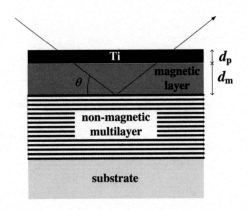

FIGURE 5.16 A multilayer-backed spin turner: nonmagnetic multilayer coated with a magnetically anisotropic layer. The functional magnetic layer with a thickness d_m is protected from oxidation by a layer (e.g., Ti) with a thickness d_p.

defined by the Bragg peak width, and the efficiency of the multilayer-backed spin turner is increased for multilayers with smaller d-spacings [4].

According to the Larmor precession law, the spin rotates about $\boldsymbol{B}$ by an angle

$$\varphi = CB\lambda \frac{2d_m}{\sin\theta}. \tag{5.45}$$

($C = 4\pi m_n|\mu_n|/h^2 \cong 4.63 \times 10^{-4}\ \mathrm{T^{-1}nm^{-2}}$) proportional to B, λ, and the total path length in the magnetic layer with a thickness d_m. At the Bragg peak ($q = q_B$), the Bragg condition is satisfied

$$4\pi \sin\theta = \lambda q_B, \tag{5.46}$$

and the precession angle at q_B is

$$\varphi = 8\pi CBd_m / q_B. \tag{5.47}$$

The Bragg reflectivity is maximum for quarter-wavelength layers at $q = q_B$:

$$d_{a,b} = \frac{D/2}{\sqrt{1-(q_{a,b}/q_B)^2}} = \frac{\pi}{\sqrt{q_B^2 - q_{a,b}^2}}, \tag{5.48}$$

where refraction in each layer in the bilayer with an optical thickness $D = 2\pi/q_B$ is taken into account ($q_{a,b}$ are the critical momentum transfers for materials a and b, alternating in the periodic multilayer).

The Larmor precession angle for neutrons reflected at the Bragg peak should be equal to π for a flipper. The required thickness of the magnetic layer can be found from Eq. (5.47):

$$d_m(\pi) = q_B / (8CB). \tag{5.49}$$

Then the structure will work as a flipper for neutrons reflected with q near q_B, providing that **B** is perpendicular to the spins of the incident neutrons.

The first neutron mirror spin flipper was built as a multilayer-backed flipper. A periodic structure of 20 pairs of nonmagnetic NiMo(6.96 nm) and Ti(6.17 nm) layers was sputtered onto two glass substrates at the sputtering machine LUNA (PNPI, Gatchina, Russia). The quarter-wavelength thicknesses of the samples correspond to $q_B = 0.5$ nm^{-1}. Note that the critical momentum transfer of Ti with a negative scattering length is an imaginary number. Due to a high aspect ratio of the CoFe target, the easy axis is introduced into the magnetic layer during sputtering. Owing to pronounced easy-axis anisotropy, the residual magnetization of the layers is quite high. One can find from the neutron data fitting in Ref. [39] that $B = 2.0$ T. One finds from Eq. (5.49) that the thickness of the magnetic $Co_{70}Fe_{30}$ layer is 71.1 nm. Two layers, CoFe(71.1 nm) and Ti(30 nm), were sputtered onto one of the samples. The Ti layer on top protects the functional layer from oxidation and prevents deterioration of its magnetic properties [18].

Experiments with the mirror flipper were carried out at the neutron reflectometer NR-4M (beam 13, reactor WWR-M, Gatchina). The maximum spectral density of the polarized beam was at the wavelength 0.130 nm. The spin-up (+) and spin-down (−) neutron reflectivities $R_{\pm}(q)$ for the mirror flipper in a field sufficiently strong to magnetically saturate the CoFe layer are represented in Figure 5.17. One can see from Figure 5.17c that the precession condition $R_+ = R_-$ is well fulfilled at the Bragg peak with $\Delta q/q = 7.5\%$ (FWHM). Note also that the peak maximum at a value 0.48 nm^{-1} corresponds to thicker layers in the NiMo/Ti multilayer.

The low q region, where the spin-up and spin-down reflectivities are quite different (cf. Figure 5.17a and Figure 5.17b), can be used to study the reversal of the CoFe layer magnetization by measuring integral reflectivities at a small glancing angle. Then the integral polarizing efficiency

$$P_{int}(B_0) = [R_{off}^{int}(B_0) - R_{on}^{int}(B_0)] / [R_{off}^{int}(B_0) + R_{on}^{int}(B_0)], \qquad (5.50)$$

measured as a function of the applied field **B**$_0$ represents the hysteresis loop; R_{off}^{int} and R_{on}^{int} are the integral reflectivities measured with the flipper switched off and on to

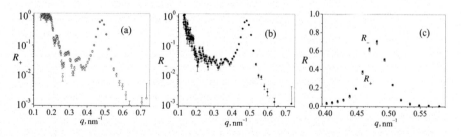

FIGURE 5.17 The spin-up (a) and spin-down (b) neutron reflectivities $R_{\pm}$ for the mirror flipper are represented as functions of q. The TOF data are obtained at a glancing angle $\theta = 4.95$ mrad in a field 47 mT. (c) $R_{\pm}$ are represented in the linear scale for a q-range near the Bragg peak. (Modified from Ref. [6].)

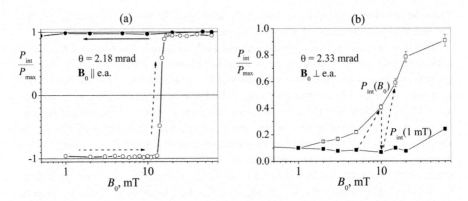

FIGURE 5.18 The dependence $P_{int}(\mathbf{B}_0)$, normalized to $P_{max}= \max(P_{int})$, obtained with a white beam of neutrons reflected from the mirror flipper (a) in the field $\mathbf{B}_0 \parallel$ easy axis in the CoFe layer, $\theta = 2.18$ mrad and (b) in the field $\mathbf{B}_0 \perp$ easy axis, $\theta = 2.33$ mrad. Previously the sample was magnetized in $\mathbf{B}_0 = -60$ mT along the easy axis; in the case (b) it was turned in $\mathbf{B}_0 = 0$ about its surface normal by 90° and then the field $\mathbf{B}_0$ changed as shown with the arrows: each rise from 1 mT is followed by a drop to 1 mT. (Reprinted from Ref. [6], with permission from Elsevier.)

change the sign of the incident beam polarization. The hysteresis loop with the field $\mathbf{B}_0$ along the easy axis (e.a.) is represented in Figure 5.18a. Note that the magnetization reversal starts near 12 mT.

The sample was magnetized in a field 60 mT, then the field was switched off and the sample turned about its surface normal by 90°. The measurements obtained with the field $\mathbf{B}_0 \perp$ e.a. are represented in Figure 5.18b. The field $\mathbf{B}_0$ changed as follows: $0 \rightarrow 1$ mT $\rightarrow 2$ mT $\rightarrow 1$ mT $\rightarrow 3$ mT $\rightarrow 1$ mT $\rightarrow \ldots$ As a consequence, the lower points (filled squares) in Figure 5.18b correspond to measurements in a field 1 mT after having applied the field with the value given in the abscissa axis. One can infer that no oppositely magnetized domains form in the fields up to 20 mT. The layer behaves as a single-domain magnetic spring. It means that the mirror spin flipper, in principle, can work in guide fields up to 20 mT.

What about its efficiency? In the standard method of measuring the flipper efficiencies, two flippers are switched off and on, and the four intensities measured with an analyzer give the efficiencies of both. However, a mirror spin flipper changes both intensity and direction of the beam, violating the requirement of the standard method that the source and the analyzer in the measurements be the same. So the scheme was modified. Both nonmagnetic multilayer and mirror flipper samples were used, one imitating the "flipper-off" state, and the other flipping spins (Figure 5.19). The flipper after the sample unit was found [40] to be more efficient with small fields at the sample, than that before the sample unit. Therefore, it was used for testing the mirror flipper, and the mirror flipper served as the first flipper in the modified scheme. The same glancing angles in each set of measurements meant the same neutron paths, therefore the same efficiencies of flipper 2 and the analyzer.

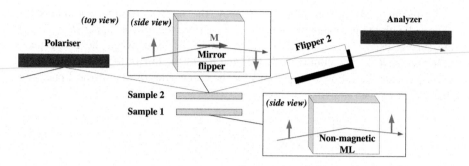

FIGURE 5.19 The scheme to measure the flipper efficiencies of the mirror flipper includes polarizer, flipper, analyzer, and two multilayers, one imitating the "flipper-off" state (non-magnetic sample 1) and the other flipping spins (magnetic sample 2). A low guide field $\mathbf{B}_0$ is perpendicular to the magnetization $\mathbf{M}$ in the mirror flipper.

Normalized TOF intensities measured for the nonmagnetic sample at $\theta = 4.95$ mrad in a field 1 mT with the analyzer and flipper 2 switched off and on show (Figure 5.20a) that the Bragg-reflected neutrons are sufficiently polarized, even though the polarization guiding is not optimal with low fields at the sample unit of the reflectometer. Normalized TOF intensities measured for the mirror flipper at the same glancing angle show (Figure 5.20b) that the polarization of the Bragg-reflected neutrons is reversed. The mirror works as a flipper, indeed.

A narrow slit on the single detector window guaranteed the same angle for the reflection from the analyzing supermirror (the same efficiency of the analyzer). Because of the additional reflection and the slit on the detector window, the normalized Bragg peaks in Figure 5.20 are noticeably lower than the Bragg reflectivities in Figure 5.17. However, only the ratios of the intensities matter in estimation of the flipping efficiency.

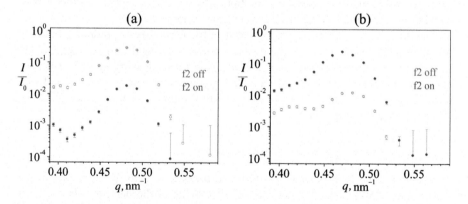

FIGURE 5.20 The TOF neutron intensities I measured near the Bragg peak for (a) the non-magnetic sample and (b) the mirror flipper at $\theta = 4.95$ mrad in a field 1.0 mT with flipper 2 switched off and on are represented as functions of q. The intensity in each TOF channel is normalized with the direct beam intensity I_0 in the respective channel. (Modified from Ref. [6].)

To evaluate the mirror flipper efficiency, the integral Bragg peak intensities obtained with the nonmagnetic ML ($I_{off,off}$, $I_{off,on}$) and the mirror flipper ($I_{on,off}$, $I_{on,on}$) are further used. The standard equations cannot be used, when the Bragg peak reflectivities R_0 for the nonmagnetic ML and R_F for the mirror flipper are not equal ($R_0 \neq R_F$). Then the following equations for the efficiencies of the mirror flipper and flipper 2 have been shown [6] to hold:

$$f_1 = \frac{1}{2}\left[1 + \frac{|\sigma_F|/\sigma_0}{1 + (P_0 P_A / \sigma_0 - 1)(|\sigma_F| + \sigma_0)/(1 + \sigma_0)}\right], \tag{5.51}$$

$$f_2 = \frac{1 + (P_0 P_A)^{-1}}{1 + \sigma_0^{-1}}, \tag{5.52}$$

where $P_0 P_A$ is the product of the efficiencies of the polarizer and the analyzer, the reflectivities R_0 and R_F cancel out in the quantities

$$\sigma_0 = \frac{I_{off,off} - I_{off,on}}{I_{off,off} + I_{off,on}}, \quad \sigma_F = \frac{I_{on,on} - I_{on,off}}{I_{on,on} + I_{on,off}}. \tag{5.53}$$

Measuring the intensities J_{off} and J_{on} in the absence of a mirror sample, one determines the product

$$P_0 P_A = [(1 + \sigma^{-1})f - 1]^{-1}, \quad \sigma = (J_{off} - J_{on}) / (J_{off} + J_{on}), \tag{5.54}$$

where f is the efficiency of the flipper used in the measurements. Each of two flippers of the reflectometer can be used. The quantity f is measured with the same guide fields in the standard scheme with two flippers and analyzer.

In the manner described the efficiencies of the mirror flipper and flipper 2 were found: $f_1 = 0.975 \pm 0.005$, $f_2 = 0.995 \pm 0.005$. It means that the spins of about 2.5% neutrons within the Bragg peak are not flipped by the mirror flipper. In other words, the NSF portion of the Bragg-reflected neutrons is about 2.5%.

The test of a neutron mirror flipper verified the feasibility of producing not only mirror flippers, but also mirror spin turners. Indeed, rotating the mirror flipper about its surface normal by $\pi/4$, we obtain a $\pi/2$ spin turner leaving spins in the mirror plane (Figure 5.15b). With magnetic layers thinner by half we obtain a $\pi/2$ spin turner that may turn initially in-plane spins to the direction perpendicular to the mirror surface (Figure 5.15c).

5.9   PROSPECTS OF DEVELOPING NSO

Theoretically, the NSF portion should be 10 times less, than that obtained in the experiment (2.5%). Indeed, calculated NSF portions in the neutron reflectivity at $q = q_B$ (dashed curve) and those in the integral Bragg peak reflectivity (solid curve) are represented in Figure 5.21 as functions of the magnetic layer thickness d_m, the other parameters being as designed. Due to spin-dependent reflection and a subtle interplay of the waves reflected from numerous interfaces the dashed curve does

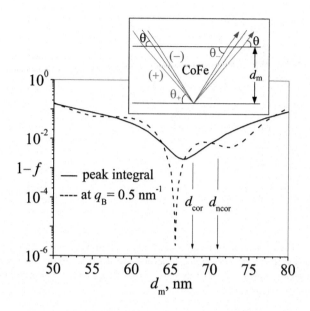

FIGURE 5.21 Calculated NSF portions in the neutron reflectivity at $q = q_B = 0.5 \text{ nm}^{-1}$ (dashed curve) and in the integral Bragg peak reflectivity (solid curve) for the spin flipper as functions of the CoFe layer thickness d_m. The arrows point to the thicknesses d_{ncor} and d_{cor} obtained with Eqs. (5.49) and (5.55), respectively. Insert: Account of the spin-dependent refraction in the magnetic layer gives a better estimation of the CoFe thickness optimum for spin flipping. Non-refracted rays are represented with dashed lines. (Reprinted from Ref. [6], with permission from Elsevier.)

not look like the square of a sine, as expected for a purely Larmor precession. The NSF integral (solid curve) is minimum (0.21%) at a thickness $d_m = 66.5$ nm, which is noticeably less than the thickness 71.1 nm, designed according to (5.49). Account of the spin-dependent refraction in the magnetic layer (see the insert in Figure 5.21) yields the precession angle in terms of $q_\pm$, the critical momentum transfers for neutrons with the spin up (+) and down (−) the field **B** in the CoFe layer:

$$\varphi(q) = d_m \left(\sqrt{q^2 - q_-^2} - \sqrt{q^2 - q_+^2} \right). \tag{5.55}$$

The correction for the spin-dependent refraction improves estimation of the thickness optimum for spin flipping. It follows from $\varphi(q_B) = \pi$, that $d_m = d_{cor} = 67.8$ nm. The thicknesses not corrected (d_{ncor}) and corrected (d_{cor}) for refraction are shown in Figure 5.21 with the arrows. One can see that the thickness d_{cor} corresponds to the NSF integral equal to 0.3%. The thickness d_{ncor} corresponds to 1% for the NSF integral calculated for $q_B = 0.5 \text{ nm}^{-1}$. The shift of q_B to 0.48 nm^{-1} increases the NSF integral up to 2.5%, matching the experimental value. One may hope that the efficiency of the mirror flipper will approach the theoretical maximum 99.8%, when all thicknesses are matched (see the respective reflectivity and the NSF portion of neutrons in Figure 5.22).

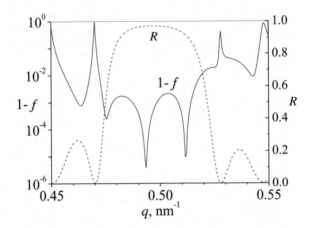

FIGURE 5.22 The neutron reflectivity R (dashed curve) and the NSF portion, 1-f, calculated as functions of q with the exact numerical method (solid curve) and in the approximation of purely Larmor precessions in a CoFe (66.5 nm) layer (dotted curve). (Reprinted from Ref. [6], with permission from Elsevier.)

Multilayer-backed flippers can combine monochromatization of a polarized beam with flipping spins of the monochromatized neutrons. When the polarized beam is initially monochromatic, supermirror-backed flippers with reflectivities close to 1 in a wide q-range can be used. Due to a wider working q-range, the resultant intensity may be increased up to 4 times. No precise adjustment of the magnetic layer thickness is needed. According to the calculations with a nonmagnetic NiMo/Ti ($m = 3$) supermirror backing a CoFe (66.5 nm) layer, the theoretical efficiency exceeds 99% in a q-range about 10% [6].

Such NSO devices can be used only with narrow and well-collimated beams, as used in neutron reflectometry. Further developments of multichannel designs and supermirror-based spin turners may appreciably lift the restrictions. Then NSO can be made broad-band, which is of special interest for numerous applications, especially at impulse reactors and spallation sources.

5.10 INNOVATIVE NSO DEVICES: CONCEPTION

Spin-turning reflectors give new solutions for manipulations with neutron spins. In particular, compact $\pi/2$-turners and flippers can be used as elements for neutron reflectometry with 3D-polarimetry, in NSE schemes, etc. Moreover, combining spin-turning reflectors, one can design compact spin manipulators such as a 3D polarizer (Figure 5.23a) for producing beams with polarization in any desired direction defined by the polar angle γ and the azimuthal angle δ. It consists of a spin selector (polarizer with pronounced remanence) and a $\pi/2$-turner. The required angles γ and δ are set by rotating the reflectors about their surface normals to change the orientation of the in-plane magnetizations M_1 and M_2. A spin rotator (Figure 5.23b) for rotating spins from any direction in the reflecting plane to any desired direction can be designed with a π-turner (mirror flipper) and a $\pi/2$-turner. A spin manipulator (Figure 5.23c)

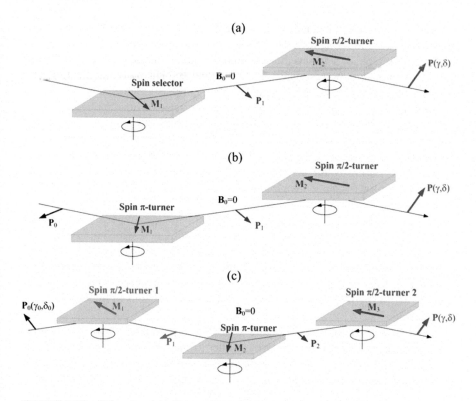

FIGURE 5.23 Schemes of (a) 3D polarizer, (b) spin rotator, and (c) spin manipulator. The neutron reflectors rotate about their reflecting surface normals.

for rotating spins from any direction defined by the angles γ_0 and δ_0 to any desired direction can be designed with a π-turner (mirror flipper) between two $\pi/2$-turners. The elements in the schemes in Figure 5.23 are supposed to be in zero field.

A mirror flipper can be used in a spin echo scheme to guide polarization from the spin selector to the sample in a uniform field $\mathbf{B}_0$ (Figure 5.24). The sample and the spin selector are at the same distance from the mirror flipper. All reflecting surfaces are parallel. The spin selector forms a beam with precessing polarization, the precession front being parallel the selector surface. The polarization at the sample surface

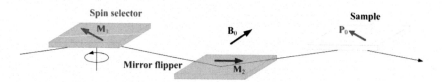

FIGURE 5.24 The mirror flipper used in a spin echo scheme to guide polarization from the spin selector polarizer to the sample in a uniform field $\mathbf{B}_0$. The selector can be rotated about its surface normal. All reflecting surfaces are parallel.

is parallel to the selector magnetization, independently of the neutron wavelength (spin echo effect). Rotating the selector about its surface normal, one can change the direction of the spin of neutrons incident onto the sample surface. The mirror flipper magnetization should be perpendicular to the field $\mathbf{B}_0$. Therefore, the field may be either parallel (Figure 5.24) or perpendicular to the flipper surface plane.

Installing an additional mirror flipper and a spin selector functioning as an analyzer, one can measure the reflected beam polarization components lying in the sample surface plane. To this end it is again sufficient to fulfill the spin echo conditions and to carry out measurements of the intensities for four selector magnetization orientations differing by 90°. To measure the component perpendicular to the sample surface, a $\pi/2$-turner is put in the place of the selector, and the selector is placed after the $\pi/2$-turner.

New possibilities from using mirror flippers include hyperpolarization of neutron beams. Hyperpolarizers are conceived as devices, which not only separate neutrons with the opposite spins, but also flip the wrong spins. In full accordance with the Liouville theorem, the polarized neutron flux can be doubled only at the expense of increased beam divergence (hyperpolarizers of type I) or beam width (hyperpolarizers of type II). Different designs of hyperpolarizers are possible.

In a hyperpolarizer of type I in Figure 5.25a ("hyperpolarizing cavities"), the spin-up neutrons are reflected from a polarizing filter (truncated supermirror sequence or periodic multilayer on a substrate transparent for neutrons); the spin-down neutrons are reflected from a mirror flipper in the spin-up state and traverse the polarizing

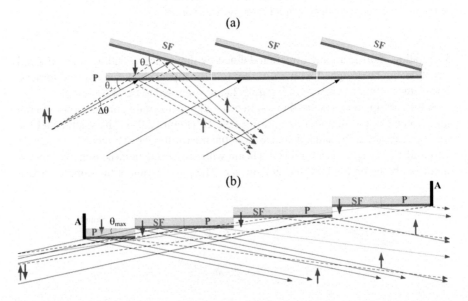

FIGURE 5.25 Hyperpolarizers built (a) by multiplication of hyperpolarizing cavities and (b) as a "hyperpolarizing staircase": P is a polarizing coating, SF is a spin-flipping coating, A is a neutron absorber.

coating at larger glancing angles without being reflected. In a hyperpolarizer of type II in Figure 5.25b ("hyperpolarizing staircase"), the polarizing supermirror coatings shadow the spin-flipping coatings so that the spin-up neutrons are reflected from the polarizing coatings and the spin-down neutrons are reflected from the spin-flipping coatings. Substrates are made from materials (silicon, quartz, etc.) quite transparent for neutrons.

Thus, hyperpolarizers double the polarized neutron flux by increasing either the divergence or the width of the beam. In most cases such a beam, with increased divergence or width, can be prepared with a conventional polarizer. Hyperpolarization may be useful, when the beam to be polarized has been restricted (for some reasons, e.g., by a neutron guide) in divergence or width.

5.11 CONCLUSION

Nanolayers with a negative neutron potential at the interfaces in polarizing coatings may suppress reflection of neutrons with the undesired spin to $10^{-3} - 10^{-4}$. The method was experimentally verified with Ti interlayers. Other factors enhancing reflection should also be addressed to develop superpolarizing coatings. Building polarizers and analyzers of new generation imposes new requirements on magnetic systems as shown in Ref. [41].

NSO will provide new possibilities and solutions in polarized neutron instrumentation. They include 3D spin manipulations and hyperpolarization. Neutron spin-turning reflectors (particularly, $\pi/2$-turners and π-turners) may be used directly or combined. Some examples were given. The test of a neutron mirror flipper verified the possibility of producing neutron mirror spin turners, including $\pi/2$- and π-turners (flippers). Calculations demonstrated that performance of the spin turners can be improved.

NSO devices may play an important role in developing alternative schemes of measurements, especially compact schemes of measurements with small samples, which are often of special interest for neutron studies. Therefore, the probing of the basic principles and elaboration of the basic elements for NSO are quite essential. The potential of NSO is still to be revealed by the studies of spin precession phenomena with real structures. Coercivity and remanence of the magnetic layers in NSO coatings are of decisive importance. The magnetic layers for mirror spin turners are too thin to produce significant stray fields. Anyway, the stray fields from remanent coatings can efficiently be compensated [29]. Surface oxidation is also to be addressed as an important issue for the work of the spin turners based on total reflection phenomenon.

NSO will contribute to development of tools for obtaining information about the objects under study. In particular, spin-turning reflectors with surfaces parallel to the sample surface can be used to implement spherical polarimetry in neutron reflectometry. Polarimetry enhances sensitivity to the phases of the reflection matrix elements. Therefore, more detailed information about magnetic mirror samples can be extracted. Polarimetry with very cold and even ultracold neutrons can be realized with NSO as a counterpart of ellipsometry, but with extremely high sensitivity to surface magnetism.

ACKNOWLEDGMENTS

The work was supported by the Federal target program of Ministry of Education and Science of Russian Federation (project No. RFMEFI61614X0004).

REFERENCES

1. O. Schärpf, "Properties of beam bender type neutron polarizers using supermirrors," *Physica B*, 156&157, 639–646 (1989).
2. A. F. Schebetov et al., "Multichannel neutron polarisers produced in PNPI," *J. Phys. Soc. Japan*, 65 (Suppl. A), 195–198 (1996).
3. P. Böni et al., "Applications of remanent supermirror polarizers," *Physica B*, 267–268, 320–327 (1999).
4. N. K. Pleshanov, "Neutron spin manipulation optics: basic principles and possible applications," *J. Phys.: Conf. Ser.*, 528, 012023 (2014).
5. N. K. Pleshanov, "Neutron spin-turning reflectors," *J. Surf. Invest.: X-ray, Synchrotron Neutron Tech*, 9, 24–34 (2015).
6. N. K. Pleshanov, "Neutron spin optics: fundamentals and verification," *Nucl. Instrum. Methods A*, 853, 61–69 (2017).
7. W. G. Williams, *Polarized Neutrons*. Oxford: Clarendon Press, 1988.
8. G. M. Drabkin et al., "Polarization of a neutron beam on reflection from a magnetized mirror," *JETP*, 42, 972–977 (1975).
9. G. M. Drabkin et al., "Multilayer Fe-Co mirror polarizing neutron guide," *Nucl. Instrum. Methods*, 133, 453–456 (1976).
10. B. P. Schoenborn, D. L. D. Caspar, and O. F. Kammerer, "A novel neutron monochromator," *J. Appl. Cryst.*, 7, 508–510 (1974).
11. V. F. Sears, "Theory of multilayer neutron monochromators," *Acta Cryst. A*, 39, 601–608 (1983).
12. J. W. Lynn et al., "Iron-germanium multilayer neutron polarizing monochromators," *J. Appl. Cryst.*, 9, 454–459 (1976).
13. C. F. Majkrzak and L. Passell, "Multilayer thin films as polarizing monochromators for neutrons," *Acta Crystallogr. A*, 41, 41–48 (1985).
14. F. Mezei and P. A. Dagleish, "Corregendum and experimental evidence on neutron supermirrors," *Commun. Phys.*, 2, 41–43 (1977).
15. M. Kreuz et al., "The crossed geometry of two super mirror polarisers—a new method for neutron beam polarisation and polarisation analysis," *Nucl. Instrum. Methods A*, 547, 583–591 (2005).
16. N. K. Pleshanov, "Superpolarizing neutron coatings: theory and first experiments," *Nucl. Instrum. Methods A*, 613, 15–22 (2010).
17. N. K. Pleshanov et al., "Observation of difference in nuclear and magnetic roughness in CoFe/TiZr multilayers by polarized neutron reflectometry," *Physica B*, 397, 62–64 (2007).
18. V. A. Matveev and N. K. Pleshanov, "On using Ti nanofilms in neutron spin optics," *J. Neutron Research*, 20, 107–111 (2018).
19. N. K. Pleshanov et al., "Specular reflection of thermal neutrons from Gd-containing layers and optimization of antireflective underlayers for polarizing coatings," *Nucl. Instrum. Methods A*, 560(2), 464–479 (2006).
20. N. K. Pleshanov, A. P. Bulkin, and V. G. Syromyatnikov, "A new method for improving polarizing neutron coatings," *Nucl. Instrum. Methods A*, 634, S63–S66 (2011).
21. N. K. Pleshanov, A. P. Bulkin, and V. G. Syromyatnikov, "Experimental verification of the applicability of a new method for improving polarizing neutron coatings," *Physics of the Solid State*, 52, 1018–1020 (2010).

22. N. K. Pleshanov, "On precessing spin neutron reflectometry," *J. Surf. Invest: X-ray, Synchrotron Neutron Tech.*, 12, 8–19 (2019).

23. N. K. Pleshanov, "Quantum aspects of neutron spin behavior in homogeneous magnetic field," *Physica B*, 304, 193–213, (2001).

24. N. K. Pleshanov, "Quantum nutation of the neutron spin," *Phys. Rev. B*, 62, 2994–2997 (2000).

25. N. K. Pleshanov, "Neutrons at the boundary of magnetic media," *Z. Phys. B*, 94, 233–243 (1994).

26. V. M. Pusenkov et al., "Study of domain structure of thin magnetic films by polarized neutron reflectometry," *J. Magn. Magn. Mat.*, 175, 237–248 (1997).

27. E. Fermi and W. H. Zinn, "Reflection of neutrons on mirrors," *Phys. Rev.*, 70, 103–110 (1946).

28. N. K. Pleshanov, "Observation of the phase shift of the neutron wave function under total reflection," *Physica B*, 198, 70–72 (1994).

29. N. K. Pleshanov et al., "Neutron reflectometry with vector polarization analysis: first steps," *J. Surf. Invest.: X-ray, Synchrotron Neutron Tech.*, 2, 846–855 (2008).

30. N. K. Pleshanov, A. Menelle, and V.M. Pusenkov (unpublished).

31. V.-O. deHaan et al., "Observation of the Goos-Hänchen shift with neutrons," *Phys. Rev. Lett.*, 104, 010401 (2010).

32. T. Ebisawa et al., "Quantum precession of cold neutron spin using multilayer spin splitters and a phase-spin-echo interferometer," *Phys. Rev. A*, 57, 4720–4729 (1998).

33. B. P. Toperverg, H. J. Lauter, and V. V. Lauter-Pasyuk, "Larmor pseudo-precession of neutron polarization at reflection," *Physica B*, 356, 1–8 (2005).

34. S. Tasaki et al., "Development of a modified neutron spin echo spectrometer using multilayer spin splitters," *Physica B*, 335, 234–237 (2003).

35. N. K. Pleshanov, "Spin particles at stratified media: operator approach," *Z. Phys. B*, 100, 423–427 (1996).

36. N. K. Pleshanov and V. M. Pusenkov, "Application of generalized matrix method to neutrons in magnetically non-collinear stratified media," *Z. Phys. B*, 100, 507–511 (1996).

37. A. Rühm, B. P. Toperverg, and H. Dosch, "Supermatrix approach to polarized neutron reflectivity from arbitrary spin structures," *Phys. Rev. B*, 60, 16073–16077 (1999).

38. V. A. Matveev et al., "The study of the oxidation of thin Ti films by neutron reflectometry," *J. Physics: Conf. Ser.*, 340, 012086 (2012).

39. N. K. Pleshanov et al., "Interfacial roughness growth and its account in designing neutron CoFeV/TiZr supermirrors with m=2.5," *Physica B*, 369, 234–242 (2005).

40. N. K. Pleshanov and V. G. Syromyatnikov, "Testing the first neutron mirror flipper," *Nucl. Instrum. Methods A*, 837, 40–43 (2016).

41. A. G. Gilev et al., "Magnetic systems for wide-aperture neutron polarizers and analyzers," *Nucl. Instrum. Methods A*, 833, 233–238 (2016).

6 Applications of Neutron Optics to Biomedicine: BNCT

Ignacio Porras

CONTENTS

6.1 INTRODUCTION

In addition to the well-known application of neutron scattering to structural biology, there is a very relevant application of neutrons to health sciences: the so-called Boron Neutron Capture Therapy (BNCT), a promising experimental form of radiotherapy, which is facing a new era with renewed projects and under active research. BNCT, among all forms of radiotherapy, has the special feature of delivering the radiation dose selectively at the cellular level. It is based on the production of heavy charged particles inside tumor cells by means of a nuclear reaction between the incident neutrons with ^{10}B nuclei that are previously introduced in the tumor cells by means of boron compounds, selectively uptaken by the malignant cells. The reaction that takes place more frequently at the tumor cells is

$$n + {}^{10}\text{B} \rightarrow {}^{7}\text{Li} + \alpha + 2.79\,\text{MeV}. \tag{6.1}$$

This reaction is chosen because the cross section is huge (3800 barns for thermal neutrons). Apart for a gamma ray of 478 keV that is produced in 94% of the captures, the most part of the energy is delivered by means of an alpha particle and recoil ^{7}Li nucleus, which have a very short range, smaller than the cell diameter, not affecting the surrounding cells. In addition to this, these particles produce a strong biological damage, as being radiation with a high-linear energy transfer (LET) to the medium, greater than the one corresponding to conventional radiation (photons or electrons). They are able to produce double-strand breaks in DNA, which are difficult to repair. For this reason, BNCT is considered a form of therapy, which has both physical (more energy is delivered at the tumor cells than to healthy ones) and biological (the biological effect is greater for a given energy delivered) advantages.

Then, BNCT is a binary therapy in which two nontoxic agents meet toward the selective cell destruction. This application is illustrated in Figure 6.1.

With these features, BNCT is especially suited for the treatment of tumors which are disseminated in critical organs, or for local recurrences already treated with radiation, reaching the dose limit of conventional radiation. Precisely, these cancers are among those of worst prognosis, and conventional therapies are not effective for their treatment.

In this chapter, the state of the art of this therapy and some advances that are expected in the near future are reviewed. These include some basic research toward new boron compounds in which facilities with neutron guides are being used. Finally, the potential use of guided thermal or cold neutrons for BNCT is discussed.

6.2 THE HISTORY OF BNCT

6.2.1 THE BEGINNING OF BNCT AND THE FIRST CLINICAL TRIALS IN THE UNITED STATES

The history of BNCT dates back to the decade in which the neutron was discovered. In 1936, just four years after the appearance of Chadwick's paper

BNCT: Effective Treatment In One Day

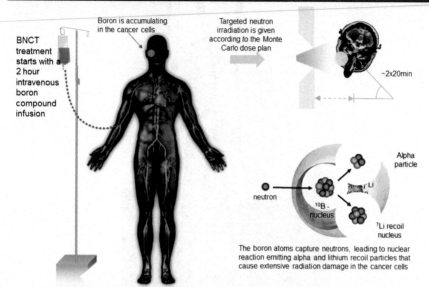

BNCT treatment starts with a 2 hour intravenous boron compound infusion

Boron is accumulating in the cancer cells

Targeted neutron irradiation is given according to the Monte Carlo dose plan

~2x20min

neutron

Alpha particle

^{10}B - nucleus

^{7}Li recoil nucleus

The boron atoms capture neutrons, leading to nuclear reaction emitting alpha and lithium recoil particles that cause extensive radiation damage in the cancer cells

(Courtesy of L. Kankaanranta & M. Kortesniemi, Helsinki University Central Hospital)

FIGURE 6.1   Illustration of the BNCT procedure.

"Existence of a Neutron", Gordon L. Locher (Figure 6.2) thought up [1] that, if a boron compound were able to be selectively accumulated in a tumor, irradiation with neutrons would produce tumor cell killing by the boron capture reaction.

The clever idea of Locher was the combination of two agents that are not expected to show toxicity by themselves: a boron compound and the irradiation with low-energy neutrons. It is interesting to note that in 1937 with Mary Nagai he published at *Nature* [2] an observation of strong mutations produced in *Drosophila melanogaster* when irradiated by a handmade source of fast neutrons (a radium-beryllium isotopic source) when compared to these same neutrons thermalized, what show the relative harmless of the latter as compared to the former, whose effect he attributed to the recoil protons produced in the elastic collisions with hydrogen by the fast neutrons.

It was until 1950 that the first clinical trial of BNCT took place. It was conducted by the Massachusetts General Hospital neurosurgeon William Herbert Sweet (Figure 6.3), who aimed for sparing normal brain tissues during radiotherapy of brain tumors [3]. In his activities, he got the participation of a Japanese Fullbright Scholar, Hiroshi Hatanaka (Figure 6.3), who started a strong BNCT program when he came back to Japan that lasts to the present dates.

The Brookhaven Graphite Research Reactor (BGRR) of the Brookhaven National Laboratory (BNL) was the neutron source used for the first clinical trials performed by Sweet and coworkers. This was the first peace-time reactor built in the United States following World War II, and its primary mission was to produce neutrons for

FIGURE 6.2 Gordon Lee Locher, scientist at the Bartol Research Foundation of the Franklin Institute, Swarthmore, Pennsylvania, demonstrating in 1936 a Geiger-Müller counter developed by himself and manufactured by Herbach and Rademan. (Image from http://national-radiation-instrument-catalog.com/press_photos.htm).

William H. Sweet   Hiroshi Hatanaka

FIGURE 6.3 W. H. Sweet and H. Hatanaka, the pioneers of the BNCT clinical trials. From R. F. Barth and A. H. Soloway. *Int. J. Radiation Oncology Biol. Phys.*, 28, 5, 1057–1058 (1994). Copyright (1994) Elsevier Science Ltd.

scientific experimentation and to refine reactor technology. In 1959, a second reactor Brookhaven Medical Research Reactor (BMRR) built specifically for medical applications, including BNCT, and equipped with a treatment room started operations and was used for clinical trials. Also, the Massachusetts Institute of Technology Reactor (MITR) was used during these years. Different boron compounds were tested, such as boric acid, sodium pentaborate, and derivatives of phenylboronic acid. A detailed description of these clinical trials can be found in Ref. [4]. The results were not successful, and this was addressed to the low selectivity of the boron compounds and to the poor quality of the neutron beam (excessive presence of gamma rays in the beam). BNCT clinical trials were stopped in the 1960s in the United States, but these efforts opened the way to the continuation of the BNCT progress, which expanded to Japan.

6.2.2 Development of BNCT in Japan

H. Hatanaka, after participating in the US clinical trials, came back to Japan and started a BNCT program. In 1968, they began a set of clinical trials based on the administration of sodium mercaptoundecahydrododecaborate ($Na_2B_{12}H_{11}SH$, commonly known as BSH), a new boron compound whose applicability to BNCT had been studied by Soloway et al. [5]. They treated malignant brain tumor patients at different reactors: Japan Research Reactor No. 1 (JRR-1), the Hitachi Training Reactor (HTR) and later, the Kyoto University Research Reactor (KUR), starting on 1974. KUR was a heavy water reactor and a high-purity thermal beam was used. However, the poor penetrability of the thermal neutrons made necessary the use of craniotomy for the irradiation in order to reach deep-seated tumors, a process that was called intraoperative BNCT. Hatanaka published in 1990 very promising results from his clinical trials with grade III and IV gliomas (Figure 6.4), in which he states, "any patient who arrived asking for the treatment was accepted without regard to the severity of the patient's condition" [6].

The work of Hatanaka showed the limitations of BNCT with thermal neutrons for very deep-seated tumors. An ingenious attempt to improve the neutron penetrability was the partial replacement of water by heavy water in the brain (taking advantages of the smaller scattering cross section of deuterium with respect to hydrogen) that was performed in his patients of Group 4.

From this stage of BNCT, I personally would like to mention also the cases of children reported by Y. Nakagawa in the 13th International Congress on Neutron Capture Therapy held in Florence, 2008 [7]. It was the first of these conferences that I attended and was impressed by the illustrative cases reported, like a one-year old child whose diagnosis was an anaplastic ependymoma with a high tumor mass (Figure 6.5). She received intraoperative BNCT and we were informed that then, 12 years after the treatment, magnetic resonance imaging (MRI) demonstrated neither tumor recurrence nor brain atrophy. She had good clinical course and was going to junior high school without neurological deficits. Conventional radiotherapy is very difficult to apply to very small children with brain tumors due to the risk of severe damage, like mental retardation, hormonal insufficiency, or brain atrophy.

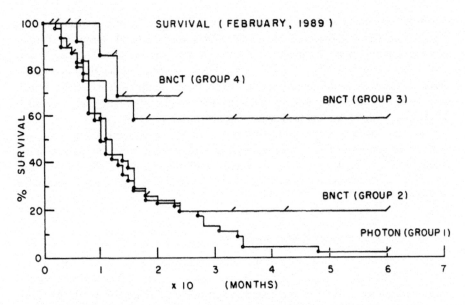

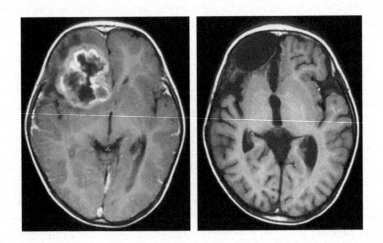

FIGURE 6.4 Survival Kaplan-Meier plot from the paper of 1990 of Hatanaka (Ref. [6]) of the outcome of treatments for grade III–IV glioma. Group 1 corresponds to 46 patients receiving conventional therapies (including irradiation with photons, hence the tag). Group 2 corresponds to all BNCT patients treated between 1968 and 1985, from which those who had the tumor within 6 cm from the brain surface (12 patients) are sorted out in Group 3. Group 4 corresponds to BNCT treatments performed in the last three years. (Plenum Press, New York, e-ISBN-13:978-1-4684-5802-2).

FIGURE 6.5 NMR images of the one-year-old child at the time of diagnosis (left) and one year after BNCT (right). (*Appl. Radiat. Isot.* 67 (2009) S27-S30. Elsevier.)

In the 1990s, with the development of epithermal neutron sources, the need of performing craniotomy to the patient (under general anesthesia) was finally removed.

6.2.3 MODERN BNCT: EPITHERMAL NEUTRON SOURCES AND THE USE OF BORONOPHENYLALANINE—CLINICAL TRIALS EXPANDED TO OTHER COUNTRIES

Although the therapeutic effect at the tumor is done by thermal neutrons, their poor penetrability can be overcome by the use of more energetic neutrons that are thermalized in the body, thus reaching thermal energies in a few centimeters' depth, as illustrated in Figure 6.6.

Consequently, by using an epithermal neutron beam, no craniotomy is needed for neutron irradiation at the research reactor site when BNCT is applied to malignant brain tumors. For this reason, strong efforts for modifying available reactors for obtaining neutron beams for BNCT started in the 1990s. The first BNCT irradiation without craniotomy with an epithermal beam was performed in 1994 at the BNL research reactor. In Japan, KUR heavy-water thermal neutron facility was modified so that it could also use epithermal neutrons from 1995. These modifications were followed by similar changes at facilities, as the JRR reactor at Tokai, Ibaraki. Other countries succeeded on the obtainment of epithermal beams and started clinical trials, mainly for glioblastoma multiforme (GBM; see Table 6.1).

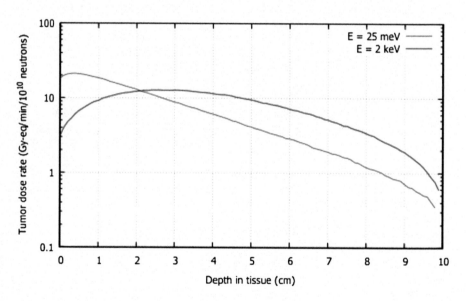

FIGURE 6.6 Depth-dose profile produced in boron-loaded tissue for 2 keV neutrons compared to thermal ones, from a Monte Carlo simulation assuming a boron concentration of 35 ppm.

TABLE 6.1

Modern BNCT Clinical Trials (since 1994) with Epithermal Beams for Brain Tumors and Head and Neck Cancers under Protocols (many other patients not included here have been treated out of protocol). GBM, glioblastoma multiforme; OBT, other brain tumors; HNC, head and neck cancers. Those trials published up to 2012 are included in the list, although KURR (Japan) and THOR (Taiwan) facilities are still active doing BNCT clinical trials. In addition to these facilities, a hyperthermal reactor, RA-6, in Argentina, is used for clinical trials of cutaneous melanoma conducted by the Instituto de Oncología Angel H. Roffo

Medical Center	Reactor	Years	No. Patients	Compound
Broohkaven Natl. Lab., Upton, NY, USA	BMRR (NY, USA)	1994–1999	53 GBM	BPA
Beth Israel Deaconess Center, Harvard Medical School, Boston, MA, USA	MITR-II (Boston, MA, USA)	1996–2003	26 GBM 2 OBT	BPA BPA
Universitätsklinikum Essen, Germany	HFR (Petten, NL)	1997–2002	26 GBM 4 OBT	BSH BPA
University of Tsukuba, Ibaraki, Japan	JRR-4, JAEA (Tokai, JP)	1998–2007	20 GBM 4 OBT	BSH + BPA BSH + BPA
University of Tokushima, Tokushima, Japan	JRR-4 or KURR (Kyoto, JP)	1998–2008	23 GBM	BSH + BPA
Helsinki University Central Hospital, Finland	FiR-1 VTT (Espoo, FI)	1999–2008 2003–2010	50 GBM 2 OBT 31 HNC	BPA BPA BPA
Faculty Hospital of Charles University, Prague, Czech Republic	LVR-15 (Rez, CZ)	2000–2002	5 GBM	BSH
Nyköping Hospital, Sweden	R2-0 Studsvik (Nyköping, SE)	2001–2005	41 GBM 2 OBT	BPA BPA
Osaka Medical College, Osaka, Japan	KURR	2002–2007 2005–2008	40 GBM 10 OBT 6 HNC	BSH + BPA BSH + BPA BPA
Osaka University, Osaka, Japan	JRR-4 or KURR	2001–2007	26 HNC	BPA + BSH
Kyoto University Research Reactor Institute (KURRI), Osaka, Japan	KURR	2001–2007	62 HNC	BPA + BSH
Kawasaki Medical School, Kurashiki, Japan	JRR-4 or KURR	2003–2008	30 HNC	BPA
Taipei Veterans General Hospital, Taipei, Taiwan	THOR (Hsinchu, TW)	2010–2011	10 HNC	BPA

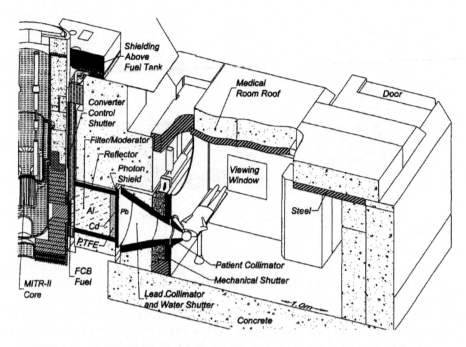

FIGURE 6.7   The BNCT irradiation facility of the MITR in the United States.

Figure 6.7 illustrates the setup built at the MITR. From a thermal column at the reactor they used a fission converter (FCB in the figure), using ^{235}U-enriched fuel, that converted thermal neutrons in fast fission neutrons, which were partially moderated to the epithermal region.

A second major breakthrough was the discovery of a boronated amino acid, *boronophenylalanine* (BPA), for different malignancies. This compound, an analog of tyrosine, a precursor of melanin, was synthesized by Mishima et al. [8] with the aim to apply BPA in BNCT as melanoma-specific boron compound. As a result, there was a large difference in BPA concentration between malignant melanoma cells and normal cells. But it was discovered that the compound accumulated not only in malignant melanoma, but also in various malignant tumors, such as brain tumors and malignant head and neck cancers. BPA is uptaken usually more than three times by tumor than by normal cells. It is believed that this is due to an increased level of amino acid transport in malignant tumor cells, and since then, BPA is the main compound used for BNCT, sometimes combined with BSH for brain tumor BNCT.

Since then, in the 1990s, new clinical trials started of what can be called "modern BNCT" in different countries, specially Japan, the United States, Finland (see Figure 6.8), the Netherlands, Sweden, Argentina, Italy, the Czech Republic, and Taiwan (see reviews of clinical trials in Refs. [9] and [10]). They have been applied for cancers of bad prognosis. Most of the protocols have been focused on brain tumors such as GBM [11–25], and recurrent head and neck cancers [26–30], for which the results were even better. In addition to this, there have also been performed trials for malignant melanoma [31] and for the treatment of multiple metastases in the liver by extracorporeal

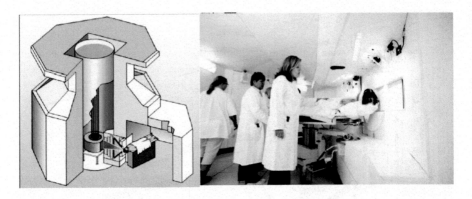

FIGURE 6.8   Sketch of the BNCT facility at the FiR1 in Finland, and a photo of the treatment room.

irradiation [32]. This clinical trial was a challenging procedure performed in Pavia (Italy): after the BPA infusion, the liver of the patient was extracted and irradiated at the thermal column of the reactor of the Laboratory of Applied Nuclear Energy (LENA), and implanted again after the irradiation. This BNCT procedure under autotransplantation was performed in two patients and the multiple tumor masses resulted completely necrotic and unknown metastasis too appeared radically treated [32].

Generally, the clinical results have been quite promising. In a NUPPEC report of 2014 [33], it has been shown a better survival of patients after BNCT than those receiving conventional therapies (see Figure 6.9), especially for head and neck cancers, for which in many cases a complete tumor response follows BNCT. The adverse effects are usually of a low grade, while the symptoms of the disease itself are reduced drastically [27].

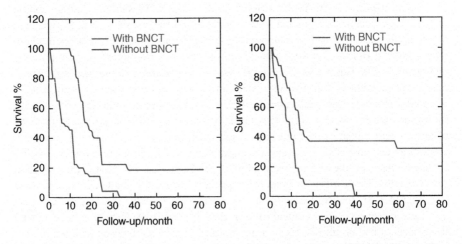

FIGURE 6.9   Kaplan-Meier survival plot of BNCT compared to conventional therapies for glioblastoma (left) and head and neck cancers (right) as reported in the NUPPEC report *Nuclear Physics for Medicine* of 2014 (http://www.nupecc.org/pub/npmed2014.pdf).

In spite of these promising results of BNCT, there is a line of active research throughout the world for improvements. This field is a paradigm of interdisciplinary research, where physicists, engineers, biologists, chemists, pharmacologists, and medical scientists (oncologists, surgeons, and other specialists) work in close connection in a problem, which have different aspects, and there is research open problem for improvement in most of these areas. The International Society for Neutron Capture Therapy (www. isnct.net) promotes the research in the field and organizes an international biannual congress (International Congress on Neutron Capture Therapy, ICNCT).

6.2.4 THE PRESENT AND FUTURE OF BNCT: ACCELERATOR-BASED NEUTRON SOURCES

One of the main limitations of BNCT is related to the neutron source. In the last century, the only neutron beams with enough intensity for BNCT available were research reactors. These are nuclear facilities outside the hospital environment and difficult to adapt to clinical trials. However, in the last years there have been developed high-current, low-energy particle accelerators which by means of nuclear reactions with a specific target can produce a neutron beam suitable for BNCT. This will allow in-hospital BNCT facilities. Current projects will be shown in Section 6.3.3, but we say now that this will open a new era for BNCT, in which the number of patients treated in clinical trials is expected to increase sharply. There are different projects toward an accelerator-based BNCT throughout the world (Figure 6.10), and it is remarkable that the number of them exceed the number of reactors used for clinical BNCT in all its history.

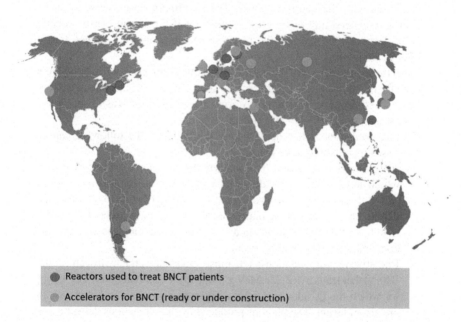

● Reactors used to treat BNCT patients

● Accelerators for BNCT (ready or under construction)

FIGURE 6.10 Places where BNCT clinical trials have been performed with research reactors (dark) and planned accelerator-based neutron sources for BNCT (lighter).

As of May 2019, there is only one facility that has started clinical trials using an accelerator-based neutron source, the C-BENS (Cyclotron-Based Epithermal Neutron Beam) at Kyoto University Research Reactor Institute [34]. A cyclotron from Sumitomo Heavy Industries, Ltd. is used to accelerate protons to 30 MeV that collide with a beryllium target for producing the neutron beam. The clinical results are expected in the next year.

6.3 NEUTRON SOURCES FOR BNCT

6.3.1 REQUIREMENTS FOR A NEUTRON SOURCE FOR BNCT

Neutrons do not have electric charge and therefore do not ionize directly the medium, they deliver energy by means of the secondary particle produced by their interactions. The interactions of a low-energy neutron beam with the body are mainly dominated by the following processes, which deliver radiation dose to the tissues:

- Elastic scattering: This is the dominating process before the neutron is thermalized. Taking into account the composition and the cross sections of the different elements in the tissue (mainly H, O, C, and N), the dominating process is the scattering with hydrogen nuclei. The energy delivered to the body per unit mass at each point by the recoiling nuclei (mostly by protons) is called fast dose (although it is produced by epithermal neutrons) and denoted by D_f. This dose component, for a given initial neutron energy, decreases with depth as the recoil proton deposits an energy, which decreases with the neutron energy. The maximum dose, delivered at the skin, can exceed any other dose component if the initial neutron energy is too large, for this reason the beam spectrum should not exceed tens of kiloelectron volts.

- Inelastic scattering: Only the highest energetic neutrons can leave the recoil nucleus in the collision in an excited state. In this process neutrons lose energy and produce a gamma ray from the excited nucleus. Their role is almost negligible for epithermal neutron beams.

- Capture processes: When neutrons are thermalized, the following capture processes increasingly take place.

  - $^{14}N(n,p)$ capture reaction: This leads to the emission of a 485 keV proton, which delivers energy locally to the tissue. Its contribution to the dose is called thermal neutron dose (sometimes it is called nitrogen dose) and denoted by D_t, as it is the main contribution from indirect charged particle from thermal neutrons in absence of boron, and it is an important contribution to the healthy tissue.

  - Radiative capture reactions: The most important is the hydrogen capture $^1H(n,\gamma)$ that leads to a 2.224 MeV gamma ray. Other reactions like $^{35}Cl(n,\gamma)$ for which the Q-value is 8.5 MeV and which leads to a photon cascade are also relevant. The photons produced by these reactions deliver a dose nonlocally. This dose, plus the dose produced by gamma ray contamination from the beam, is called gamma dose and denoted by D_γ.

- $^{10}B(n,\alpha)$ reaction: The reaction in which BNCT is based. The dose
  delivered locally by means of an alpha particle and a recoil ^7Li nucleus,
  with high LET, is called boron dose and denoted by D_B. This is the
  dominant dose component at the tumor. The 478 keV gamma ray emitted in the reactions where ^7Li reaches, its excited state is included in D_γ
  instead of in D_B.

Therefore, the total absorbed dose is given by

$$D = D_f + D_t + D_\gamma + D_B. \tag{6.2}$$

In Figure 6.11 the different dose components, as a function of depth, obtained from
a Monte Carlo (MC) simulation are displayed for neutrons of 10 keV, in order to
illustrate their different role [35]. It has been assumed a typical boron concentration
in tumor (35 ppm) and in healthy tissue (10 ppm), and a standard tissue: the four-
component International Commission on Radiation Units (ICRU) soft tissue (ICRU4
tissue), which contains 10.1172% H, 11.1000% C, 2.6000% N, and 76.1828% O (mass
proportions).

These different dose components are of different LET. It is well known that radiation of different LET may have different relative biological effectiveness (RBE),
which is taken into account by means of weighting factors, representing the equivalent photon dose that produces the same effect that the actual component. Therefore,
for comparing to a conventional radiotherapy irradiation, it is evaluated the so-called
weighted or biological dose:

$$D_W = w_f D_f + w_t D_t + w_\gamma D_\gamma + w_B D_B. \tag{6.3}$$

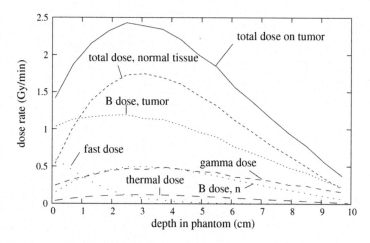

FIGURE 6.11 Dose components, as described in the text, as a function of tissue for 10 keV
monoenergetic neutrons (from Ref. [35]).

TABLE 6.2

Weighting Factors Recommended by Coderre and Morris [36]

Weighting Factors w_i	Normal Tissue	Tumor
Thermal dose factor	3.2	3.2
Fast dose factor	3.2	3.2
Gamma dose factor	1	1
Boron dose factor	1.3	3.8

The weighting factors w_i are defined as the ratio of a reference photon irradiation dose, D_p, and the value of the physical dose component D_i needed to produce the same effect:

$$w_i = \frac{D_p}{D_i}, \qquad i = t, f, \gamma, B. \tag{6.4}$$

The weighting functions have been obtained by means of radiobiology measurements. The values most commonly used in BNCT clinical trials [36] are displayed in Table 6.2.

By evaluating the weighted dose in a phantom, assuming a given concentration of boron in tumor and in normal tissue we can investigate what is the range of optimal neutron energies for BNCT. If we chose a representative general tissue (ICRU4) we can obtain, from a MC simulation, the ratio between the tumor dose and the maximum dose at healthy tissue, which is illustrated in Figure 6.12 for different

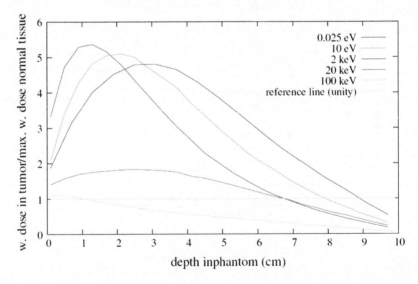

FIGURE 6.12 Ratio of the weighted dose at tumor and the maximum dose in normal tissue for different neutron energies, from Ref. [34]. The optimal energy is of 2 keV.

TABLE 6.3

Recommendations of the International Atomic Energy Agency (IAEA) for the Neutron Beam Parameters. Bounds Mean the Minimum Bounds Acceptable for Therapy and Target Means the Desired Ones for Optimal Therapy

	Bounds	Target
Flux of epithermal neutrons (n/cm²s)	$> 5 \times 10^8$	$\geq 10^9$
Ratio neutron current/neutron flux (divergence)	> 0.7	any > 0.7
Ratio thermal flux/total flux	< 0.05	any < 0.05
Fast neutron dose per epithermal neutron (Gy/epi n)	$< 1.3 \times 10^{-12}$	$\leq 2 \times 10^{-13}$
Gamma dose per epithermal neutron (Gy/epi n)	$< 1.3 \times 10^{-12}$	$\leq 2 \times 10^{-13}$

monoenergetic neutron beams of various energies. The advantage depth (AD) is an in-phantom figure of merit, which is defined as the depth at which the tumor dose equals the maximum dose in healthy tissue and gives an idea of the depth of tumors treatable with BNCT. This value is maximum for 2 keV neutrons, reaching an AD near 9 cm for 10 ppm in normal tissue, 35 ppm at tumor (for higher boron concentration at the tumor, AD increases).

Therefore, as the neutron beams either from a reactor or from an accelerator are not monoenergetic, what is recommended for the spectrum of a neutron source for BNCT of deep-seated tumors is to lie mostly in the epithermal region, which is defined as the range between 0.5 eV and 10 keV.

The International Atomic Energy Agency published a technical document [37] with some recommendations for the desired neutron beam parameters. These are illustrated in Table 6.3. The minimum epithermal flux is required in order to perform treatments in a reasonable time (in the range 20–30 minutes for an epithermal flux of 10^9 n/cm²s). The second restriction gives an upper bound to the divergence of the beam and guarantees that the beam will be directed mostly to the zone of interest. The others try to minimize the contribution of thermal, fast neutrons, and gamma (these last two are really undesirable for BNCT).

These parameters were established for reactor-based neutron sources. For example, the upper limit for the energy of a neutron to be considered epithermal has been discussed recently [38]. It is rather arbitrary as it is considered as an order of magnitude for cutting the high-energy neutron tail from a reactor, which may include neutrons up to 1 MeV.

6.3.2 Neutron Sources from Reactors

As it has been mentioned previously, the BNCT progress in the previous decades has been possible with the conversion of existing research reactors. This has meant modifying or adding components such as the reflector, a beam port or thermal column, shielding, collimators, and filters in order to try to obtain a beam of the intensity and quality needed.

From the broad neutron spectrum of a reactor (containing fast undesirable neutrons to several MeV), a more suitable beam can be achieved (with a large fraction of neutrons in the epithermal range) by shifting or filtering the spectra from the core. Although it has been demonstrated that the displacement of the spectrum with a moderator provides greater efficiency in the production of an epithermal radius than filtering, the technique of choice is determined by the design of the existing reactor. The shifting method requires the availability of a large opening in the shield, such as is often used for a thermal column. If the reactor does not have a space of this type, a more powerful reactor (> 10 MW) with a beam port has the possibility to filter the beam. Alternatively, part of the shield can be opened or removed to provide space for a spectrum changer.

Adequate moderator or filter materials chosen must not degrade in a high radiation field. Any neutron activation products from the materials should be short lived. Some suitable candidates are Al, C, S, Al_2O_3, AlF_3, D_2O, and $(CF_2)_n$. Combinations of Al followed by Al_2O_3 or AlF_3 are very efficient because the O and F cross sections fill in the valleys between the energy resonance peaks of Al. Fluental™ was developed by the VTT Centre of Finland and behaves well under radiation. Using this material as a moderator, a suitable beam for BNCT was obtained from the 250 kW reactor FiR1 at Espoo, Finland [39].

For these spectrum shift facilities, a challenge is to obtain the desired recommended epithermal flux of 10^9 n/cm^{2}s. This is achieved by placing the moderator as close to the reactor core as possible to maximize the input of fast neutrons, and fitting all beam-conditioning components and shutters within the existing shielding dimensions (Figure 6.13).

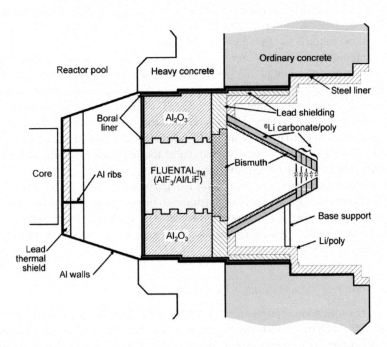

FIGURE 6.13 Sketch of the materials used for producing an epithermal beam from a research reactor. (IAEA TECDOC-1223 [37]).

6.3.3 ACCELERATOR-BASED NEUTRON SOURCES

As it has been mentioned previously, the main factor that has restricted the application of BNCT has been the need of moving the patients to a research reactor for the irradiation. Is it believed that the development of neutron sources that could be placed inside a hospital will increase the number of patients treated.

For obtaining neutron beams according to the IAEA recommendations, different neutron-producing reactions have been studied [40]. In Figure 6.14 some of the most popular reactions that produce a high yield at low energy are displayed.

Choosing a particular reaction is a compromise between yield, neutron energy, energy of the accelerated charged particle, intensity of the particle beam, and target properties. The most popular is $^7Li(p,n)^7Be$ because of the sharp increase of the yield near the threshold, which allows to obtain a high-neutron flux at moderate proton energies with minimal neutron energy, which implies less moderation in order to reach an epithermal beam. The drawbacks are the low melting point of Li (170°C), which makes necessary an effective target refrigeration system or dealing with a liquid target, and the production of radioactive ^7Be. The reaction $^9Be(p,n)^9B$ requires higher proton energies, depending on the current available. Examples are the C-BENS system of Kyoto (Japan), in which a 1 mA, 30 MeV is used [41], or the Tsukuba project, iBNCT, which is based on a 10 mA, 8 MeV accelerator [42]. The larger the proton energy, the greater moderation of neutron energies is needed. In Table 6.4 a list of the facilities constructed or foreseen for next years can be found.

As the neutron energies resulting from the abovementioned particle-induced reactions exceed the epithermal range, some moderation is required. Materials

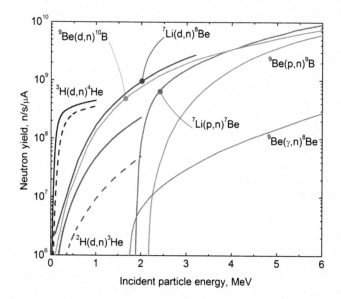

FIGURE 6.14 Neutron yield curves for several reactions from Ref. [40]. The solid plots refer to 100% atomic beams (except for the $^9Be(\gamma,n)^8Be$ reaction), while the two dashed plots refer to 100% molecular ion beams.

TABLE 6.4

Initiatives Toward an Accelerator-Based BNCT Facilities Completed or Projected. As of May 2019, only the C-BENS facility is currently performing clinical trials. Updated from Ref. [35].

Institute	Accelerator	Beam energy	Intensity	Reaction	Max. n energy
Kyoto Univ, Japan (in clinical trials)	Cyclotron	30 MeV	1 mA	$^9Be(p,n)$	28 MeV
Helsinki Univ. Cent. Hospital, Finland	Electrostatic (Hyperion)	2.6 MeV	30 mA	$^7Li(p,n)$	0.89 MeV
Budker Institute, Novosibirsk, Russia	Vacuum insulated Tandem	2 MeV	2 mA	$^7Li(p,n)$	0.23 MeV
IPPE Obninsk, Russia	Cascade generator KG-2.5	2.3 MeV	3 mA	$^7Li(p,n)$	0.57 MeV
Birmingham Univ., UK	Electrostatic (Dynamitron)	2.8 MeV	1 mA	$^7Li(p,n)$	1.1 MeV
Tsukuba Univ., Japan	RFQ-DTL[a]	8 MeV	10 mA	$^9Be(p,n)$	6.1 MeV
CNEA Bs. As., Argentina	Tandem	1.4 MeV	30 mA	$^9Be(d,n)$	5.7 MeV
	Electrost. Quadrupole	2.5 MeV	30 mA	$^7Li(p,n)$	0.79 MeV
INFN, Italia	RFQ[a]	5 MeV	50 mA	$^9Be(p,n)$	3.1 MeV
SOREQ, Israel	RFQ-DTL[a]	4 MeV	2 mA	$^7Li(p,n)$	2.3 MeV
LBNL, USA	Electrostatic	2.5 MeV	50 mA	$^7Li(p,n)$	0.79 MeV
National Cancer Center, Japan	RFQ[a]	2.5 MeV	20 mA	$^7Li(p,n)$	0.79 MeV
Xiamen Humanity Hospital, China	Electrostatic (VITA)	2.5 MeV	10 mA	$^7Li(p,n)$	0.79 MeV
Nagoya Univ., Japan	Electrostatic (Dynamitron)	2.8 MeV	15 mA	$^7Li(p,n)$	1.1 MeV
Granada Univ., Spain	Electrostatic (Hyperion)	2.1 MeV	30 mA	$^7Li(p,n)$	0.35 MeV
Gachon Univ. Gil Medical Center, South Korea	RFQ-DTL[a]	10 MeV	8 mA	$^9Be(p,n)$	8.1 MeV
Southern Tohoku Hospital, Fukushima, Japan	Cyclotron	30 MeV	1 mA	$^9Be(p,n)$	28 MeV

[a] RFQ stands for Radio Frequency Quadrupole and DTL for Drift Tube Linac.

such as MgF_2, Al, LiF, AlF, D_2O, BeO, MgO, Al_2O_3, or combinations of them have been proposed for this purpose. This also implies using reflectors surrounding the moderator for reducing loss of flux, either Pb or Bi, as high mass number nuclei scatter nuclei at higher angles. In addition to this, filters of materials like Fe can increase the spectrum at certain epithermal energies where its total

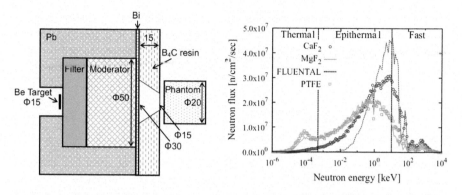

FIGURE 6.15 BSA proposed by Inoue et al. [43] and spectra obtained from different choices for the moderator.

cross section has a minimum (transmission window). After the moderator, layers of materials for absorbing thermal neutrons and gammas can be used. Finally, neutron-absorbing materials like lithiated polyethylene (Li-poly) or boron-containing resins can be used for collimating the beam. All this structure is called beam shape assembly (BSA). In Figure 6.15 an example of BSA proposed by Inoue et al. [43] is displayed, as well as the spectrum obtained by different choices of the material for the moderator. The performance of MgF_2 to produce a pure epithermal beam is remarkable.

Finally, it should be mentioned that the spectra of accelerator-based neutron sources will differ significantly from the spectra from reactors. In Figure 6.16

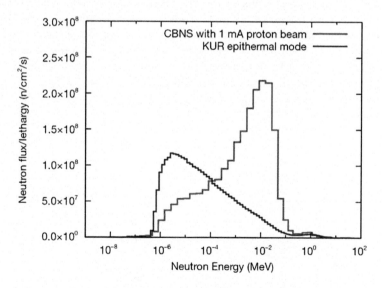

FIGURE 6.16 Comparison of the spectrum of a ABNS and a reactor (Image from Ref. [44] http://tro.amegroups.com/article/view/4713/html#B4)

we illustrate a comparison between the reactor- and the accelerator-based sources of KURRI [44]. The spectrum is closer to the optimal energies for the second, so it can be expected an even better clinical performance from the new sources.

6.4   THE BORON COMPOUNDS

6.4.1   REQUIREMENTS FOR A BORON COMPOUND FOR BNCT

The most important requirements for a BNCT delivery agent are a low intrinsic toxicity, a selective high tumor uptake (> 20 ppm) with low normal tissue uptake (desired tumor:normal tissue and tumor:blood boron concentration ratios greater than 3:1), and a faster clearance from blood and normal tissues than from tumor (persistence at tumor enough to perform the irradiations).

In this section, it would be briefly discussed the main compounds for BNCT and some trends for future agents. The interested reader is referred to very recent reviews [45, 46].

6.4.2   THE COMPOUNDS USED IN CLINICAL TRIALS FOR BNCT

As it has been previously mentioned, there are two main boron compounds that have been used in clinical trials:

- BSH (Figure 6.17, left): Sodium mercaptoundecahydroclosododecaborate ($Na_2B_{12}H_{11}SH$), is a polyhedral borane anion. BSH was used in Japan, and in the clinical trials in the JRC Petten, the Netherlands, to treat patients with malignant gliomas of high grade. The reason to use BSH lies on the fact that, as a small hydrophilic molecule, it does not cross the intact blood-brain barrier (BBB), but can penetrate into the brain passively when the BBB is disrupted [5], as in the GBM. Although BSH has been applied for the treatment of GBM in infusions with no toxic effects, the efficacy has been limited, what is address to a not greater than 3:1 selectivity ratio [47].

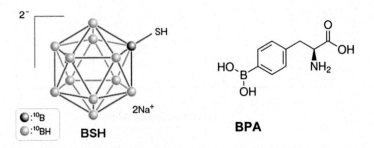

FIGURE 6.17   Molecules of BSH and BPA.

- BPA (Figure 6.17, right): (L)-4-dihydroxy-borylphenylalanine, a boron-containing amino acid, known as boronophenylalanine or BPA, first synthesized by Snyder et al. in 1958 [48], and introduced by Mishima in Japan, it is the currently used boron compound for BNCT. Because of the similarity with a precursor of melanin, it was assumed that BPA would selectively be taken up by melanin-synthesizing cells, and it was used to treat several patients with cutaneous melanomas by injecting it perilesionally [8]. Experimental data of Coderre et al. at the BNL in the United States demonstrated that BPA also was taken up by other types of tumors [49]. Based on this observation, BPA was increasingly used clinically for patients of high-grade gliomas. BPA, due to its poor solubility, is infused by means of a fructose complex (BPA-F), which improves its water solubility [50]. As it has been observed in Table 6.1, it has been the compound used in many clinical trials in number the United States, Finland, Sweden, and Japan for BNCT treatments of patients with high-grade gliomas and recurrent tumors of the head and neck region.

The major problem with both BSH and BPA is the variability, especially in the case of brain tumors, in the tumor uptake by different regions of the tumor, and between different patients receiving the same dose of the boron compound. This has been observed in clinical studies [51, 52]. This variability in the tumor uptake of BPA and BSH is addressed to the complex intratumoral histologic, genomic, and epigenomic heterogeneity within high-grade gliomas and to intertumoral heterogeneity between patients [45].

6.4.3 New Boron Compounds under Study, Including Boron-Rich Nanoparticles

It is believed that the most critical aspect of BNCT lies on the boron compound, and the search of a new carrier that improves the concentration and selectivity of BPA is a very active research problem, as it would be a breakthrough for this therapy.

Some low-molecular weight agents, as BSH and BPA, have been studied. An interesting compound is GB-10 (sodium decaborate, $Na_2B_{12}H$), which has been used in only a few animal studies; although it has an approved US Food and Drug Administration (FDA) Investigational New Drug designation, but it never has been used clinically [45].

Higher-molecular weight compounds that may carry much more boron atoms to the tumor cells are being under investigation. The progress in the very active research field of bionanotechnology, where a myriad of nanostructures are studied for both passive and active targeting of tumors can be translated to BNCT. While for chemotherapy, a nanostructure that contains inside the antitumor agent must reach the cell and release its content, which has to act on the nucleus of the cell, for BNCT this last requisite may not be so critical, as the neutron may react with the boron inside the nanoparticle and the products of the reaction may reach the DNA. Liposomes are a very interesting candidate for this nano-targeted BNCT. Interested readers are referred to the review of H. Nakamura and M. Kirihata on the subject [53].

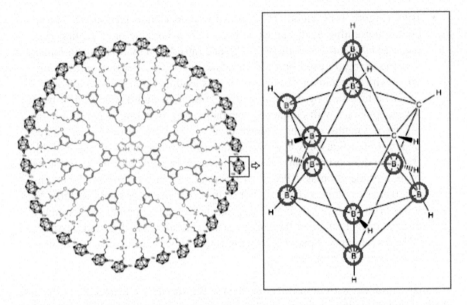

FIGURE 6.18 A poly(aryl ether) dendrimer, synthesized by Cabrera-González et al. [54], with tetraphenylporphyrin as the core and 32 peripheral closo-carborane clusters. Each cluster, shown on the right, has ten boron atoms (circled in dark).

Another example of the capability of carrying many atoms to the tumor is boron-rich dendrimers. They can accumulate many boron clusters around a structure, which depending on its nucleus may have tumor affinity. Examples of promising structures of this kind are those based on porphyrin-type nucleus, like the one illustrated in Figure 6.18.

6.5   THE TREATMENT PLANNING

6.5.1   The Role of Monte Carlo Simulation for BNCT

Due to the complex interactions of a neutron with tissue and the variability of the relative importance of the different process and the energy delivered with the neutron energy, which varies through its travel inside the medium, no accurate simple algorithms based on tabulated coefficients can be applied for the estimation of the dose delivered. Therefore, all treatment planning is done by means of computer codes that are based on MC simulations. For example, all dose profiles shown in this chapter have been obtained from MC simulations performed with the code MCNPX, v.2.5.0 [55].

All MC codes perform neutron transport calculations and are able to evaluate the neutron spectral fluence $\phi_n(E)$ at any elemental voxel inside the medium. The basic scheme of an MC calculation for the neutron transport is described here.

The trajectory of a neutron starts with a given position (x, y, z), kinetic energy (E), and the three-component unit vector (u, v, w) corresponding to its direction of movement. The distance s that travels until its next collision is sampled according to the probability distribution function

$$p(s) = \lambda^{-1} e^{-s/\lambda},$$

where λ denotes the mean free path, given by

$$\lambda = \sum_j n_j \sigma_{\text{tot},j}(E)$$

n_j denotes the atomic density of nuclei of type j, and $\sigma_{\text{tot},j}(E)$ is the total cross section (reaction plus scattering) for nuclei j and neutron energy E. Then the particle is displaced to the new position:

$$(x', y', z') = (x, y, z) + s(u, v, w).$$

Then the type of collision is decided, according to the relative probabilities. An interaction process k with a nucleus of type j is produced with the relative probability

$$p_{k,j} = \lambda n_j \sigma_{k,j}(E).$$

Then by generating a random number, the particular process can be assigned for this iteration. If this is a reaction in which the neutron is absorbed, the neutron disappears, and the next initial neutron enters the process.

If an elastic scattering takes place, we can assume, for the neutron energies of interest for BNCT (<1 MeV), that the scattering is isotropic in the center of mass frame. Then the exit angle θ_{CM} is sampled from a random number ξ in the interval $[0,1]$ as

$$\theta_{\text{CM}} = \arccos(1 - 2\xi).$$

In the laboratory system, the angle with respect to the incident direction is given by

$$\cos\theta = \frac{A\cos\theta_{\text{CM}} + 1}{\sqrt{A^2 + 2A\cos\theta_{\text{CM}} + 1}},$$

while the azimuthal φ angle is randomly distributed in the interval $[0,2\pi]$. The new direction of the neutron is then obtained by a rotation of the vector according to these angles, which is given by

$$u' = u\cos\theta + \frac{\sin\theta(uw\cos\varphi - v\sin\varphi)}{\sqrt{1 - w^2}}$$

$$v' = v\cos\theta + \frac{\sin\theta(vw\cos\varphi + u\sin\varphi)}{\sqrt{1 - w^2}}$$

$$w' = w\cos\theta + \sqrt{1 - w^2}\,\sin\theta\cos\varphi.$$

Except for $w = \pm 1$ (particle initially traveling in the z-direction). In this case, the following formulas should be applied:

$$u' = \pm \sin\theta \cos\varphi$$

$$v' = \pm \sin\theta \sin\varphi$$

$$w' = \pm \cos\theta.$$

The kinetic energy is reduced according to

$$E' = \frac{A^2 + 2A\cos\theta_{CM} + 1}{(A+1)^2} E.$$

And then with the new parameters, the neutron is simulated again. All the magnitudes intrinsically depend on the medium via the cross sections, therefore if an interphase is found, the particle must be stopped at the interphase and is simulated in the new medium according to its composition.

During the simulation process, as the particle position, kinetic energy, and direction are known, it is possible to store, at a particular voxel inside the medium, the number of particles that pass through its boundaries and the neutron fluence $\phi_n(E)$ can be obtained.

If secondary particles are generated, like photons, they have to be transported according to their own process of interactions. Heavy charged particles, as protons or recoil nuclei do not need to be transported as they can be assumed to be absorbed locally.

In reference MC BNCT dose calculations, the absorbed dose rate is calculated with the formula (7.2). Each component can be calculated from the spectral neutron fluence following the expression:

$$D_f = \int_{0.5\,eV}^{\infty} F_n(E)\phi_n(E)\,dE$$

$$D_t = \int_{0}^{0.5\,eV} F_n(E)\phi_n(E)\,dE,$$

where $F_n(E)$ denotes the kerma factor, defined by

$$F_n(E) = \sum_{j,k} \sigma_{kj}(E)\frac{x_j}{M_j} N_A \epsilon_{kj},$$

where the index j runs for the different nuclei in the medium (excepting boron, which is taken into account in other dose term), present in a mass fraction x_j (M_j denotes atomic mass), and k denotes the interaction processes. ϵ_{kj} denotes the kinetic energy released by charged particles in the interaction process k with nucleus j, σ_{kj} is the cross section, and N_A is Avogadro's number.

The Boron dose is given by

$$D_B = \int_0^\infty \Gamma_B(E)\phi_n(E)\,dE,$$

where

$$F_B(E) = \sum_k \sigma_{k,B}(E)\frac{x_B}{M_B}N_A\epsilon_{k,B}.$$

Finally, the gamma dose can be approximated by the photon kerma, once the photon fluence is determined by the MC simulation, by means of the mass energy absorption coefficients:

$$D_\gamma = \int_0^\infty E_\gamma\,\frac{\mu_{en}}{\rho}(E_\gamma)\phi_\gamma(E_\gamma)\,dE_\gamma.$$

The neutron kerma factors have been tabulated for different tissues by the ICRU [56]. If removing the sum in j, the individual contribution of the different elements can be compared. In Figure 6.19 we illustrate the values reported by Goorley et al. [57]

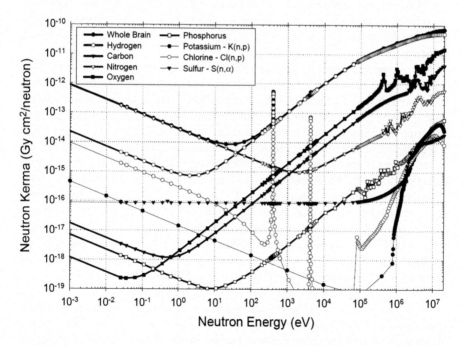

FIGURE 6.19 Values of the neutron kerma factors for the elements in adult brain reported by Goorley et al. [57].

for the elements present in adult brain tissue. It can be seen how the most important are nitrogen (because of the (n,p) capture reaction) for thermal neutrons and hydrogen (because of the elastic collisions) for high-energy neutrons. Between 10 and 100 eV, the kerma factor reaches a minimum. It is interesting to note that elements with minor presence in the tissue, as chlorine, may become important at particular energies due to resonances in their cross section.

The treatment planning in BNCT clinical trials has been performed with MC-based codes as SERA [58], JCDS [59], NCTPlan [60], and THORplan [61]. They can use, as input, geometries constructed from medical images and evaluate dose depth profiles, isodose curves, and dose-volume histograms. For the new generation of accelerator-based neutron sources, new planning codes are being developed, like the TsukubaPlan that can integrate other radiation sources such as protons [62].

6.5.2 Improving Data for Treatment Planning

In a BNCT clinical trial performed in Helsinki [21], it was observed a much better survival in one group of patients receiving a planning tumor volume (PTV) dose of 35 Gy (w) than other group that received a slightly smaller one, 31 Gy (w). The dose in BNCT is delivered with a safe margin because of uncertainties in the estimation of the dose received by the organs at risk. Reducing these uncertainties may allow to increase the irradiation time and to improve the efficacy of the treatments. The ICRU recommends for all kinds of radiotherapy keeping the uncertainties of every step of the planning below 5%. In the case of neutrons, the complexity of the problem makes this a challenging goal, but important efforts are being done toward reducing them.

It is a consensus opinion among BNCT researchers that the boron distribution achieved both at tumors and at normal cells is the major source of uncertainty, because of inhomogeneities. At the level of basic research, techniques like the neutron autoradiography can give some light to the microdistribution of boron in cells [63], and an emerging technique recently announced at the 18th ICNCT (http://www.icnct18.org/) is the single-cell ICP-MS [64], which could be a powerful tool for studying the statistics of the boron uptake in vitro cell cultures. With respect to the real patient case, in recent years the PET diagnosis tool has been applied for imaging the boron distribution in the body by the use of BPA linked to ^{18}F (^{18}F-BPA-PET) [65]. Improvements in this imaging technique, as well as the application of other real-time dose measurements like the use of Compton cameras for detecting the prompt gammas from the boron reaction [66] during the treatment, will probably improve the treatment planning.

Another source of uncertainty is the radiobiological data required (the weighting factors) for the estimation of the photon equivalent dose. The current factors have been obtained from a limited set of measurements, mostly related to the case of brain tumors, but have been used also for other treatments like head and neck cancers. In addition to this, the own formalism of Eq. (6.3) has been questioned because it assumes linearity of the dose-response relationship for all the dose components (and even for the reference photon dose to compare with) and additivity of them when it is known that the dose response is better described by a linear quadratic model. For improving the formalism, a more accurate formalism based on the concept of the photon iso-effective dose has been proposed [67]. In addition to the renovation of the formalism, new accurate radiobiological data for different tumors and normal tissues are required in order to improve the treatment planning.

With respect to the nuclear data for BNCT, although the uncertainties are smaller than the abovementioned previous sources, it is interesting to improve the knowledge of the cross sections of importance for the BNCT dose calculation. The range of energies of relevance goes from thermal to hundreds of keV, and for some nuclei as key for BNCT as ^{14}N, there are some discrepancies between evaluations and experimental data. Also, ^{35}Cl, as it have been mentioned above, may have an important role for some neutron energies. For these reasons at the neutron time of flight facility (n_TOF) at CERN, the cross sections of neutron capture reactions of ^{35}Cl and ^{14}N have been measured recently in order to improve the accuracy of these values [68].

6.6   THE APPLICATION OF NEUTRON GUIDES AND LENSES TO BNCT

This section is related to the aspects of BNCT most related to the scope of this book, these are the applications of neutron optics to BNCT research by means of pure low-energy neutron beams and the potential application of neutron guides and lenses to clinical BNCT.

6.6.1   Dose Produced by Thermal and Cold Neutrons

It has been discussed that the therapeutic effect of BNCT on tumors is due to thermalized neutrons in the tissue. This is an obvious consequence of the $1/v$ behavior of the boron capture cross section, which is maximum at the lowest energy possible. For cold neutrons this cross section is even larger, which means that the number of captures would be enhanced. However, their penetrability is very small, so the effect will be produced at very short depths, as it is illustrated in Figure 6.20.

Therefore, cold neutrons can be used for effective irradiation of small layers of cells in in vitro experiments. These are interesting for determining the survival curves after irradiation, from which the thermal neutron weighting factor can be obtained (the RBE of cold and thermal neutrons should be equal, as the energy delivered by the capture products is the same) and for testing new boron compounds in vitro. A specific setup for these purposes will be discussed next.

6.6.2   Application of Neutron Guides to BNCT Research: The PF1B Line at ILL

In vitro experiments for obtaining neutron radiobiological data for BNCT have been historically performed at reactors providing thermal neutron beams from tangential beam tubes facing the water moderator, called thermal columns. These are beam

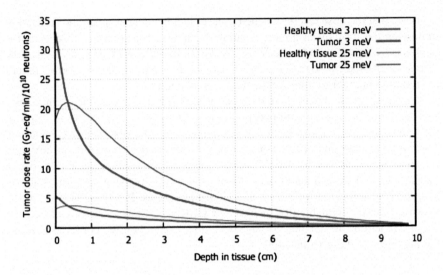

FIGURE 6.20 Depth-dose profile, estimated with Monte Carlo simulation, produced by a pencil beam of monoenergetic neutrons in ICRU four-component tissue for two energies in the cold-thermal range. The boron concentration assumed for the normal tissue is 10 ppm, while for tumor it is 35 ppm. The enhancement at the surface increases with decreasing energy, but the advantage depth is reduced (about 3 cm for the 3 meV neutrons).

tube inserts that provide a thermalized neutron spectrum. Nevertheless, such beams are contaminated by energetic gamma rays due to neutron captures in the moderator and Compton-scattered photons from the reactor core. Therefore, the biological effect in the samples will be produced by both neutrons and photons, and the measurement of the pure neutron effect is intrinsically affected by large uncertainties from the background deduction. As an example, in the experiments of Coderre et al. [69] performed at BMRR, from where the current weighting factors were obtained, the reported gamma dose reported was greater than the pure neutron dose.

Some reactors and spallation neutron sources are equipped with so-called neutron guides where neutrons are efficiently transported over large distances by total reflection on a vacuum-matter interface. The critical angle for total reflection depends on the wavelength of the neutrons (i.e., the energy, so that lower energy neutrons have larger critical angles) and on the material and structure of the coating surfaces (neutron mirrors and supermirrors). The neutron guides are coated with these mirrors and are slightly bent to prevent a direct view of the source. Due to this curvature, neutrons with low energy and low divergence are transported efficiently using sequential garland reflections (never touching the inner wall of the guide), while larger divergence neutrons produce zig-zag reflections (inner to outer walls). On the design of such guides, the curvature has to be chosen such that the transition from garland to zig-zag reflections is allowed, and therefore the guide is efficiently filled. After several reflections, neutrons with a critical angle above the limit (i.e., high divergence and/or high energy) are completely suppressed from the guide. Only thermal and slower neutrons can follow the curvature by total reflections, while faster neutrons and gamma rays go straight and are stopped in the shielding of the neutron guide. Therefore, at the exit of such guides the initial components of fast neutrons as well as gamma rays from the reactor are suppressed. Even in that case, some additional background exists, since it is produced locally by interactions of the neutron beam. Nevertheless, an appropriate selection of the beam windows (separating the vacuum of the neutron guide from the experimental setup), beam stop (downstream of the setup), and surrounding material allows to minimize the total background from gamma rays and fast neutrons. Certain reactors are equipped with a so-called cold source where a vessel containing a material at low temperatures, e.g. liquid deuterium at 25 K, is introduced for the purpose of moderating neutrons to a lower energy. Beam pipes can be arranged to extract cold neutrons from these cold sources, corresponding to average neutron energies of few meV, i.e., lower than thermal energies. Lower energy neutrons can be more efficiently transported through guides than thermal neutrons, since they have larger critical angles for total reflection at the inner walls of the guides, thus increasing the beam intensities available for experiments.

One of these beams can be found at Institut Laue-Langevin (ILL) in Grenoble (sketched in Figure 6.21). In this line, called PF1b [70], scientists perform experiments that study nuclear reactions induced by neutron captures. The cross section of neutron capture reactions follows a perfect $1/v$ behavior (where v is the neutron velocity) at low energies (i.e., thermal and below) in the vast majority of the cases. For this reason, a cold neutron beam can fully replace a thermal beam, provided that the thermal equivalent capture flux is used.

As said above, the bent guide prevents fast neutrons and gamma rays to reach the samples placed at the experimental hall. Due to these appropriate features, our group has started a campaign of measurements of the thermal neutron weighting factor at ILL for different cultures of tumor and normal cell lines, in order to reduce uncertainties on the evaluation of the photon iso-effective dose. In addition to this, we plan to perform tests of new boron compounds by irradiation with this cold neutron source. For both types of experiments, the following setup has been proposed: cells are placed in quartz cuvettes with a minimum inner space (2 mm), which is filled with the culture medium containing the cells. In this way, we can minimize the secondary gamma production by hydrogen. These cuvettes are placed horizontally before the irradiation so the cells (adherent cells) attach to one of the inner faces of

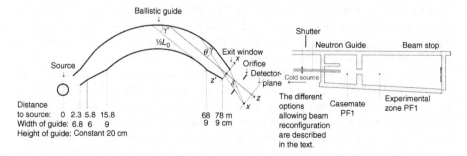

FIGURE 6.21 Ballistic guide of PF1b cold neutron beam line at Institut Laue-Langevin (ILL) and sketch of the experimental area.

FIGURE 6.22 Teflon holder with the cell-containing quartz cuvettes (left). Setup top view (center). Setup with a sample placed (right).

the cuvette. For the irradiation they are placed vertically, with the cell layer close to the beam (see Figure 6.22 for the setup). The first results from these measurements will be published soon [71].

6.6.3 A FUTURE FORM OF RADIOSURGERY WITH NEUTRONS?

Low-energy neutrons, from thermal to cold, can be guided and focused, not only for obtaining pure neutron beams like PF1b at ILL, but also for producing neutron beams at the millimeter or submillimeter scale. The use of focused neutron beams could locally increase the dose with an even smaller damage to close tissues. This strategy can be useful for treating internally the surrounding tissues after the resection of a tumoral region. In this way, these procedures would provide the capacity to completely eliminate the cancer cells that could remain after surgery and degenerate in local recurrencies.

Neutron guides, elliptic or parabolic, focusing supermirrors for several neutron technologies are widely extended, and many known devices as these are in use or under development currently. Some of them are able to focus neutrons to spots at the millimeter range.

In the following example, we study the focusing of either a thermal or a cold neutron beam in a 2-m long guide containing four inner concentric parallel conduits that close from a total section of 10×10 cm to 5×5 cm, as displayed in Figure 6.23.

Although the intensity is reduced, the flux is increased. It can be evaluated with the use of McStas code, from which it is found that the thermal flux in increased by a 1.65 factor, and the cold one by 3.5. The lateral and depth dose distributions produced in tissue for both original and focused beams are shown in Figure 6.24, where a strong concentration of the dose delivered can be observed, especially for cold neutrons, in a very local region.

Furthermore, polycapillary fibers allow neutron transmission and focusing, and examples of neutron lenses made of such materials are available in the literature [72]. These devices are able to increase efficiently the neutron density by several factors (from twice to tens or even more), at a cost of reducing the total neutron intensity (interpreted as the total number of neutrons per second). In these devices, neutrons can pass through hundreds of hollow micrometer wide channels. These channels are perforated in glass-like materials whose smooth inner surface enables multiple total reflections, as in the neutron guides described above. By bending and converging several of such channels,

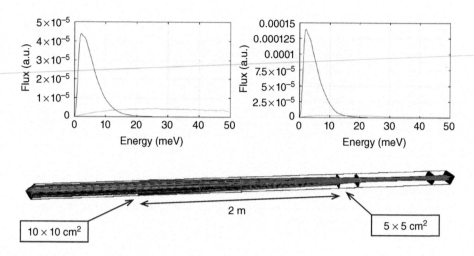

FIGURE 6.23   A neutron-focusing guide.

focusing neutron lenses can be produced. Neutrons that do not enter to the channels or do have high divergence are absorbed within the surrounding material to avoid the degradation of the focused beam spot. At the focusing point, spot sizes near or below a millimeter can be achieved, thus high precision in neutron beam localization can be attained.

Therefore, in principle, a millimeter-focused (thermal or cold) neutron beam can be produced. This is of the order of the width of a catheter that can be introduced in the body. This beam produces an intense dose deposition along the axial line for few millimeters (up to a centimeter) with low dispersion of the neutron beam. This is due to the strong absorption in tissue (mainly from the boron compounds in the

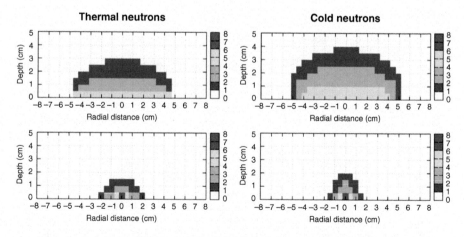

FIGURE 6.24   Dose map from a neutron beam at the entrance of the guide and at the exit. The dose delivered is calculated for a tumor assumed loaded with a concentration of boron of 35 ppm and is normalized to the maximum value of the dose produced at normal tissue (for which a boron concentration of 10 ppm is assumed).

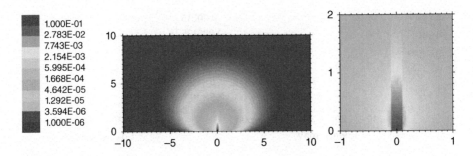

FIGURE 6.25 Average energy deposition due to a thermal neutron beam of 1.0 mm radius into an ICRU4 tissue phantom. Horizontal and vertical axes indicate distance in centimeter. The color scale is logarithmic. The picture on the right is an enlargement of the central-bottom part of the figure on the left.

tumor, but also from nitrogen and hydrogen captures). Additional but less intense dose deposition is present in the nearby tissue due to secondary gamma radiation. This is illustrated by means of an MC simulation in Figure 6.25.

These results suggest the possibility of applying neutron guides to treat internally the surrounding tissue after resection of a tumor, but more research is required in order to determine a practical use.

ACKNOWLEDGMENTS

The author wish to thank Pablo Torres-Sánchez for his help with some of the calculations and figures here shown and for revising the text.

REFERENCES

1. G. L. Locher, Biological effects and therapeutic possibilities of neutrons. *Am. J. Roentgen. Radium Ther.*, 36, 1 (1936).
2. M. Nagai and G. L. Locher, Production of mutations by neutrons. *Nature*, 140, 111–112 (1937).
3. W. H. Sweet and M. Javid, The possible use of slow neutrons plus boron-10 in therapy of intracranial tumors. *Trans. Am. Neurol. Assoc.*, 76, 60–63 (1951).
4. D. N. Slatkin, A history of boron neutron capture therapy of brain tumours. *Brain*, 114, 1609–1629 (1991).
5. A. H. Soloway, H. Hatanaka, and M. A. Davis, Penetration of brain and brain tumor. VII. tumor-binding sulfhydryl boron compounds. *J. Med. Chem.*, 10, 714–717 (1967).
6. H. Hatanaka, Clinical results of boron neutron capture therapy, in *Neutron Beam Design, Development, and Performance for Neutron Capture Therapy*, edited by O. K. Harling, J. A. Bernard, and R. G. Zamenhof, Basic Life Science series, vol. 54, pp. 15–22 (Plenum Press, New York, 1990).
7. Y. Nakagawa, T. Kageji, Y. Mizobuchi et al., Clinical results of BNCT for malignant brain tumors in children. *Appl. Radiat. Isot.*, 67, S27–S30 (2009).
8. Y. Mishima, C. Honda, M. Ichihashi et al., Treatment of malignant melanoma by single thermal neutron capture therapy with melanoma-seeking 10B-compound. *Lancet*, 2, 388–389 (1989).

9. R. F. Barth, M. G. H. Vicente, O. K. Harling et al., Current status of boron neutron capture therapy of high grade gliomas and recurrent head and neck cancer, *Radiat. Oncol.*, 7, 146 (2012).

10. R. F. Barth, Z. Zhang, and T. Liu, A realistic appraisal of boron neutron capture therapy as a cancer treatment modality. *Cancer Commun. (Lond).*, 38, 36 (2018).

11. M. Chadha, J. Capala, J. A. Coderre et al., Boron neutron-capture therapy (BNCT) for glioblastoma multiforme (GBM) using the epithermal neutron beam at the Brookhaven National Laboratory, *Int. J. Radiat. Oncol. Biol. Phys.*, 40, 829–834 (1998).

12. W. Sauerwein and A. Zurlo, The EORTC boron neutron capture therapy (BNCT) group: achievements and future projects. *Eur J Cancer*, 38, S31–34 (2002).

13. V. Dbaly, F. Tovarys, H. Honova et al. Contemporary state of neutron capture therapy in Czech Republic. *Cesaslov Neurol Neurochir.*, 66, 60–63 (2002).

14. H. Joensuu, L. Kankaanranta, T. Seppälä et al., Boron neutron capture therapy of brain tumors: clinical trials at the Finnish facility using boronophenylalanine. *J. Neuro-Oncol.*, 62, 123–134 (2003).

15. T. Yamamoto, A. Matsumura, K. Nakai et al., Current clinical results of the Tsukuba BNCT trial. *Appl. Radiat. Isot.*, 61, 1089–1093 (2004).

16. S-I. Miyatake, S. Kawabata, Y. Kajimoto et al., Modified boron neutron capture therapy for malignant gliomas performed using epithermal neutron and two boron compounds with different accumulation mechanisms: an efficacy study based on findings on neuro-images. *J. Neurosurg.*, 103, 1000–1009 (2005).

17. R. Henriksson, J. Capala, A. Michanek et al., Boron neutron capture therapy (BNCT) for glioblastoma multiforme: a phase II study evaluating a prolonged high-dose of boronophenylalanine (BPA). *Radiother. Oncol.*, 88, 183–191 (2008).

18. S. Kawabata, S-I. Miyatake, T. Kuroiwa et al., Boron neutron capture therapy for newly diagnosed glioblastoma *J. Radiat. Res. (Tokyo)*, 50, 51–60 (2009).

19. S-I. Miyatake, S. Kawabata, K. Yokoyama et al., Survival benefit of boron neutron capture therapy for recurrent malignant gliomas. *J. Neuro-Oncol.*, 91, 199–206 (2009).

20. S. Kawabata, S-I. Miyatake, R. Hiramatsu et al., Phase II clinical study of Boron neutron capture therapy combined with X-ray radiotherapy/temozolomide in patients with newly diagnosed glioblastoma multiforme. *Appl. Radiat. Isot.*, 69, 1796–1799 (2011).

21. L. Kankaanranta, T. Seppälä, H. Koivunoro et al., L-boronophenylalanine-mediated boron neutron capture therapy for glioblastoma or anaplastic astrocytoma progressing after external beam radiation therapy. *Int. J. Radiat. Oncol. Biol. Phys.*, 80, 369–376 (2011).

22. K. Nakai, T. Yamamoto, H. Aiyama et al., Boron neutron capture therapy combined with fractionated photon irradiation for glioblastoma: a recursive partitioning analysis of BNCT patients. *Appl. Radiat. Isot.*, 69, 1790–1792 (2011).

23. T. Yamamoto. K. Nakai, T. Nariai et al., The status of Tsukuba BNCT trial: BPA-based boron neutron capture therapy combined with X-ray irradiation. *Appl. Radiat. Isot.*, 69, 1817–1818 (2011).

24. H. Aiyama, K. Nakai, T. Yamamoto et al., A clinical trial protocol for second line treatment of malignant brain tumors with BNCT at University of Tsukuba. *Appl. Radiat. Isot.*, 69, 1819–1822 (2011).

25. S-I. Miyatake, M. Furuse, S. Kawabata et al., Bevacizumab treatment of symptomatic pseudoprogression after boron neutron capture therapy for recurrent malignant gliomas. Report of 2 cases. *Neuro Oncol.*, 15, 650–655 (2013).

26. I. Kato, K. Ono, Y. Sakurai et al., Effectiveness of BNCT for recurrent head and neck malignancies. *Appl. Radiat. Isot.*, 61, 1069–1073 (2004).

27. I. Kato, Y. Fujita, A. Maruhashi et al., Effectiveness of boron neutron capture therapy for recurrent head and neck malignancies. *Appl. Radiat. Isot.*, 67, S37–S42 (2009).

28. L. Kankaanranta, T. Seppälä, H. Koivunoro et al., Boron neutron capture therapy in the treatment of locally recurred head-and-neck cancer: final analysis of a phase I/II trial. *Int. J. Radiat. Oncol. Biol. Phys.*, 82, e67–75 (2012).

29. L. Kankaanranta, K. Saarilahti, A. Mäkitie et al., Boron neutron capture therapy (BNCT) followed by intensity modulated chemoradiotherapy as primary treatment of large head and neck cancer with intracranial involvement. *Radiother. Oncol.*, 99, 98–99 (2011).

30. L. W. Wang, Y. W. Chen, C. Y. Ho et al., Fractionated BNCT for locally recurrent head and neck cancer: experience from a phase I/II clinical trial at Tsing Hua Open-Pool Reactor. *Appl. Radiat. Isot.*, 88, 23–27 (2014).

31. P. R. Menéndez, B. M. Roth, M. D. Pereira et al., BNCT for skin melanoma in extremities: updated Argentine clinical results. *Appl. Radiat. Isot.*, 67, S50–53 (2009).

32. A. Zonta, T. Pinelli, U. Prati et al., Extra-corporeal liver BNCT for the treatment of diffuse metastases: what was learned and what is still to be learned. *Appl. Radiat. Isot.*, 67, S67–75 (2009).

33. Nuclear Physics European Collaboration Committee (NUPECC), Nuclear Physics for Medicine Report, (2014) http://www.nupecc.org/pub/npmed2014.pdf.

34. T. Mitsumoto, S. Yajima, H. Tsutsui et al., Cyclotron-based neutron source for BNCT. *AIP Conf. Proc.*, 1525, 319–322 (2013).

35. I. Porras, J. Praena, F. Arias de Saavedra et al., Perspectives on Neutron Capture Therapy of Cancer, in *Proceedings of the 15th International Conference on Nuclear Reaction Mechanisms*, edited by F. Cerutti, A. Ferrari, T. Kawano, F. Salvat-Pujol, and P. Talou, *CERN-Proceedings-2019-001* (CERN, Geneva, 2019), pp. 295–304.

36. J. A. Coderre and G. M. Morris, The radiation biology of boron neutron capture therapy. *Radiat. Res.*, 151, 1–18 (1999).

37. International Atomic Energy Agency, *Current Status of Neutron Capture Therapy*. IAEA-TECDOC-1223. (IAEA, Vienna, 2001).

38. P. Torres-Sánchez, I. Porras, F. Arias de Saavedra et al., On the upper limit for the energy of epithermal neutrons for boron neutron capture therapy, *Radiat. Phys. Chem.* 156, 240–244 (2019).

39. I. Auterinen, P. Hiismäki, P. Kotiluoto et al., Metamorphosis of a 35 years old Triga reactor into a modern BNCT facility, in *Frontiers in Neutron Capture Therapy*, edited by M. F. Hawthorne, K. Shelly, R. J. Wiersema, 1, pp. 267–275 (Plenum Press, New York, 2001).

40. D. L. Chichester, *Production and Applications of Neutrons Using Particle Accelerators*, INL/EXT-09-17312 (Idaho National Laboratory, Idaho Falls, 2009).

41. H. Tanaka, Y. Sakurai, M. Suzuki et al., Experimental verification of beam characteristics for cyclotron-based epithermal neutron source (C-BENS). *Appl Radiat Isot.*, 69, 1642–1645 (2011).

42. H. Kumada, A. Matsumura, H. Sakurai et al., Project of development of the linac based NCT facility in University of Tsukuba. *Appl. Radiat. Isot.*, 88, 211–215 (2014).

43. R. Inoue, F. Hiraga, and Y. Kiyanagi, Optimum design of a moderator system based on dose calculation for an accelerator driven boron neutron capture therapy. *Appl. Radiat. Isot.*, 88, 225–228 (2014).

44. Y. Kiyanagi, Accelerator-based neutron source for boron neutron capture therapy. *Ther. Radiol. Oncol.*, 2, 55 (2018).

45. R. F. Barth, P. Mi and W. Yang, et al., Boron delivery agents for neutron capture therapy of cancer. *Cancer Commun.*, 38, 35 (2018).

46. H. Cerecetto and M. Couto, Medicinal chemistry of boron-bearing compounds for BNCT-glioma treatment: current challenges and perspectives, in *Glioma—Contemporary Diagnostic and Therapeutic Approaches*, edited by I. Omerhodžić and K. Arnautović (IntechOpen, London, 2018). DOI: 10.5772/intechopen.76369.

47. T. Kageji, Y. Nakagawa, K. Kitamura et al., Pharmacokinetics and boron uptake of BSH (Na2B12H11SH) in patients with intracranial tumors. *J. Neuro-Oncol.*, 33, 117–130 (1997).

48. H. R. Snyder, A. J. Reedy and W. J. Lennarz, Synthesis of aromatic boronic acids. Aldehydo boronic acids and a boronic acid analog of tyrosine. *J. Am. Chem. Soc.*, 80, 835–838 (1958).

49. J. A. Coderre, J. D. Glass, R. G. Fairchild et al., Selective delivery of boron by the melanin precursor analogue p-boronopheny-lalanine to tumors other than melanoma. *Cancer Res.*, 50, 138–141 (1990).

50. K. Yoshino, A. Suzuki, Y. Mori et al., Improvement of solubility of p-boronophenylalanine by complex formation with monosaccharides. *Strahlenther Onkol.*, 165, 127–129 (1989).

51. J. H. Goodman, W. Yang, R. F. Barth et al., Boron neutron capture therapy of brain tumors: biodistribution, phar-macokinetics, and radiation dosimetry sodium borocaptate in patients with gliomas. *Neurosurgery*, 47, 608–621 (2000).

52. H. Koivunoro, E. Hippelainen, I. Auterinen et al., Biokinetic analysis of tissue boron (10B) concentrations of glioma patients treated with BNCT in Finland. *Appl Radiat Isot.*, 106, 189–194 (2015).

53. H. Nakamura and M. Kirihata, Boron compounds: new candidates for boron carriers in BNCT, in *Neutron Capture Therapy: Principles and Applications*, edited by A. Sauerwein, A. Wittig, R. Moss, and Y. Nakagawa, pp. 99–116 (Springer, Berlin, 2012).

54. J. Cabrera-González, E. Xochitiotzi-Flores, C. Viñas et al., High-boron-content porphyrin-cored aryl ether dendrimers: controlled synthesis, characterization, and photophysical properties. *Inorg. Chem.*, 54, 5021–5031 (2015).

55. D. B. Pelowitz, *MCNPX™-User's Manual, Version 2.5.0. Publication LACP-05-0369* (Los Alamos National Laboratory, Los Alamos, 2005).

56. International Commission on Radiation Units and Measurements, Nuclear Data for Neutron and Proton Radiotherapy and for Radiation Protection, ICRU Report 63.

57. J. T. Goorley, W. S. Kiger III and R. G. Zamenhof, Reference dosimetry calculations for neutron capture therapy with comparison of analytical and voxel models. *Med. Phys.*, 29, 145–156 (2002).

58. R. Zamenhof, E. Redmond II, G. Solares et al., Monte Carlo-based treatment planning for boron neutron capture therapy using custom designed models automatically generated from CT data. *Int. J. Radiat. Oncol. Biol. Phys.*, 35, 383–397 (1996).

59. D. W. Nigg, C. A. Wemple, D. E. Wessol et al., SERA—an advanced treatment planning system for neutron therapy and BNCT. *Trans. Am. Nuc. Soc.*, 80, 66 (1999).

60. H. Kumada, K. Yamamoto, A. Matsumura et al., Verification of the computational dosimetry system in JAERI (JCDS) for boron neutron capture therapy. *Phys. Med. Biol.*, 49, 3353 (2004).

61. T. Y. Lin and Y. W. Liu, Development and verification of THORplan—a BNCT treatment planning system for THOR. *Appl. Radiat. Isot.*, 69, 1878 (2011).

62. H. Kumada, K. Takada, K. Yamanashi et al., Verification of nuclear data for the Tsukuba plan, a newly developed treatment planning system for boron neutron capture therapy. *Appl Radiat Isot.*, 106, 111–115 (2015).

63. A. Portu, A. J. Molinari, and S. I. Thorp, Neutron autoradiography to study boron compound microdistribution in an oral cancer model. *Int J Radiat Biol.*, 91, 329–335 (2015).

64. R. Merrifield, L. Amable, and C. Stephan, Single-cell ICP-MS analysis: quantifying the metal concentration of unicellular organisms at the cellular level, *Spectroscopy*, 33, 33–39 (2018).

65. L. Menichetti, L. Cionini, and W. A. Sauerwein, Positron emission tomography and [18F.BPA: a perspective application to assess tumour extraction of boron in BNCT. *Appl Radiat Isot.*, 67, S351–354 (2009).

66. C. H. Gong, X. B. Tang, D. Y. Shu et al., Optimization of the Compton camera for measuring prompt gamma rays in boron neutron capture therapy. *Appl Radiat Isot.*, 124, 62–67 (2017).

67. S. J. Gonzalez and G. A. Santa Cruz, The photon-iso-effective dose in boron neutron capture therapy. *Radiat Res*, 178, 609–621 (2012).

68. J. Praena, I. Porras, M. Sabaté-Gilarte et al., The 14N(n,p)14C and 35Cl(n,p)35S reactions at n_TOF-EAR2: dosimetry in BNCT and astrophysics. CERN Document Server: CERN-INTC-2017-039/INTC-P-510, (2017) https://cds.cern.ch/record/2266484

69. J. A. Coderre, M. S. Makar, P. L. Micca et al., Derivations of relative biological effectiveness for the high-LET radiations produced during boron neutron capture irradiations of the 9L rat gliosarcoma in vitro and in vivo. *Int. J. Radiat. Oncol. Biol. Phys.*, 27, 1121–1129 (1993).

70. H. Abele, D. Dubbers, H. Häse et al., Characterization of a ballistic supermirror neutron guide. *Nucl. Inst. Meth. Phys. Res. A*, 562, 407–417 (2006).

71. I. Porras, M. Pedrosa, J. Praena et al., Radiobiology in vitro measurements for boron neutron capture therapy, ILL Experimental Report 3-07-376 (2018) http://userclub.ill.eu

72. H. H. Chen-Mayer, D. F. R. Mildner, G. P. Lamaze et al., Neutron focusing using capillary optics and its applications to elemental analysis. *AIP Conf. Proc.*, 475, 718 (1999).

Index

Note: *Italicized* page numbers refer to figures,
bold page numbers refer to tables.

3D magnetic field, 177, *178*

A

Absolute temperature, 6–7, 32, 35
Absorbed dose (D), 251
Absorbing boundaries, 110
Absorption cross sections, 172
Accelerators
    BNCT, 249, *249*, **256**
    DONES, 10
    linear, 12
    medium, 12
    neutron beams, 253
    neutron sources, 127, 156, 250, 257, 264
    particle, 249, xv
    pulsed proton beams, 127
    small, 12
Acceptance angle, thermal neutrons, 130–131
Activation, 9, 15–16
Activation reactions, 9
Advantage depth (AD), 253
Alpha particles, 240, 251
Alternating materials, 207
Amplitudes, 47–49
    coherent, 135
    Gaussian, 88
    of hologram, 61
    Laue diffraction, 52–53
    probability, 127–128, 213
    transmission, 86
    X-ray, 54
Analysis of matter, 126, *126*
Analyzers, 177, 205, 206, *206*, 228–230,
    229, 235
Angles, *see* Critical angle; Glancing angles
Antibarrier layer, 210
Antireflective, underlayer, 206, *207*, 208, 210, *211*
Aperture function, 86
Approximate collimators, 13
Approximate multiple scattering equations, 21
Approximation(s)
    with one amplitude, 45–47
    quasi-classical, 137
    slow neutron optical potential, 20–24
    for thermal neutrons, 20
    with two amplitudes, 47–49
Archaeometry, 16

Artificial sources for neutrons, 10–13;
    see also Neutrons
    accelerator-based, 127
    medium accelerators, 12
    nuclear reactors, 11–12, 126–127
    small accelerators, 12
    spallation source, 12–13, 127, **128**
Atomic nuclei, 6
    absorption of thermal neutrons, 54
    cross sections by individual, 18–20
    fissile, 11
    incoherence effects, 23
    of materials with low atomic mass, 13
    moderation, 13
    neutron beams interaction with, 13
    in thin crystal lattice, 50
Atomic-resolution slow neutron holography, *59*,
    59–62
Attenuation coefficient, 170–172
Attenuation contrast, 169–172
    linear coefficient, 170
Average energy deposition, neutron beam, *270*

B

Bacon, G. E., 4
Ballistic guides, 267
Bartol Research Foundation, *242*
Basis sets
    Frozen-Gaussian, 100–105
    optimal, 105
    pseudospectral, 107–110
    spectral, 105–106
BBB, *see* Blood-brain barrier (BBB)
Beam(s)
    area, 179
    collimated/collimator, 13–14, 181–182, *182*
    contamination, 266
    cross sections, 182–184, *183*
    geometry, *168*, 168–169, 180–184, *181*
    neutron, *see* Neutron beams
    neutron beams interaction with
        atomic nuclei, 13
    pipes, 267
    requirements, 177–180
    slow neutrons, 125
    X-ray, 125
Beam adjusted collimator, 181–182, *182*
Beam geometry
    direct, 180–184, *181*
    spatial resolution, *168*, 168–169